UPROOTING AND DEVELOPMENT

Dilemmas of Coping with Modernization

Current Topics in Mental Health

Series Editors: Paul I. Ahmed
U.S. Office of International Health
Department of Health and Human Services
and
Stanley C. Plog
Plog Research, Inc.

STATE MENTAL HOSPITALS: What Happens When They Close
Edited by Paul I. Ahmed and Stanley C. Plog

COPING WITH PHYSICAL ILLNESS
Edited by Rudolf H. Moos

THE PRINCIPLES AND TECHNIQUES OF MENTAL HEALTH CONSULTATION
Edited by Stanley C. Plog and Paul I. Ahmed

TOWARD A NEW DEFINITION OF HEALTH
Edited by Paul I. Ahmed and George V. Coelho

THE YEAR 2000 AND MENTAL RETARDATION
Edited by Stanley C. Plog and Miles B. Santamour

UPROOTING AND DEVELOPMENT: Dilemmas of Coping with Modernization
Edited by George V. Coelho and Paul I. Ahmed

UPROOTING AND DEVELOPMENT

Dilemmas of Coping with Modernization

Edited by

George V. Coelho

National Institute of Mental Health
Rockville, Maryland

and

Paul I. Ahmed

U. S. Office of International Health
Department of Health and Human Services
Rockville, Maryland

With the assistance of

Ying-Ying T. Yuan

Lawrence Johnson and Associates, Inc.
Washington, D.C.

Springer Science+Business Media, LLC

Library of Congress Cataloging in Publication Data

Main entry under title:

Uprooting and development.

(Current topics in mental health)
Includes bibliographical references and index.
1. Mental health. 2. Child mental health. 3. Minorities – Mental health. 4. Social change – Psychological aspects. 5. Industrialization – Psychological aspects. I. Coelho, George V. II. Ahmed, Paul I. III. Yuan, Ying-Ying T. [DNLM: 1. Stress, Psychological. 2. Adaptation, Psychological. 3. Social environment. WM172 U68]
RA790.5.U67 303.4'3 81-16539

ISBN 978-1-4684-3796-6 ISBN 978-1-4684-3794-2 (eBook)
DOI 10.1007/978-1-4684-3794-2

Originally published by Plenum Press, New York in 1980
Softcover reprint of the hardcover 1st edition 1980

Dedicated to

our children, whose future promises change, resilience, and growth

George Arjun, Jr., Susie Sumangali,
and Victor Anand Coelho;

Paul Farzand, Jr., Irene Michelle,
and Rochelle Yvonne Ahmed;

and Nicole Patricia Rui-Zhen Yuan

Contributors

Paul I. Ahmed, M.A., L.L.B., Public Health Advisor, U.S. Office of International Health, Office of the Assistant Secretary for Health, Department of Health and Human Services, Rockville, Maryland

Lorna Rhodes AmaraSingham, Ph.D., Social Anthropologist, Bethesda, Maryland

Kurt W. Back, Ph.D., Chairman and Professor, Department of Sociology, Duke University, Durham, North Carolina

Inge Bretherton, Ph.D., Research Psychologist, Department of Psychology, University of Colorado, Boulder, Colorado

Richard Harvey Brown, Ph.D., Associate Professor of Sociology, Division of Behavioral Sciences, University of Maryland, College Park, Maryland, and Visiting Lecturer, Faculté des Sciences Sociales, Département de Sociologie, Université d'Ottawa, Canada

John H. Bryant, M.D., Deputy Assistant Secretary of International Health, Office of International Health, Office of the Assistant Secretary for Health, Department of Health and Human Services, Rockville, Maryland

George V. Coelho, Ph.D., Senior Social Scientist, Office of the Assistant Director for Children and Youth, National Institute of Mental Health, Rockville, Maryland

Lucy M. Cohen, Ph.D., Professor of Anthropology, Department of Anthropology, Catholic University, Washington, D.C.

Marc Fried, Ph.D., Professor of Psychology and Human Sciences, and Director, Laboratory of Psychosocial Studies, Boston College, Boston, Massachusetts

Elaine Ruth Goldberg, M. Ed. Educational Consultant and Writer, Boston, Massachusetts

James S. Gordon, M.D., Research Psychiatrist, Center for Studies of Child and Family Mental Health, National Institute of Mental Health, Rockville, Maryland

Otto Klineberg, Ph.D., Emeritus Professor of Psychology, Columbia University, and Visiting Professor, Ecole des Hautes Etudes en Sciences Sociales, Paris, France

Aliza Kolker, Ph.D., Associate Professor of Sociology, George Mason University, Fairfax, Virginia

Peter Marris, Ph.D., Professor, School of Architecture and Urban Planning, University of California, Los Angeles, California

Murali Nair, D.S.W., Associate Professor and Director of Admissions, Graduate School of Social Work, Marywood College, Scranton, Pennsylvania

Paul B. Pedersen, Ph.D., Research Fellow, Culture Learning Institute, East–West Center, Honolulu, Hawaii

Maria Pfister-Amende, M.D., M.P.H., Psychiatrist and President, Swiss National Committee for Mental Health, Zurich, Switzerland

Parmatma Saran, Ph.D., Assistant Professor, Department of Sociology, Baruch College, The City University of New York, New York, New York

Seth Spaulding, Ph.D., Professor, International and Development Education Programs, University of Pittsburgh, Pittsburgh, Pennsylvania

Jane J. Stein, Medical Science Writer, Washington, D.C.

Margaret A. Thomson, M.A., Subcommittee Staff Director, Subcommittee on Forests, Committee on Agriculture, U. S. House of Representatives, Washington, D.C.

Frank M. Tims, Ph.D., Program Analyst, National Institute of Drug Abuse, Rockville, Maryland

Edward A. Tiryakian, Ph.D., Professor of Sociology, Department of Sociology, Duke University, Durham, North Carolina

Joseph E. Trimble, Ph.D., Associate Professor of Psychology, Western Washington University, Bellingham, Washington

Sidney Werkman, M.D., Professor of Psychiatry, University of Colorado School of Medicine, Denver, Colorado

Ying-Ying T. Yuan, Ph.D., Division Manager and Senior Scientist, Lawrence Johnson and Associates, Inc., Washington, D.C.

Foreword

Uprooting has to do with one of the fundamental properties of human life—the need to change—and with the personal and societal mechanisms for dealing with that need. As with the more general problems of change, uprooting can be a time of human disaster and desolation, or a time of adaptation and growth into new capacities.

The special quality of uprooting is that the need to change is faced at a time of separation from accustomed social, cultural, and environmental support systems. It is this separation from familiar supports that either renders the uprooted vulnerable to the destructive consequences of change, or creates freedoms for their evolution into new and constructive patterns of life. Whether the outcomes will be destructive or constructive will be determined by the forces at work: the nature and power of the uprooting forces versus the personal and societal capacities for coping with them.

Uprooting events are so widespread as to be compared with the major rites of life, but with the difference that dislocation is involved. Uprooting reaches from self-imposed movements such as rural-to-urban migration, running away, and traveling abroad for schooling, to natural and man-made disasters such as earthquakes, political oppression, and war. The impacts vary from the need to adapt to a new culture for an interim period of study to the desolating consequences of the total loss of family, friends, home, and country.

The variants of uprooting apply to all societies and all countries, rich and poor, north and south, east and west. No society is spared the threats and the possibilities of uprooting.

Thus, uprooting is everywhere to be seen; and this volume explores it in its great variety. It documents the forms of uprooting, describes the social and personal dynamics of coping or failing to cope with the stresses of uprooting, and, importantly, sets out the possibilities of organized societal response that can ease the stresses and promote the supporting structures that will assist the uprooted in coping with changes. In so doing, this book serves the heuristic purpose of stimulating our thinking about the uprooted, pushes us to a better understanding of their problems, and leads us toward more constructive actions to assist them. For these reasons, we are indebted to the editors and authors of this stimulating work.

John H. Bryant, M.D.

Preface

As our physical and social environments change with a momentum and massiveness that is unparalleled in history, vital interpersonal attachments and institutional supports are being disrupted. The foundations for maintaining personal well-being and community mental health are being continually challenged. At a study-group meeting of the International Association for Child Psychiatry and Allied Professions held in 1977 in India it was suggested, for example, that in traditional and slowly developing societies today a "good enough environment" for child development is usually provided by the intergenerational nexus of significant people in the child's social world. Specifically, it was hypothesized that the best defense against mental illness is the continuity of emotional and social support that is built into the three-generation family.[1] Today, however, this type of support group is under severe strain due to many environmental and cultural changes. Hence the challenge today is not only for individuals to learn coping skills, but also for institutions to anticipate and avert or alleviate the new stressful situations that are increasing the emotional hazards for many vulnerable groups in various societal transitions.

Uprooting, the crucial metaphor of modernization, may be considered to be the dominant trend underlying much social change. Indeed, the World Health Organization has identified uprooting as

> the common factor in a number of psychological high-risk situations, such as migration, urbanization, resettlement, and rapid social change. Uproot-

[1] J. Anthony & C. Chilland (Eds.), *Children and parents in a changing world*. International Year Book of the International Association for Child Psychiatry and Allied Professions. New York: Wiley-Intersciences, 1978.

> ing occurs in most countries of the world and is often associated with meaningless violence, the abuse of alcohol and drugs, criminality, and reactive mental disorders.[2]

Rapid technological, economic, social, and political changes are challenging traditional family structures and roles. New tasks of adaptation call for new strategies in changing environments as young people strive to learn models of competent adult roles and responsibilities.

Now, as never before, humankind is becoming a single global ecological system.[3] A catastrophic change in one habitat has repercussions in others, thousands of miles away. What are some of the coping mechanisms that individuals use for managing the stressful situations of uprootings which are becoming commonplace in a complex changing world? What collaborative coping strategies can communities collectively develop for moderating the pace and scale of stressful change?

We need to know more about the psychosocial environments of uprooted peoples and the high-risk situations to which the most vulnerable younger and older groups are exposed. As uprootings of different kinds proliferate in our time—many of them positively valued—we need new concepts and data for defining the parameters of change, the limits of our tolerance, and the costs to health and emotional well-being of individuals and communities exposed to these difficulties.

Learning to be modern is a constant challenge in collaborative coping. We need to take a long-term perspective forward and backward in time, and across different cultures, in order to understand the ancient roots of behavior and the development of adaptive mechanisms in individual development and social organization as changes occur within the self and in the environments that are discontinuous and that become unfamiliar. The environmental conditions that once were "good enough" to facilitate the learning of competence from generation to generation are now fast disappearing. Ancient patterns of behavior persist while culture patterns rapidly change. "We do not know," as Philip Handler writes, "how personal unhappiness and social distress is a consequence of man's biological nature in conflict with an unnatural and essentially nonhuman environment."[4]

The aim of this volume is to stimulate social scientists, mental-

[2] World Health Organization, Psychosocial factors and health. In P. I. Ahmed & G. V. Coelho (Eds.), *Toward a new definition of health*. New York: Plenum, 1979, pp. 87–111.

[3] N. Reeves, The ultimate selection. In A. M. M'Bow *et al.* (Eds.) *Suicide or survival: Round table on the challenges of the year 2000*. Paris: United Nations Educational, Scientific, and Cultural Organization, 1978.

[4] P. Handler, *Biology and the future of man*. New York: Oxford University Press, 1976.

health practitioners, and urban policy planners (a) to focus on emotional and interpersonal aspects of environmental stress that is associated with uprooting and rapid change in our time; (b) to plan multidisciplinary cooperative research on issues of coping behavior and adaptation of high-risk groups, viewed in cross-cultural perspectives; (c) to devise national and international institutional mechanisms for practical action designed to alleviate the health burdens on the most vulnerable groups, who are at high risk and most in need of support in the face of stressful change. The global impact of uprooting calls for policies of collaborative action of communities, government agencies, and international associations.

The volume does not attempt to cover all kinds of stressful stituations, or all dimensions of environmental change. We need data and insights from cross-cultural research to describe and to explain the variety of adaptive behavior patterns developed by those who are uprooted in different ecological settings. The editors hope to stimulate more detailed study of variations by ethnicity, age, sex, dominant culture, or minority subculture, from the point of view of behavioral adaptation to stress. The special value of this volume is that it provides a general paradigm of coping behavior in stressful situations of change.[5] The articles focus on adaptation issues in diverse uprooting situations that have been treated in isolation and under disparate topics associated with rapid modernization, such as migration and cross-cultural education, acculturation, forced relocation, refugee resettlement, marginality, and so on.

The theme and theoretical thrust of the volume is that, though high-risk conditions are produced by rapid physical and socioenvironmental changes due to uprooting, they produce different effects on different population groups, depending on the developmental status, personality, and cultural predispositions of the individuals concerned. Conceptual tools are provided for diagnosing stressful situations and assessing their impact on different population groups that may be vulnerable to emotional and health hazards.

The format and style of communication is designed to appeal to mental, behavioral, and social scientists who are concerned with the stressful changes of various kinds of uprooting that affect individual and group functioning. It is also expected that urban designers, international health-planning officials, and development investors will recognize the importance of behavioral and institutional factors in

[5] D. A. Hamburg, G. V. Coelho, & J. E. Adams, Coping and adaptation: A synthesis of biological and social perspectives. In G. V. Coelho, D. A. Hamburg, & J. E. Adams (Eds.), *Coping and adaptation*. New York: Basic Books, 1974, Chapter 13.

mediating the environmental impact on quality of life in communities that are being transformed through their enterprises. Several articles provide useful tools for identifying change parameters that need to be included in planning models and measures of development and cost–benefit assessments. The volume is organized into six sections, as follows:

Part I, *Coping with the Inner and Outer Worlds of Change*. This section defines the stress situations of global scope and consequence that challenge the adaptive capacities of individuals and groups who are exposed to uprooting in space and time. The first two articles present historical and cross-cultural perspectives on the contemporary *problematique* of change. They highlight the worldwide ecological changes that in unique and unprecedented ways have tended to increase our vulnerability as a species, and the coping challenges for groups as well as institutions.

A major theme of the first section, elaborated in the second and recapitulated in the last three, is that coping as a response to rapid change and crisis cannot be understood as an individual personal enterprise. Rather, coping involves a process of community engagement and collaborative action on the economic, political, and societal levels.

Part II, *Meanings and Impacts of Uprooting*. This section examines the dimensions of change in various forms of uprooting from the vantage point of diverse disciplines: behavioral biology and anthropology, urban planning and architecture, sociology, developmental and social psychology, psychiatry, and public health. The articles in this section present diagnostic tools for analyzing social functioning and adaptive behavior in stress at the personal and the social-system levels.

Positive self-image, self-esteem, and role-adaptation are used as key link concepts for articulating the interaction of developmental functions in personal systems within the larger institutional systems.

The analytic frameworks provide tools for assessing variations in group adaptive behavior under stressful change within and across cultures, as well as in individual coping responses; for some individuals hurt more easily than others, some disintegrate more rapidly than others, and some recover and learn from severe crises of change more efficiently than others.

Part III, *Stressful Situations of Children and Adolescents in Transition: The Role of Attachments and Social Supports*. This section examines the experience of loss that is brought home by change and the reaching for new bonds. Examples are given from early developmental periods to illustrate the adaptive behaviors of children and adolescents, who are exploratory and stress-seeking creatures, but are also immature, dependent, and vulnerable to changes that threaten their feelings of self-worth, self-identity, and relatedness to caring persons in their familiar environment.

The authors in this section examine the dynamics of coping in stressful situations of uprooting of young people who move both within and across cultural boundaries. They highlight the emotional and interpersonal impacts of environmental change, especially as it affects children living in nontraditional families or institutions, runaway youth, and adolescent dependents who are uprooted when the family is relocated to a post overseas.

Part IV, *Stressful Situations of Foreign Students: The Challenges of Cross-Cultural Education*. This section focuses on adaptive strategies of students who are challenged to learn new social roles and cultural rituals during the various phases of studying abroad. The literature on foreign students is reviewed for cross-cultural generalizations and hypotheses. Guidelines are provided for improving exchange programs of international education that often produce the side effects of uprooting young students who undergo prolonged training and study abroad. Recommendations are made for improving the procedures and techniques of selection, orientation, cross-cultural counselling, and follow-up support on reentry in the home country.

Part V, *Stressful Situations of New Settlers: Coping Strategies of Immigrant Women and New Ethnic Minorities*. This section deals with the tasks of adaptation of new settlers in North American society and the process of coping and acculturation within a pluralistic nation. The articles selected emphasize the role of cultural predispositions that shape the diverse coping patterns of new settlers, especially women and families from new ethnic minorities that have come to work in North America during the last few decades.

Part VI, *Stressful Situations of Uprooted Populations: The Role of Public and Government Bodies*. This section deals with the catastrophic impact of the uprooting of population groups forced to migrate due to war emergency, political or ideological exile, or the pressure of technological and economic enterprises that are congruent with rapid modernization.

Although the data and hypotheses presented in the articles are derived from population groups and environmental settings in the United States, they should stimulate cross-cultural testing and validation. Interdisciplinary research is needed on the macro- and micro-aspects of these stressful situations associated with uprooting. Policy analysis is also needed regarding the role of preventative measures, at individual and institutional levels, that can ameliorate the health hazards that affect the most vulnerable groups and populations uprooted in different ways.

The authors were selected with a view to including a diversity of ethnic and linguistic backgrounds, namely, Armenian, Chinese, German, Hebrew, Hindi, Indonesian, Urdu, Yiddish, and several modern Romance languages. All have experienced uprooting in the course of their personal growth and professional development. The editors are grateful to the authors, who carried out numerous revisions with diligence and grace. A special debt of gratitude is due to Ying-Ying Yuan, who joined the editors in the last phase of the project and brought a keen anthropological sensitivity to the cross-cultural examination of major themes and issues presented in the volume. Dr. Yuan shared with the editors the responsibility of preparing the introduction to each section and planning the final organization of the substantive contents.

Acknowledgments

Our thanks are also due to several individuals who helped at different times to prepare various portions of the manuscript for publication, especially Patti Jackson, Jane Montgomery, Cathi Bruder, Rani Coelho, and Kathy Fitzpatrick.

The editors wish to express their deep gratitude to their families for providing the supportive environment and inspiration without which this work would not have been possible, especially to Nancy, Rochelle, Irene, and Paul Ahmed Jr., and to Rani, Victor, Susie, and George Coelho, Jr.

George V. Coelho
Paul I. Ahmed

Contents

Part IV: Stressful Situations of Foreign Students: Challenges of Cross-Cultural Education

Part V: Stressful Situations of New Settlers: Coping Strategies of Immigrant Women and New Ethnic Groups

Chapter 17

Chapter 18

Chapter 19

I

Coping with the Inner and Outer Worlds of Change

Introduction

The authors consider several aspects of social development within the larger context of rapid social change. The articles focus on the cognitive, emotional, and psychosocial issues of uprooting of family groups and populations exposed to the impacts of modernization.

The opening chapters emphasize the serial drama of human adaptive behavior: the uprooting of groups driven by expectations and environmental demands. Unique to our contemporary phase of evolution, however, new patterns of coping are emerging with great intensity and urgency. Adaptive behavior developed under conditions that facilitated group-care of the young in small communities. These conditions of primary-group socialization have been drastically changed because of accelerated global urbanization. This section exposes the increasing difficulty of developing merely individual coping mechanisms in conditions of rapid change, where the social systems that facilitate the learning and use of these mechanisms are also in flux. New collaborative coping strategies, therefore, must be learned for groups to meet unprecedented environmental challenges.

In Chapter I, Coelho, Yuan, and Ahmed examine uprooting as a behavioral and institutional phenomenon. Although uprooting has always been one human way of coping with change in terms of symbolmaking and of physical movement, the conditions and character of uprooting in our contemporary world have greatly changed. Changes because of uprooting occur today often without warning, and with greater frequency, scope, and scale than in the past. The changes are accumulating and accelerating at an increasingly rapid rate. Although physical movement across political boundaries speeds up technological diffusion, basic human needs of mutual aid, self-esteem, and sense of community persist, but are increasingly fragmented, and fail to be fulfilled. Groups adapt by devising culturally patterned ways of problem solving in the face of crises and emergencies. Although modern means of transportation and communication expand a popular consciousness of change, much present day cross-cultural mobility occurs without the proximate support networks of family, friends, or community members. In addition, as competition increases for scarce energy sources, arable land, food, and a minimum standard of living, uprooted people are challenged to adapt more rapidly than they can learn to change.

The authors identify stress situations that affect vulnerable populations, but argue that uprooting also provides an opportunity for growth and increased competence in coping with new situations. It is argued that individual coping mechanisms developed by persons experiencing the stresses of uprooting require cultural reinforcement and institutional support in order to be effective. Practical guidelines and recommendations are provided for research and action programs that will facilitate coping behavior under the conditions of uprooting.

In Chapter 2, Coelho and Stein review the available data on trends of urbanization by the year 2,000, and hypothesize that integration of communities that was possible at the primary-group level becomes more problematic in times of rapid environmental and social change. The authors emphasize that the conjunction of overcrowding and uprooting poses major emotional and health hazards to populations. An assessment of the quality of life in three urban centers is presented, to suggest strategies for adapting to the changes due to uprooting. In addition to being a physical relocation, uprooting is experienced as a subjective reality. The continuity of shared traditions, social relations, and valued interpersonal linkages, for example, can overcome the psychological distress associated with noise and overcrowding in urban areas. The disruption of such social supports and meaningful attachments is experienced as a traumatic loss. Several hypotheses are presented to indicate the role of specific factors that increase the stresses of uprooting. Especially stressful are dislocations which occur during important developmental transitions in the life course, or during high-risk situations associated with the loss of specific attachments. Specific population groups that are particularly vulnerable to stress under the circumstances of rapid change include young children, adolescents, the disabled, and the aged, as well as the marginally affiliated, who are most in need of institutional support and social services.

In Chapter 3, Brown examines the relationship between individual coping behavior and institutional means by which individuals can determine their own future when large-scale events threaten their forms of life. Brown first discusses what might be called "the politics of identity"—the constraints and resources that persons discover in their efforts to create a viable self. Brown then analyzes the planning process from this point of view, and argues that the role of "citizen" requires that members of a society collaborate "in constructing meanings that strengthen both their personal identities and their collective self-concept as a political community." The coping citizen is one who acts publicly as a moral agent in affirming personal and collective choices, rights, and duties. Brown maintains that, just as the individual must no longer remain passive, so corporate and governmental actions can no longer be considered morally neutral. He compares the elitist approach to social planning with a self-directed, humanistic approach. In the humanistic planning process the planner becomes a propman rather than a scriptwriter, a catalytic presence that helps citizens reconceptualize their roles and enact their performances. Planning in this collaborative mode becomes a process of mutual learning and collective development, rather than an exercise in technocratic domination.

In Chapter 4, Fried approaches the issue of the relationship between changes that induce stress and the resulting strain. He defines the stress–strain ratio as indicating the level of strain generated by the same number of units of stress. After considering means of measuring stress, he analyzes the involvement of the individual in the condition of social change by considering the definition of roles, role activities, role functions, role relationships, and role arrays. Psychological adaptation is thus defined as "the process of modification of role behavior in response to changes in psychological or physiological functioning, or to changes in sociocultural, politicoeconomic, or environmental processes." For example, the degree of role investment by an individual in a residential area influences the degree of strain inflicted by residential relocation. Resistance to role alteration in a changed environment can also result in discordance between the individual and the social system. After analyzing the constraints on role adaptation in relation to macrolevel societal stresses, he concludes that individual achievements cannot solve the adaptational problems of a mass society.

1

Contemporary Uprootings and Collaborative Coping: Behavioral and Societal Responses

George V. Coelho, Ying-Ying T. Yuan, and Paul I. Ahmed

Introduction

Rapidly changing societies are resulting in increasingly stressful situations for their populations, and are accentuating the inevitably stressful transitions which take place during the life span of each person. Changes which occur on a societal scale, or on a global scale, disrupt the interpersonal bonds and the institutional supports which provide the foundations for the development of individual participation in the community. The theme of our argument is that the challenge of modern adaptation lies not only in the individual learning of coping skills, but also in the functioning of societal arrangements for preserving and recreating institutions which ameliorate the management of life tasks.

The goal of this article is an heuristic one. We seek to stimulate behavioral research, and planning and policy analysis. The mental-

The views expressed in this chapter are the authors' own, and do not necessarily represent the position of the institutions with which they are associated.

George V. Coelho • Office of the Assistant Director for Children and Youth, National Institute of Mental Health, Rockville, Maryland 20857. **Ying-Ying T. Yuan** • Laurence Johnson & Associates, Inc., Washington, D.C. **Paul I. Ahmed** • Office of International Health, Department of Health and Human Services, Rockville, Maryland 20857.

health and sociocultural sciences have identified several classes of behavior indicative of stressful situations which are associated with uprootings. In addition, practitioners and public-policy officials, who are concerned with the provision of human services, are beginning to recognize that there are behavioral parameters to coping with rapid and intense changes. In this paper we attempt to integrate the studies of individual and group behavior under stress caused by uprooting, and the policy objectives of those who plan and provide health and social services.

The Precursors of Uprooting

Throughout the long history of human development, peoples around the world have relocated in order to survive. During the thousands of years of hunting and gathering, human groups learned to explore more hospitable environments and/or to use more varied resources in order to build cultural meanings and social definitions. The last 10,000 years of agriculture brought a fundamental transformation in the social-technical way of life. The social organization of the extended family and clan in a community organization provided an environment conducive to the learning of coping behavior. These major population uprootings have occurred using a diversity of strategies for physical and cultural survival. These strategies have become part of the coping repertoire of *Homo sapiens*. During the past 200 years, however, the human dominance over the natural ecological system has rapidly and radically transformed relationships between *Homo sapiens* and the environment. As increasing numbers of people uproot themselves in time and space, they transform their environment on a scale that exceeds their own expectations and their control based on past learnings. Coping skills are no longer merely learned and applied in ongoing social environments. Social environments change and grow with unprecedented rapidity and scale.

Population Growth and Urban Expansion

In the forefront of major contemporary concerns is the rapid increase in population. Population growth has been the major precursor of one form of uprooting: the movement from rural to urban settings. In 1930 the world's population was estimated to be two billion. Within the next 25 years the world's population, now nearly four billion, will double. The International Year of the Child Report indicates that 90% of that increase will occur in developing countries, and that children under 15 will comprise over a third of that increase. Such population growth will intensify rapid urbanization, as many millions of the rural poor, illiterate, and unemployed migrate into the cities of these countries. India alone, for example, adds over 11 million people each year to its population of 600 million people. The population of Brazil has also grown at a phenomenal rate. In 1950, Brazil had a population of 52 million; in 1980, it will have a population of over 120 million, and in twenty years it is expected to double. About one-tenth of its population will reside in Rio de Janeiro. In addition to the inability to maintain development at a reciprocal rate, the rapid population growth poses a great threat to the mental and emotional health of the people.

Speed and Scale of Technological Change

In the past, the impact of uprooting was limited to specific populations, localities, and periods. In the contemporary world, however, the uprooting of populations from their habitats is occurring with unprecedented speed and abruptness in both the environmental and the social arenas. These changes are accelerated and amplified by the diffusion of technology to the point that the impact of change is massive in scale and global in scope (Platt, 1972).

A student from Melanesia can be transported to Cambridge, Massachusetts, a distance of some 20,000 miles, within 24 hours. As supersonic jet travel and mass transportation become widely available, families can be shipped from one continent to another within a few hours. The logistic preparation for such moves is often easily arranged, but the physiological and psychological adaptation to the move may be of greater complexity and difficulty. When the uprooting involves moving at short notice into a vastly different culture, the disruption of significant social supports tends to increase. Uprooting under such circumstances can often result in an inability to commit oneself to the social expectations and rituals of members of the host community. The uprooted individual invariably becomes an observer and an outsider, rather than a participating agent capable of shaping the environment to meet his or her needs.

Sources of Uprooting Stress and Tasks of Adaptation

It is important to recognize that technological changes often offer the hope of meeting adaptive requirements more effectively. However, social systems must also change in order to utilize these technological opportunities. In turn, this process provokes changes in socialization practices as individuals are prepared for effective roles. A very rapid rate of change poses a distinctive problem for the culture and its children, because socialization practices must be modified in order to meet the emerging conditions. These culturally valued patterns of behavior are difficult to change, because sense of identity, self-esteem, and interpersonal relationships may be dependent upon them. Nevertheless, the demands for behavioral changes are intense when environmental changes are occurring (Hamburg, Coelho, & Adams, 1974, p. 410).

Uprootings are stressful because dislocation results in an increased difficulty in communicating, both verbally and nonverbally. Local school systems are frequently unprepared to assist a newcomer who speaks a foreign language. Other social institutions also lack bilingual personnel. Thus, immigrants become functionally illiterate in the new country, and face additional problems of loss of self-confidence and economic self-sufficiency. In Israel, for example, the immigrants who had attained a prior competence in Hebrew were able to integrate with greater ease into Israeli society (Ahmed, Tims, & Kolker, Chapter 22, this volume).

Another source of uprooting stress is the loss of sensory contact with a familiar physical environment. The stimulus from aesthetic and physical associations is important to most individuals. Time and space perceptions also vary from one society to another (Hall, 1959, 1966). Uprooted persons may suffer cultural shock when they are abruptly deprived of familiar environmental associations that are within their learned cultural expectations. However, for many population groups, the change in the ecological environment is of intensity rather than of kind. Noise and overcrowding in urban settlements produce an increasing variety of environmental stresses to both new and old urban dwellers. (Coelho & Stein, Chapter 2, this volume).

Uprooting also requires the learning of new behavioral patterns. Uprooting disrupts the familiar social networks which provide mutual exchanges, emotional support, and self-identity. Individuals who are uprooted often experience a loss of continuity in the performance of their functional roles. The psychological and technical coping skills that are functional in one setting may hinder adaptation in another. For example, the Chemahawin Indians could no longer continue to fish as their main subsistence activity when they were transported to the mainland of Canada (Trimble, Chapter 20, this volume). Foreign students may be assigned multiple roles as foreign student, ambassador, cultural representative, and the like when they move to different academic settings (Pedersen, Chapter 14, this volume).

Arenas of stress are further exacerbated by the rate of adaptation.

The process of adaptation has been compared to the process of primary socialization from birth to adolescence. However, in most cases the uprooted person cannot be compared to an infant. The uprooted suffer from a loss of their sustaining cultural roots. They are required to relearn or modify existing competencies, and to mesh several cultural traditions. Indeed, if we consider that a person has not been completely resocialized until complete assimilation or integration occurs, socialization may continue over several generations. Within a short period of time, cultural adaptation may occur. Cultural adaptation which is the primary task of the uprooted involves adopting everyday social rituals and external conventions which allow for social and economic participation within the society (Gordon, 1978). Structural or complete assimilation necessitates the acceptance into primary and secondary groups. This degree of adaptation may continue at a slower pace because of the individual's inability to cope, or the inability or unwillingness of the society to facilitate coping. Nevertheless, the rate of social adaptation may insure a continuity of being and meaning which is necessary for individual self-development.

Changing Social Interaction Patterns: Independence and Dependence

A related source of stress is the loss of opportunities for intimate and informal interactions that are based on close friendships. These interpersonal relations involve the qualities of trust, spontaneity, and native humor that provide a sense of belonging to the local community. The loss of such relationships may have the impact of bereavement (Marris, Chapter 5, this volume). Under such circumstances, it is difficult to compensate for the loss of the mutual care associated with the close bonds of friendship.

Even under stable conditions, children are reluctant to transfer their attachment from one caretaker to another. Once their bonds of attachment and support are developed, children retain their loyalty in the face of separation and substitutions (Bretherton, Chapter 9 this volume). Similarly, when families of officials on overseas assignments relocate in societies which are viewed as full of hardships and hazards, adolescents tend to have increased conflicts of dependency with their parents, because the latter are themselves insecure in the new environment, and so try to restrict the striving of their children for independence (Goldberg, Chapter 10, this volume).

The problems of renegotiating relationships of dependence and independence are not limited to crises within the family. The basic human needs that individuals and groups have for both dependence and independence are not easily balanced under the unstable situation of transition (Dubos, 1968). Migrants from traditional, hierarchical societies tend to depend on government agencies when they have no access to familial support systems. Emergency assistance is often necessary for the alleviation of the immediate and painful sense of dislocation, and the practical needs of uprooted groups. Nevertheless, bureaucratic agencies may foster long-term dependency on the part of those seeking help, which increases the burden to the recipient as well as to the giver of aid. Under such conditions, both government bureaucracy and privately owned agencies need to establish new social structures that provide health and economic supports without obstructing the coping initiatives of the uprooted (Ahmed, Tims, & Kolker, Chapter 22, this volume; Chapter 21, this volume; Chapter 23, this volume).

Tasks of Adaptation Facing the Uprooted

The crucial task of adaptation for the uprooted individual is to be able to maintain self-image and competency (Back, Chapter 6, this volume). Although a major motivation of rural people migrating to cities is the prospect of improved economic conditions, their hopes are often far from realized. In some cases, the change means increased material gain. Yet these monetary gains are only in the lower sector of the economy. Thus, gains in economic means may be accompanied by a decreased sense of worth and of belonging.

Foreign students who have completed their studies abroad are doubly uprooted because the skills that they have learned in Western technological societies are not easily transferred to less capital-intensive economies. New social inventions that support the adaptation of those skills to labor-intensive and illiterate rural societies are needed (Pedersen, Chapter 14, this volume). For the uprooted person, the task of adaptation includes preserving attained coping skills, and learning new skills. This double task creates a dilemma for the uprooted, especially for the mature adult. It is extremely stressful to have to learn new rules and rituals, where achieved competences are unrecognized. The uprooted person is challenged to seek new positive role models, and to incorporate new value orientations into previously learned patterns of behavior. Uprooted people face the task of deciphering new social codes of interaction which are partly linguistic and symbolic and which affect the expectations and behavioral responses of the people concerned.

Many migrant groups experience psychological disorders when the sociocultural conditions of uprooting do not permit their coping with change. Developmental and personality variables may further aggravate the emotional burdens of uprooting. If uprooting occurs during a major developmental transition, stressful life-change events produce a high-risk situation for young people. Mobile American adolescents who have lived abroad report that they remember the tension of separation for several years after they return home (Goldberg, Chapter 10, this volume). Participation levels of returning American adolescents decrease initially, since so many changes have occurred during the period that they were out of the country. They tend to remain observers on the periphery, rather than participating in the social rituals of their school and peer groups (Werkman, Chapter 11, this volume).

Personal goals, motivations, and attitudes are, of course, crucial in defining whether tasks of adaptation are perceived positively or negatively. Individuals differ in their abilities to appraise situations and meanings of the stressful experience. Stressful situations are less threatening to people whose driving sense of purpose transcending personal risk is a paramount incentive.

Tasks of adaptation may also change over time. For example, in the temporary and most often benign form of uprooting which foreign students experience in the United States, different coping issues become crucial at different stages, from the time of departure to the reentry into the home culture. (Klineberg, Chapter 13, this volume). Student and community perceptions may change over time, depending on historical, political, and situational factors. These perceptions will then influence the selection and the use of particular coping strategies. Furthermore, social policy and public attitudes influence how host communities prepare for and respond to foreign students in their midst.

Behavioral Strategies in Managing Essential Tasks of Adaptation

There are certain basic minimum requirements that must be met for effective social adaptation of a given population that is exposed to the stresses of uprooting. These requirements apply to a wide variety of situations, including the relatively benign planned moves of foreign students or American adolescents abroad, or the life-threatening situations of refugees dislocated by war or natural disaster. We discuss three behavioral strategies that are commonly used both in catastrophic circumstances of uprooting and in less stressful dislocations. These coping strategies are: (1) acquiring and using information; (2) participating in a valued support system; and (3) anticipating and planning for contingencies.

Acquiring and Using Information

In uprooting situations, acquiring relevant information is in itself a coping challenge. First, one needs to know where to go to gain information. Second, to gain access to this code involves selecting messages from multiple stimuli, coding them for storage and communication with others, and retrieving them as the new situation demands new skills for coping with change. Before relocation, uprooted peoples need to acquire behavioral information that enables them to operate on a day-to-day basis.

Foreign students, migrant workers, dependents of families posted overseas, and others, need to be assisted in locating information sources. These information sources may involve instrumental aspects, such as the location of various facilities, housing and employment opportunities, and recreational and community resources. They may also include historical and cultural aspects relevant to the social mores of the people. In addition, there is need for behavioral information, that is, how the newcomer responds to cues of dress, hospitality, social idiom, and, more generally, how one learns the expressive grammar of exchange that makes the person feel at home.

Acquiring information is, however, a two-way process. It involves mutual facilitation of communication between the newcomer and the native resident. Acquiring relevant information is not sufficient to enable an individual to mitigate all the stresses of uprooting. There must be opportunities for utilizing the information in culturally supportive ways. Participatory planning on the part of the host community is recommended as a way of enlisting the newcomer in the decision-making process that affects his acceptance of the move (Brown, Chapter 3, this volume).

Participation in Social Support Networks

In most situations of contemporary migration, individuals or small family groups have to depend on their own resources and coping skills. Adaptation to new situations is greatly facilitated when the new settlers have access to networks of social support. These networks are usually face-to-face primary relationships that provide emotional reassurance and social support to the individual. Uprooting means losing access to the network of family and friendship supports that help to sustain the meaning of a person's social identity as parent or child, spouse or friend. Losing connections with this network means being deprived of the opportunity to share symbolic meanings and social relations.

Losing friends when one moves may, of course, open up opportunities for gaining new friends and exploring new relationships with congenial groups. In the initial stages of foreign-student adaptation, it has been shown that voluntary association with compatriots may provide an important buffer against culture shock, and in some cases provide assistance in solving everyday practical problems of shelter, food, clothing, etc. The primary function of these associations of compatriots, however, is to reinforce support for the individual's cultural heritage in an accepting atmosphere. But sharing recreational and religious cultural settings (though temporarily useful) is not enough for functional adaptation. Families living overseas in a diplomatic enclave or military compound may be shielded from the chronic stresses of culture shock, surrounded as they are by a familiar language and social system with its own rituals and rules. On the other hand, they may also be vulnerable to feelings of boredom, if they are not exposed to environmental diversity as well as to novel situations. Learning from culture shock can promote growth in our world of change.

Thus, the transfer of coping skills from one stressful situation to another setting can aid in facilitating the development of competence. In addition to learning a new language or new techniques for moving and settling, effective coping includes the learning to be flexible and open to change. Young people who have coped effectively in a new culture may strengthen not only their own confidence in new situations, but also that of their parents and those who are responsible for successfully relocating people. Thus there is a continuous feedback, in which the individuals develop new coping resources for managing the situation that, in an earlier developmental phase, may have been experienced as stressful.

Anticipation, Rehearsal, and Contingency Planning

Compatriots can also assist newcomers in anticipating certain common adjustment problems and providing living role models for functioning competently in a new culture. Having been "in the same boat," old timers can provide practical means for the rehearsal of coping strategies. Just as, in serious illness, constructive "work of worry" is important as a coping mechanism for the individual patient (Janiś, 1974), so vigilant appraisal of the threatening situation, and constructive anxiety, can be generally adaptive.

Lack of information can inhibit the planning process at several stages. On the one hand, rural residents who are forced to relocate and make way for industrial power enterprises are hesitant to react to relocation proposals, since they lack the detailed information necessary to assess the company plans for moving them (Thomson, Chapter 23, this volume). Local government leaders, on the other hand, may lack the technical and cultural expertise necessary for spelling out a comprehensive plan for community development. The uprooted, therefore, face a twofold dilemma. They cannot locate the levers of change, and they lack the power to determine the directions of change, since decision-making authority is vested in other sectors of the society. Those who control the economic sources of the communication system may grant some opportunity for involvement to uprooted groups, but at the same time they may limit the extent of that involvement. Thus, many coping strategies may be most effective in certain areas, as for example, in intrafamilial relationships, but less successful in economic sectors (Pearlin & Schooler, 1978). Coping strategies may not be directly transferred unless the settings are interconnected through community institutions that have broad cultural validity and authority.

For uprooted persons or population groups, contingency planning or rehearsal of options is likely to be most effective under the following conditions:

1. When there is access to information sources that are relevant to their daily tasks of adaptation in terms of culturally shaped expectations.

2. When individuals can participate in viable social organizations that can plan changes.

3. When the cultural matrix enables individuals to move across social settings and to experience an interconnected structure of meanings, codes, and rituals that are associated with cognitive and communicative behavior.

Collaborative Coping: The Societal Response

When uprooting is too rapid or arbitrary, individuals suffer psychological and health hazards. Coping behavior is a function not only of learned personal strategies, but also of the social situation of the uprooted. The effectiveness of coping strategies depends not only on intrapsychic mechanisms, but also on the institutional supports, social provision, and restoration of practical means and ways of problem solving in transition situations. Under such conditions, uprooting may be the stimulus for improving the individual's quality of life as well as the economic status and social amenities of the group. Uprooting may enable individuals and groups to seek refuge from political oppression or religious persecution.

Immigration and refugee-resettlement policies which reflect national priorities and cultural prejudices affect the adaptation process. Even in countries which welcome immigrants, immigration policies vary with the historical period and the origin of immigrants (Ahmed, Tims, & Kolker, Chapter 22, this volume; Kolker & Ahmed, Chapter 21, this volume). For example, in the United States, Cuban refugees were allowed the option of resettling in geographically congenial surroundings. Cubans who settled in Florida recreated a milieu which was climatically and culturally familiar. Vietnamese refugees, on the other hand, were dispersed throughout the country without reference to the environmental and social impact of such dispersal. Although there is some indication that the Vietnamese are also gradually resettling into small ethnic communities, the initial shock of adjustment has been severe, and radically different from that of other refugees (Liu, Lamanna, & Murata, 1979).

In summary, uprooting produces stressful situations that affect the emotional and social well-being of individuals, especially during life-course transitions. Nevertheless, such stressful transition situations may also stimulate, in time, effective coping behaviors. The challenge is to manage tension and anxiety within tolerable limits, to maintain a positive self-image, to preserve a continuity of meaning and relationships with significant others. Above all, coping means exercising the creativity of individuals and groups in a social system, maintaining a coherent view of the future, and developing the ability or power to change one's situation for oneself and others. In the last analysis, coping is a collaborative enterprise of individual personalities and social support systems within a community. Government agencies and other organizations that allocate resources and regulate access to them may help or hinder the coping efforts of individuals or groups. The key issue is how community institutions influence the mobilization of "convoys of social support" (Kahn, 1979). Therefore, research on properties of networks, their size, stability, homogeneity, symmetry, connectedness, and interrelations, is likely to yield important insights into the composition and functions of convoys of social support. Policy-oriented analyses of the effects of various interventions may lead to more enlightened behavior, official and familial, and thus prevent the unnecessary disruptions of these networks by governmental actions or certain social policies that interfere with natural support systems.

Conclusion

Our essay has selected a range of stressful situations examined in the present volume in order to highlight the adaptive aspects of human behavior as manifested by people who are exposed to the crises of rapid environmental change. We recognize that the pragmatic consequences of major and minor situations of uprooting vary. There are different consequences for different populations, depending on such factors as access to institutional and familial sources of emotional and social support, opportunities for information seeking, lead time for planning for contingencies, and the host community's receptivity and readiness to cooperate with public agencies in assisting newcomers. How vulnerable people behave in different kinds of uprootings depends on their status and situation, which may be different for groups such as students, corporation executives, migrant workers often separated from their families, frequently mobile service families with their dependents, refugees exposed to catastrophic dislocations and deprivations, and victims of natural disaster or war. The coping perspective is useful in providing a framework for identifying the common tasks of adaptation as well as the diverse strategies employed by individuals and groups in stressful situations of change. The coping perspective permits social scientists to assess the differential impact of various stressful situations of uprooting on different population groups, and to test hypotheses about both positive and negative consequences of such situations on health- and mental-health-related behaviors of the people concerned. The coping perspective also emphasizes the developmental process in sociocultural and behavioral adaptation.

Furthermore, the concept of collaborative coping that is proposed in this essay will serve to highlight the crucial link and interaction between personal and communal coping systems of action. It also provides conceptual unity to the group efforts of practitioners and policymakers in the human-service fields. By recognizing and strengthening the individual and group coping initiatives of people in crises of change, the mental-health risks to which uprooted people are exposed can be alleviated.

Recommendations

As people are uprooted from one culture to another with increasing rapidity and frequency, mental-health problems of the uprooted will be of increased concern not only to the individual and families who expect to move, but also to government agencies and various helping professions, including public-health and social workers. Concerned public, voluntary, or governmental groups that are involved in resettlement planning work must assess their cultural and technical resources more carefully as they seek to help not only the uprooted, but also the communities that receive them.

Multidisciplinary studies are needed in order to identify those population groups which are threatened with relocation or exposed to rapid and frequent mobility. Especial attention needs to be paid to the vulnerabilities of children, adolescents, and the aged who are uprooted as refugees from war-torn or disaster-stricken areas, they being those most likely to be exposed to health and mental-health risks and most in need of social support and developmental assistance in their resettlement. In addition, the stressful impacts of dislocation and relocation under conditions of sudden environmental change must be brought to the attention of urban planners and developers, and personnel officials of international or multinational organizations concerned with students, families, and service officials. Special training in cross-cultural knowledge and counseling skills is recommended strongly for international resettlement officials and public-health workers, in order to increase their efficiency and effectiveness in providing developmental assistance to uprooted people, and in preparing host communities for their cooperative contribution to the humanitarian process.

Acknowledgments

For comments and suggestions received on earlier drafts of this article, the authors wish to express their appreciation to Aliza Kolker, Donellen Sogn, and Richard Brown.

References

Dubos, R. Environmental determinants of human life. In D. Glass (Ed.), *Biology and behavior: Environmental influences*. New York: Rockefeller University and Russell Sage Foundation, 1968.

Gordon, M. *Human nature, class, and ethnicity*. New York: Oxford University Press, 1978.

Hall, E. T. *The silent language*. Garden City, N.Y. Doubleday, 1959.

Hall, E. T. *The hidden dimension*. Garden City, N.Y. Doubleday, 1966.

Hamburg, D. A., Coelho, G. V., & Adams, J. E. Coping and adaptation. In G. V. Coelho, D. A. Hamburg, J. E. Adams (Eds.), *Coping and adaptation*. New York: Basic Books, 1974.

Janis, I. L. Vigilance and Decision Making in Personal Crises. In G. V. Coelho, D. A. Hamburg & J. E. Adams (Eds.), *Coping and adaptation*. New York: Basic Books, 1974.

Kahn, R. L., & Antonucci, T. C. *Convoys over the life course: Attachments, roles and social support*. Ann Arbor: Institute for Social Research, University of Michigan, June 1979.

Liu, W. T., Lamanna, M., & Murata, A. *Transition to nowhere: Vietnam refugees in America*. Nashville: Special Services for Groups, Asian American Mental Health Research Center and Charter House Publishers, 1979.

McHale, M. C., McHale, J., & Steatfield, G. F. *World of children*. Population Reference Bureau, Vol. 33., No. 6, Washington, D. C.: U.S. Government Printing Office, January 1979.

Pearlin, L. I., & Schooler, C. The structure of coping. *Journal of Health and Social Behavior*, 1978, *19* 2–21.

Platt, J. The world transformation and what must be done. In G. V. Coelho & E. A. Rubinstein (Eds.), *Social change and human behavior*. National Institute of Mental Health, HSM 72–7122. Washington, D. C.: U. S. Government Printing Office, 1972.

2

Change, Vulnerability, and Coping: Stresses of Uprooting and Overcrowding

George V. Coelho and Jane J. Stein

A rural laborer migrates to the city, with hopes of bringing his family there within a few years, or perhaps returning home with enough money to buy or rent a small parcel of farm land. Conversely, an unemployed young man from a large city is hired for a civil-service job in the provinces. Both of these workers face major disruptions in their personal lives as they leave behind families, friends, an accustomed life-style, and a physical and cultural environment to which they made adaptations.

Mechanized industries—fertilizer plants, canning factories, packaging processors—move into a rural area and radically change the agricultural village into a small city. Tourism is promoted in a relatively unpopulated area, and the subsequent resort development results in local people being denied access to their own beaches. Without even uprooting themselves, the people in these communities are subjected to rapid social changes, and the quality of their lives is often adversely affected by the impact of the altered physical and cultural environment.

Blocks of old houses are razed to make way for modern apartments, office buildings, or highways, leaving thousands of families to face

George V. Coelho • Office of the Assistant Director for Children and Youth, National Institute of Mental Health, Rockville, Maryland 20857. **Jane J. Stein** • Washington, D.C. 20013.

forced relocation. Corporation executives, military personnel, international civil servants, foreign service workers, and their dependents routinely change the cities or countries in which they live and work—another form of forced relocation with its accompanying disruptive stresses, especially for the spouses and adolescent children.

The increasingly urbanized human settlements around the world get more crowded, polluted, and noisy. Commuting to and from work gets more congested and stressful as a daily event. Resources such as adequate housing and mass transportation become scarcer, making life in the cities more precarious and competitive.

The human settlements to which the people in the above scenarios are moving, or to which they are exposed, offer something attractive and positively valued: the expectation of better standards of living. Although modern urban environments try to meet many economic and social needs, they are also primary sources of dislocation and stress. There are physical realities with which to contend—pollution and related health problems, high noise levels, ugliness and blight: and there are social realities—isolation amid crowding, lack of personal contacts, poverty, and unemployment.

Each of the scenarios presents potential stressors which characterize the future development of human settlements: dislocations, separations, lack of security, loss of personal identity or place in the community, problems of adapting to new neighborhoods or cultures, competition for limited community resources. Although the new environments promise something better, quality of life is often not commensurate with standard of living.

Can We Adapt to an Urban Planet?

As a society develops a modern life-style, it tends to acquire the apparatus of industrialization and urbanization, which in turn sets into motion new patterns of social interaction. Globally speaking, most of the human race is still rooted in a rural way of life. But by the year 2,000 this will be an urban planet, with more than two-thirds of our race living and working in towns and cities. The human consequences of such rapid change occurring within so short a time are likely to increase the hazards to health and emotional well-being, as well as impair the quality of life by aggravating social stress, group conflict, and aggression.

Are we ready and well adapted to cope with the environmental challenges that are in scope, tempo, novelty, and complexity unprecedented in our long history as *Homo sapiens?* Evidence from cultural and behavioral scientists suggests that there are two major constraints on our capacity to adapt: (1) institutionalized solutions to life's problems tend to lag behind new developments in a culture; and (2) our genetic priming for social relations requires a certain depth and continuity of intimate contact and face-to-face interaction (Bruner, 1972; Platt, 1972, Washburn, 1972). It is instructive to examine the tenacity, and yet tenuousness, of these human needs.

To grasp the behavioral challenge of human adaptation to our habitat we need to take a species-wide perspective. For 99% of the time that man has existed, human societies have flourished in small populations. For nearly 10,000 years our habitats were essentially agricultural settlements. The industrial environment became widespread, physically and symbolically, only recently: it has revolutionized our lives for barely 200 years. We are faced with conflicts between the new and the old—between the requirements of a technology that is very new and of our own making, and the predispositions of a biology that is very old and built into our survival as a species (Hamburg, Coelho, & Adams, 1974).

Are we learning to cope with the stresses and vulnerabilities that accompany our unique developmental schedule? Human beings develop from an immature and dependent organism into a functioning adult according to a program of growth that is longer than that of any other mammal on earth. Such prolonged dependency makes human beings highly vulnerable, and sensitive to the attention, care, support, and respect of significant people in their social orbit.

There are, however, some adaptive advantages to such a protracted dependency: human beings develop needs not only for a society—as do other mammals—but for a culture that protects their social relations as well as their privacy (Bruner, 1972). Under favorable conditions, this helps prepare individuals to adapt to a wide range of environments, and to develop an extensive repertoire of coping skills to face the vast array of stressful situations throughout life, to identify with significant adult models who are respected and loving caretakers, and to experiment with diverse social roles in rehearsing for future commitments in work, vocation, love, and other human relations. In the stable environments usually found in traditional cultures, guidelines for social behavior are transmitted from generation to generation through example, apprenticeship, myth, song, fable, and other cultural symbolizations of valued human experience. These guidelines, transmitted by the older to the younger generations, provide advance preparation and sources of emotional support for managing life crises.

But in times of rapid social and environmental change accelerated by technological innovation and diffusion, these guidelines become poorly suited to the new conditions. Rooted in certain codes and behavior patterns which are appropriate for the older established environment, they cease to be relevant or credible in new situations. Established guidelines in social learning, if they have worked well for long periods of time, are difficult to change, since self-esteem and close interpersonal relationships are dependent on them.

When technological change comes too fast and too soon for a society, it makes stable adaptations difficult if not impossible to achieve without severe pain, emotional stress, and conflict. Yet the need for change in behavior is actually greatest when drastic environmental changes are occurring. Failure to change behavior under such circumstances could result in considerable risk of extinction—for the person, for society, and even for the species (Hamburg *et al.*, 1974).

Health Hazards and Emotional Stresses in Urban Habitats

> These things which one has been accustomed to for a long time, although worse than things which one is not accustomed to, usually give less disturbance.

So said Hippocrates nearly 25 centuries ago, reminding his contemporaries of the health risks of social change. Although some of the metropolises of Hippocrates' time had density rates even higher than the rates in many of the cities of the world today (700 per hectare in ancient Rome versus 300 per hectare in modern-day Paris), the pace and complexity of social change in the world today is far more rapid, and is unprecedented in the history of civilization. As the environment changes more and more rapidly, physical stress reactions tend to increase in intensity and duration (Cassel, 1971).

Although developmental changes are to be expected—indeed, adapting and changing are essential for creative survival—rapid situational and environmental changes severely tax the adaptive capabilities of the human being.

How do the typical urban living and working conditions in our modern everchanging habitats affect health and the emotional quality of life? Inadequate housing may or may not be responsible for bad health, because it is a secondary effect of poverty, along with poor nutrition and several other urban-related problems, But, as we shall discuss later, good housing does not necessarily mean greater life satisfaction.

Likewise, crowding may or may not be a cause of ill health. Even though cities have been increasing in size, death rates have fallen more rapidly in crowded areas than in the more sparsely populated rural areas. Part of this phenomenon may be owing to the improved medical care and sanitation in the cities, and part to the migration of younger people to the cities, leaving an older and more susceptible population behind in rural areas. Hong Kong, with a density four times that of most U.S. cities, has one of the highest levels of physical and mental health in the world (Cassel, 1971).

Noise is an ubiquitous urban stressor which has potential impacts on health. Jets, air compressors, sirens, and automobile traffic are all common sources of uncontrollable noise to which we are all exposed. However, there is much conjecture as to whether execessive noise conditions can cause physical or mental health disorders. Researchers recently concluded that in most cases noise does not affect performance—except when a person is working on a highly complex task (Glass & Singer, 1972). And even then, only unpredictable or uncontrollable noise will disrupt performance. Other research indicates that noise disturbs quality rather than quantity of work (Cohen, 1969). For example, when astronauts were exposed to extremely high noise levels from a jet engine at full thrust, they experienced difficulty in carrying out simple arithmetical operations, and tended to put down any answer, in order to end the experiment quickly.

There is medical evidence that noise can cause heart attacks in individuals with existing cardiac injury, and that continued exposure to loud noises can cause such chronic effects as hypertension or ulcers. At issue is whether the stressful effects of noise—alone or together with other urban stress factors—can eventually overwhelm man's capability for healthy adjustment, with resulting physical or mental health problems.

So it is not simply that stressful events and crowded, noisy living conditions play a role in the occurrence of illness. It is necessary to understand *what* events influence *what* illnesses, under *what* conditions, and to know the availability and viability of coping mechansims which individuals and groups can use (Mechanic, 1974a).

In most developing countries, increased physical and mental health problems are often seen in the wake of rapid population growth and urbanization. A World Health Organization committee recently reported that one-fifth of the patients attending health centers in developing countries have a significant psychiatric disorder (Organization, 1975). Although the process of development and rapid social change produces numerous problems of adjustment and adaptation, the situation is exacerbated by additional difficulties, such as crowded living conditions and unemployment. Adaptation to social change does take place, but it is predictably the newcomers to the situation who are the segment of the population at highest risk. (Cassel, 1971). Moreover, with many of the traditional social supports missing, there is an increasing incidence of suicide in urbanized areas, especially among younger people (Brooke, 1974).

Those susceptible to mental and physical disease, it can be deduced, are persons who have a loss of status, self-esteem, and sense of community. They are deprived of meaningful group membership, exposed to ambiguous and conflicting demands for which they have had no previous experience, and frustrated in achieving their goals and aspirations. More than half the population in developing countries is under 15 years of age, and it is this group which is potentially most vulnerable to health hazards, as the children and adolescents are exposed to considerable risks in their physical and emotional development.

People at Risk

Every environmental change acts as a psychosocial stressor (Levi, 1974). This means that people who are exposed to major and sudden changes in their environment are likely to experience emotional stress and a variety of related physiological and behavioral reactions. Stress in life is inevitable, and to a certain moderate degree it is both motivating and energizing in creative problem solving. Yet, at specific times of life and under specific environmental situations, individual people as well as groups become highly vulnerable to stresses. Behavioral and mental-health research shows the complexity of factors which affect the impact of environmental change on human behavior, and specifically on people in risk situations. The following hypotheses will illustrate how these behavioral factors operate, and how they pose adaptive challenges.

Hypothesis 1. In any society, people are likely to experience psychological stress if an unexpected or unusual change occurs in the immediate social or physical environment, such as uprooting because of war or natural disaster, or forced relocation because of urban renewal. As an example, moving to new living quarters requires many new adaptations for a family. There are new friends and neighbors to make, new schools for children to adjust to, and possibly great financial and job demands. When urban renewal forced hundreds of families in a major U.S. city to relocate, many of them grieved—as in mourning—the loss of their neighborhood, despite their moving to better living conditions (Fried, 1973).

Hypothesis 2. People are likely to experience even greater stress than usual when environmental change occurs at the same time as a severe life crisis, for example, the death of a loved one, divorce, separation, or accidental injury. Grieving widows, when compared with married women, have a significantly higher morbidity rate within a year after such a radical change in their lives (Clayton, Desmarais, & Winokur, 1968).

In a recent study charting life changes, stress, and health, more than half the changes resulted in poor health effects (Holmes & Masuda, 1974). Moreover, the more serious the life situation or changes, the greater the incidence of adverse health effects reported: 79% of the people who reported having major life crises (serious events such as death of a loved one, marital separation or reconciliation, as well as a significant accumulation of less stressful events) were associated with health changes, as against only 37% of those who reported having only mild life crises. So life changes appear to be relevant factors in affecting disease patterns such as time of onset and severity.

Hypothesis 3. People are likely to experience aggravated stress and health hazards if changes in their environment and personal life situations cluster at critical developmental periods, such as adolescence, first pregnancy, menopause, retirement, and other turning points in the life cycle (Holmes & Masuda, 1974). Most of the precipitating events involve some sort of interaction of people with their environment, and encompass essentially all of the changes in life situations with which people normally deal. Even a job promotion can be a stressful change, involving an increase in status and income; and pressures to perform well in the new job can lead a person to question his or her competence, or in other ways threaten self-esteem.

Many of the developmental and environmental changes that take place in moderate degrees throughout the life cycle are desirable. But severe or catastrophic changes can overload the adaptive capacities of the human body and emotions, lowering resistance and increasing vulnerability. Yet illness is not inevitable: change often has different impacts on different people.

Hypothesis 4. In situations involving high vulnerability, people in certain high-risk groups are more likely than the rest of the population to experience severe stress and health hazards. Population groups considered to be at high risk include children and adolescents, the aged and the disabled, and poor and disadvantaged minority groups (Psychosocial, 1974).

We are not passive victims of the environments in which we live. There is a variety of ways in which individuals cope with environmental and life changes—some more effectively than others. From this perspective, our account of human vulnerability in the face of high-risk situations caused by rapid environmental change would not be complete without the following final hypothesis.

Hypothesis 5. Although people experience severe stress and potential hazards to their health in times of environmental change, they are also capable of coping competently with these challenges. The effectiveness of their coping mechanisms depends on the perception of the threatening or promising elements in the situation (Janis, 1974, Lazarus, 1975), motivation and readiness to respond creatively to the environmental challenge (Mechanic, 1974b), emotional and social supports that are available and used in such crises (Adams & Lindemann, 1974), and cultural provisions for appropriate institutionalized solutions to crises (Goldschmidt, 1974).

Although national and international planners are increasingly concerned with the quality of life, they have primarily aimed their efforts at promoting better standards of living through improved housing, roads, transportation, and other utilities. But they can no longer discount the health hazards, emotional stresses, and social pathology directly or indirectly related to the impact of massive projects that disrupt people's communities. They need to pay attention to the behavioral and cultural factors in human adaptive responses to habitat, for the impacts of rapid environmental change associated with high mobility, uprooting, and crowding can seriously affect the emotional quality of life and the health of individuals and communities.

Implications for Human Settlements: Case Studies in Modernization

In the next 25 years, more than 3.2 billion people will be living in urban areas—almost exactly double the urban population of today. The bulk of this increase will take place in developing countries. As an indication of the magnitude of this growth, London, Paris, and New York are estimated to have average annual population growth rates of about one percent. Karachi and Jakarta will grow around five percent per year, Lagos at more than six percent. There are 90 cities in the developing world today with populations of over one million; by the year 2,000 there will be close to 300 such cities (Urbanization, 1972).

The European urbanization experience occurred over centuries, and population growth and diffusion of innovation were relatively slow compared to the contemporary situation. The slow-paced changes associated with the Industrial Revolution in the West resulted in the eventual emergence of related social, economic, and political institutions to which the population could gradually adapt. But today, urbanization takes place in an unprecedented short span and with little time for preparation for long-term adaptation.

Most people all over the world are curious about the attractions of urban living. They seek out the material amenities and convenient access to public services and utilities (housing, health care, transportation, sewerage, water, gas, electricity); educational and occupational opportunities for their children; and higher standards of living and diversity of life-styles that urban living usually promises and often provides.

If these human needs are dependable motivations and attitudes, it is important to integrate them into plans for the megapolitan developments that are transforming our currently rural planet into a predominantly urban one by the year 2,000.

It is a disturbing fact of contemporary life that improvements in the standard of living through better housing, income, education, health care, and other social indicators do not necessarily guarantee improvements in the quality of life (Schneider, 1975). The Urban Institute in Washington, D. C. assessed the life conditions in 13 of the largest cities in the United States, and found no specific relationship between objective indicators of quality of life (housing, health care, income) and subjective experiences (aspirations, expectations, happiness, satisfaction) (Flax, 1972). Researchers reported that those cities that were best off objectively were not the cities where people reported being the most satisfied. Similarly, the cities that were worst off objectively were not the cities where the people were least satisfied.

How large new urban centers will become is at present unanswerable, but there are indications that many cities will double, even triple their present size. Since the Urban Institute study on the quality of life focused only on the largest cities, could it be that beyond a certain city size the level of satisfaction has little to do with how many social, economic, health, and cultural amenities the city provides? If so, what does this mean for our urban planet?

Two behavioral phenomena of human settlements prevalent around the world—uprooting and crowding—will be reviewed in this section. Particular attention will be paid to the vulnerability of high-risk groups involved in these phenomena, and what the role of coping behavior, social support systems, cultural values, and institutions are in determining whether these phenomena lead to physical and emotional illness and social pathology, or to individual and group coping competence.

- If there is an accelerated increase in the already high levels of mobility within virtually every country, high-risk groups will be further exposed to the health and emotional hazards of uprooting, and a stable basis for building competent coping skills is likely to be greatly threatened.

- If crowded living conditions increase with the size and number of human settlements, the situation will result in harsher competition and aggressive conflict over the access to limited, scarce, and valued resources. How do people within the cities seek both privacy and a sense of community in their efforts to cope with their habitats?

Assuming that there will be a series of megapolitan urban networks of several million people each, what are the implications for human settlements?

Uprooting—and the Uprooted

Except for those few of us who have grown up to be adults living entirely in modern urban areas, most of the people now living in the world remember themselves, their parents, and grandparents as rural dwellers and agricultural settlers. Throughout most of human history, people in a community came to know fewer than fifty people throughout their lifetime. Their lives were spent in a relatively unchanging culture, and a stable environment bounded by a few square miles, within which, for the most part, people grew up, loved, learned to live and play, work, marry, bring up children, suffer, and die. Essentially it was a small-world society, the kind of society that has been the human way for thousands of years in all the cultures that have evolved on our planet.

Was that a happy or harmonious kind of society? The question is purely academic: we cannot, nor would we want to, go back to naturalism and scrap our technology. But we can learn some lessons from it for improving the quality of life and emotional well-being.

Assuming a perhaps somewhat idealized model, an individual's place in that small-world society was acknowledged. Moreover, interdependence with others was visibly reinforced through face-to-face transactions in daily intimate relationships, and through an intricate network of reciprocal obligations and mutual aid. In such a society the patterns of coping behavior were institutionalized: there was support and care for individuals in crises; cultural guidelines for social behavior throughout the life cycle; and ritual celebration and ceremonial sharing which legitimized public concern and social solidarity.

Was it a creative and inventive society? Not in our terms of technological innovation. But there were artists and craftsmen of skill and ingenuity, as well as competent engineers in human relations.

Without the arts of cooperation that evolved in such a society, our species would have been extinct long ago.

What is the human price for the newly developing quality of life and life-style? What are the coping challenges faced by people under stressful conditions of high mobility and uprooting—especially those moving from rural to urban areas, or to different cultures?

Practical help is needed for uprooted people in the various stressful situations they encounter by moving to an urban environment. Consider the following as prerequisites for effective participation in alien urban environments (Inkeles, 1966).

- If you are moving from a small town or rural area to a big town or city, learn to handle pertinent information of all kinds—about time, money, schedules of transportation, etc.
- Learn skills to read instructions and fill out forms.
- Learn to filter out key messages from the multitudinous bombardment of sensory stimuli, and pick out the necessary signals without being distracted.
- Learn to develop relationships with strangers of all kinds, such as how to deal with repairmen when things go wrong that need their attention. Tolerate the superficiality of the contact with them, which is usually limited to the temporary and specific transaction.
- Develop tolerance for diversity, ambiguity, and novelty, for frustration without explosive anger that alienates others, or narcissistic dependency which imposes unusual demands on others.

There often are chronic feelings of bewilderment and homesickness, with nostalgic longings for the good old days of local gossip and humor, and melancholic ruminations with old friends over familiar associations and celebrations of the past. For many there is a loss in status, with feelings of lowered self-esteem and a gnawing sense of having lost one's place and identity, and the failure to get proper attention or recognition. (The sense of being denied the respected role and status held in home countries is common even in situations where there is an improvement in standards of living and social amenities). There is also depression and anger resulting from the frustration of being dependent and from the failure to find close friends with whom to share experiences and feelings.

The vulnerability of migrants in a rootless status is dramatized in the following report of a WHO Seminar on the Health Aspects of the Urban Environment (1974):

> Problems of psychological alienation are particularly acute in the case of the rootless person who, hoping to improve his situation, comes to the

> city, where he finds himself adrift in an anonymous and egoistic society. His hopes are dashed by shanty towns, social ghettos, underemployment, isolation, lack of assistance, and nostalgia mixed with frustration. Although there is running water, the sewers are inadequate, refuse-collection services are nonexistent or sporadic; there is no public transport, or only a very infrequent service; chemists' shops, hospitals, and schools are far away; the daily journey to work is a veritable expedition. Little by little the hoped-for well-being turns into daily frustration; family life disintegrates, health is threatened.

There are many different kinds of uprooting, with different effects on different groups or even members of the same family—depending on their situation, resources, and strategies for coping. For example, only part of a family may move to a new settlement. Perhaps the family split takes place after a move is made, as new areas of competence are developed by individuals in the family; or conversely, lack of motivation or ability to adapt to the new situation causes dissatisfaction, conflict, and ultimately the breakup.

Considerable anguish and guilt may afflict the newcomer in town as older or younger members of the family are left behind, even though the new situation *promises* benefits and improvements for them. If the whole family moves, the demands and opportunities in the new situation often condition them to new urban mores: the adolescents tend to form attachments to their peers in school or in the neighborhood. With these supports, the teens pay less and less attention and deference than was expected and traditionally reserved for the elders in the family.

As the women enter work roles or seek educational opportunities—for many this is their first life experience outside the house and neighborhood—conflicts are likely to arise in relationships between the sexes and generations. New relationships have to be worked out for sharing domestic tasks and child care, and roles adapted to the new situation have to be negotiated. For example, if the work situation demands commuting or different schedules for husband and wife, the leisurely opportunities for intimate sharing of concerns and for relaxed problem solving and pleasurable conversations are drastically reduced. In times of family crises, role conflict between spouses may be aggravated as one partner regresses to earlier patterns of dependency. However, what is needed at these times is an adaptive response, and complementarity to meet the exigency.

In the case of disadvantaged minority groups who lack occupational and educational skills and resources, the stress of uprooting is often compounded because it is accompanied by downward social mobility—a circumstance which further diminishes their status and self-esteem. In such stressful situations, there is greater potentiality for family conflict, aggression, and social pathology such as an overdependence on alcohol and other drugs for temporary relief. The lack of the accustomed communal supports and problem-solving mechanisms of their own cultural group makes them a highly vulnerable population group.

In summary, tasks of adapting to uprooting which results from rapid urbanization and high mobility patterns associated with modernization, involve the learning or modification of attitudes and behavior so that individuals and groups can cope with subjective feelings of threat to their self-esteem, and can develop a positive self-image to take advantage of the opportunities of the new situation. The objective requirements of an urban life—with its demands for new skills and strategies for dealing with the challenges of the physical and social environment—also need to be handled. Failure to adapt in time may result not only in serious emotional difficulties for the individual or family, but also in interpersonal tension, family conflict, violence, and social pathology involving alcoholism and other forms of drug dependence.

Crowding as a Psychosocial Stressor

Crowding is a psychological phenomenon not necessarily directly related to the number of people. It is experienced as a feeling of stress resulting from the disparity between individual supply and demand for space (Stokols, 1972).

- Crowds at a soccer game or street festival can be fun, exciting, and bring a sense of community and participation.
- The addition of a second person in an office may give the original occupant feelings of being crowded. Work must be accommodated to the habits of the other, and the space that was once his or hers must now be shared.
- An introvert may feel crowded in a situation that an extravert finds socially enjoyable.
- What may appear to be crowded conditions to an outsider may not to the occupants of the area, especially if their relationships with each other are friendly and cooperative, and if they have had much experience working and living under conditions of limited space.

So it is possible to feel crowded in the presence of a few people, or not crowded in the presence of many, depending on the social relations involved or developed in time.

To understand crowding in situations in which people live, it is necessary to know how many people there are inside a house, outside a house, and for how long—for years, or only temporarily, as in a jammed subway. A luxury apartment in the city has relatively few people inside, whereas there are high concentrations outside; an urban ghetto has high concentrations of people both inside and outside; and it is not uncommon to find crowding in the open expanse of rural areas, with families living together in a small shack.

Cultural values and attitudes affect tolerance levels, and have great relevance to how crowded a person may feel in a dwelling. Arab families of eight or ten persons sleep in a small adobe hut consisting of one or two rooms; the typical Chinese family of five to seven persons shares a two-room apartment; and many Asian children grow up sleeping together with their parents. Yet these situations are rarely viewed as intolerable: they are culturally acceptable and time-honored adaptations to life. Moreover, there is a vast difference between crowding with strangers and crowding with people who know each other.

Noise can affect feelings of crowdedness and frustrate desires for privacy. Most apartment dwellers are under considerable constraint if they are aware that their next-door neighbor can hear them. Living in a sound-porous fishbowl can be an invasion of one's privacy, as well as a distressing situation.

How people make use of the space they have—adapting it to fit their needs—has a great impact on how livable it is (Michelson, 1970). Take the example of providing a functional privacy amidst crowds. In a study of children living in high-density apartments, higher school grades were found among youngsters whose homework was done in rooms devoted at that time to only quiet pursuits. In cases where space was arranged so that the children had to force their attention away from more social pursuits, such as watching television or listening to family conversations, in order to concentrate on work, their grades fell.

Changes in design can make living comforable in high density areas. Different solutions are developed to correspond with the many cultural variables and needs. Consider the adaptation to small spaces and very high densities achieved by the Japanese. Faced with huge masses in a country with no room in which to expand, the Japanese have traditionally made their dwellings small. Virtually every inch is open for utilization through physically undifferentiated use of interior space. Each room potentially can be used like any other. But outside of their personal space, millions of Japanese live in cities such as Tokyo that sprawl over vast areas with little coordination of streets and much dirt and abuse in public areas. The city offers little escape from crowds into open public green spaces. Tokyo, the most populated city in the world, has less than one square yard of park area per person. There are 10 square yards per person in London, and 19 square yards per person in New York. (Park areas do, however, greatly increase the potential of another urban stress—crime).

Contrary to feeling crowded, high-density living can often promote feelings of isolation and loneliness. Living on the upper floors, in apartments that are cramped for space and at a distance from outdoor spaces where people gather together, can often create a situation of social isolation.

A group of families in Germany were recently studied, and those living in high-rise apartments (in contrast to those in private dwellings) had more incidences of physical and mental complaints. (Michelson, 1970). Researchers trace some of these problems to the fact that women in the apartments were more socially isolated: they were farther away from the outdoor spaces where they usually can spontaneously conduct social conversations with their neighbors. Even their physical health can be measured as part of the mental-health syndrome: Women who are isolated may think much more of their ills than those in the mainstream of social activity.

Adding to the discomfort of urban living are complex environmental stimuli, a rapid tempo of life, high noise levels, and multiplicity of events. The day itself becomes crowded with events. There can be too many inputs with which a person has to cope, or perhaps successive inputs come so fast that input A is not yet processed when input B is presented.

City life constitutes a continuous set of encounters with overloads. To maintain psychological integrity in these situations, strategies are needed to pace and manage interaction with others in the environment.

One approach is to allocate less time to each input; for example, to conserve psychic energy by becoming acquainted with far fewer people than rural counterparts do, and by maintaining more superficial relationships with these acquaintances. Another adaptive mechanism is to disregard or select out low-priority inputs. Urbanites, for example, routinely disregard drunks on the street.

The negative side of this is that it deprives the individual of a sense of direct contact and spontaneous integration in life. Although tuning out may protect the individual, it may also estrange him or her from the social world. Seeking anonymity to reduce stress and interaction does not necessarily provide a satisfying experience. Neither does it establish a secure place in a social environment (Esser, 1972).

Based on a series of studies comparing behavior of people in metropolitan areas and in smaller cities or towns, there is evidence that there is a pattern of behavior for adapting to the stresses of urban life. There is a greater amount of noninvolvement, impersonality, and aloofness in urban life. Sensory stimuli are screened, and a blasé attitude is often developed.

As a test, an abandoned car was left in New York and in Palo Alto, California (Milgram, 1970). The hood was put up to indicate that the car was in some difficulty. The responses in the two communities were very different. In New York, passersby began to dismantle the car within seven minutes—in broad daylight—and all the movable parts were stripped from it within the first 24 hours. In contrast, the Palo Alto car was unharmed. Indeed, when rain began to fall, a passerby lowered the hood to keep the motor dry.

Attitudes of anonymity, impersonality, and mutual obligation differed considerably. In New York, passersby tend to feel remote from the owner of the car, unlikely ever to meet the other people on the street. But in Palo Alto, passersby are more likely to have some familiarity with each other, to expect that known individuals may well be involved, and to feel some empathy with those in difficulty.

The differences in behavior are not so much the differences between two sets of people; rather they reflect the responses of similar people to different situations. The city is a situation to which the majority of individuals have specific adaptive responses.

In summary, there is a wide variety of adaptations to crowded, high-density living. There is no evidence that crowding causes breakdown of mental health, or that it interferes with customary family roles and rituals. Crowding itself does not harm people—and that provides enormous hope in our urban world. But it does invoke stress situations which call for the development of a wide variety of coping strategies.

Integration into the Social Network

The adaptational issue which is most important for those living in urban areas relates not so much to crowding and uprooting as to being integrated into a social network within that environment. Meaningful contact sustained over time with significant others in the social orbit is fundamental for coping with social and environmental stress.

The following cases in cities around the world illustrate how meaningful personal relationships help mitigate some of the overpowering sense of crowding and personal deprivation.

Boston. The West End section of Boston was an old, deteriorating neighborhood with five-story walk-up apartments, tiny back yards, and winding streets. The houses there were not kept in good repair, and the city declared it fit for renewal (Fried & Gleicker, 1972). The people who lived there, however, had adapted to that environment and were quite satisfied living there. Densities were high enough so that many related families could live near each other, and because of the close living conditions people tended to be friendly and concerned about each other. Although the housing itself was never idealized, the people who lived there valued the combination of the type and siting of buildings relative to each other. They were at home in their neighborhood—an area that went far beyond their private dwelling unit. The social life was an almost uninterrupted flow between apartment and street; children played on the streets, men and women leaned out of windows, families gathered together on front stoops. The street, local bars, religious and community centers all served as points of social contact, and as ways of alleviating impersonal crowding and unwanted isolation.

What concerned many of the residents was where to live after the neighborhood was razed for urban renewal. The suburbs were pictured as cold, dreary places which could not support their way of life. How could people meet their friends frequently and spontaneously when they might have to drive to each other's houses? The numbers of families congregated on a West End block couldn't locate so close to each other in the low densities that characterize so many suburbs.

Many of the residents resolved this conflict by seeking out neighborhoods similar to West End. They dispersed quite widely throughout the city, but most often in sections of Boston known for their high densities and mixed land uses.

Hong Kong. Urban Hong Kong has probably the highest residential densities ever known in the world (Mitchell, 1971). In individual dwelling units more than a quarter of the people sleep three or more in a bed; 13% sleep four or more in a bed. In the high-density units it is not unusual to have ten or more people in a unit, with two or more unrelated families sharing it. There is often no tap water, flush toilets, or cross ventilation.

Based on interviews with thousands of residents of high-density buildings, the feature that most adversely affected people was sharing the same dwelling with nonrelatives. Sustaining social relations among nonkinspeople is more likely to produce interpersonal conflicts, no matter how crowded the conditions are.

Floor levels are also a significant factor, because it is difficult for residents in the upper stories of multistory buildings to move about and get away from their homes. Consequently residents are forced into close relationships with others in their dwelling unit. Researchers found better mental health among those living on the ground floor—perhaps because that group of residents can move out into the street for additional social and personal space.

Kuala Lumpur squatter area. Living conditions in squatter areas are generally crowded, but they are not necessarily conditions of utter squalor (McGee, 1967). Moreover, the sense of living in a city with millions is often mitigated, since life in a squatter area usually incorporates only a few square miles around home and work. Often a sense of community can be developed within these areas.

In one squatter kampang in Kuala Lumpur, nearly all the residents had come, over the years, from rural areas, in search of a better livelihood. Their present occupations varied from office boy, driver and gardener to unskilled laborer. Most of the families in the area retained close links with their home village. Many of the men had come to Kuala Lumpur as single men, and had gone back to their home villages to marry, bringing their brides back to Kuala Lumpur. Once or twice a year they returned to their home village.

One common feature of life in the kampang is fear of being evicted and having no place to live. To avoid this, the residents elect a head of their kampang, who is responsible for contacting the authorities if they are threatened with eviction. Thus, positive, shared organized activities developed within one crowded kampang, to ensure perpetuation of the present life-style, and to adapt to the problems of the urban environment.

Lagos. Social interaction takes another form in Nigeria, where families descended from the same older man traditionally congregated around the surviving patriarch. (Michelson, 1970). These groups live in large compounds with open areas in the center for meeting, cooking, and other social activities. Each family has one section of the compound for private living quarters.

In Lagos, one such time-honored compound was razed for urban renewal, and was replaced by a series of row houses with yards separating each from the others. These houses incorporated strongly ingrained western notions of privacy—a development about which many residents complained. "The condition of the houses doesn't suit me," explained one resident. "They're just self-contained houses, and I'm used to communal living." The spontaneous interaction that went on in the compound virtually ceased in the new units.

Those people who did like the new settlement tended to be Nigerians returning from work or study abroad. They had in many ways shed more traditional life-styles, and they had adapted to ways more congruent with the environment created by the new Western-style housing.

Toward Collaborative Coping in an Interdependent World

The experiences of living in existing and to-be-developed human settlements pose many physical as well as mental stresses. But the effects of these stresses can often be countered. Although neither a return to nature nor a scrapping of technology is necessary, some sort of decomplexification is.

- On the individual level, a reordering of priorities, and determination of which strategies work best for one's own life situation and personal style.
- On the group level, increasing social participation in community efforts aimed at taming technology and harnessing it to better fit human needs of dignity and self-esteem.
- And on a global level, being aware of the fragility and the interconnectedness of the human web on the planet.

Individual Strategies

Coping refers to a wide range of effective patterns in meeting stressful encounters with the environment. Coping is not so much a matter of one short-term strategy as an array of diversified strategies developed over the long term. Most men and women are not passive when faced with the stresses of life: rather they are quite active, repeatedly searching, sifting through and evaluating options that are presented. Being able to alter the stressful person–environment relationship through some direct or indirect action often mobilizes the effort to do so.

Key to coping with all stress is developing strategies to (1) relieve the stress and keep it within tolerable limits; (2) maintain a sense of personal worth; (3) maintain personal relations with others; and (4) meet the requirement of the stressful situation—adjusting to life in a new city on a new job, living with another family (Hamburg & Adams, 1967).

For most people, meaningful contact sustained over time with significant others in the social orbit is fundamental for coping with stress. A person who has someone who cares for him or her is likely to resolve tensions more adequately than one who does not. Members of a supportive family, for example, can help mitigate the pressures of life by providing emotional as well as financial support.

How well adapting to the stress of moving to a new city is handled, for example, depends in great part on support and guidance from significant others. In many of the cities of the world, the majority of migrants have contacts when they arrive. Friends and relatives—major sources of contact—offer more than companionship: they also provide information, a valuable service in cities where job- and housing-information channels are informal. Generally people seek out others with similar experiences, or with similar ethnic or national backgrounds, so that they can talk about common problems and ways of resolving them. Even without using supportive help from others, simply knowing that it is available can often increase personal strength.

Flexibility in handling stress situations is an important aspect of fostering the ability to develop and accept alternatives. Just exposing people to stimulating environments, no matter how varied, will not prepare them adequately for possible futures that change so rapidly that everything will be novel. What is needed is finding new ways of satisfying social needs, through substitution—recreation and cultural pursuits, for example—or by increasing tolerance for anticipated stresses.

There is a limit, however, to how flexible people can be. Social elasticity is seriously threatened by mobility: with people moving constantly from place to place, the behavioral challenge is to create deep personal relationships and to cope with separation. The psychosocial states cannot be manufactured at will; they have to be learned.

Coping on the Group Level

Aristotle thought that a city should have only as many people in it as can know each other by sight. In the temple towns in South India, where the great philosophic teachers flourished in the 12th century, all the town's inhabitants lived within walking distance of the temples, within earshot of the bells. Leonardo da Vinci designed an urban environment which would serve every person—complete with houses, roads, canals, terraced buildings, and shopping places.

Recreating the interpersonal and institutional structures of the small-world society to which we have been exposed for thousands of years is a key way of decomplexifying modern life, as well as effectively achieving group coping efforts. The family unit is likely to continue as the universal social organization, even though it will change in size, composition, and style to fit different societies and cultures within a society.

Decomplexifying life also involves increasing the range of reciprocal bonds between individuals and between groups. Children will have to learn to socialize not primarily for dependence or independence, but rather for interdependence. Family bonds must be harnessed in such a way as to build a larger unit—a process which requires the development of very elaborate forms of organization. It is the social group or community which must provide the necessary supports and resources to help individuals cope.

According to the Blueprint for Survival, a document drawn up by the editors of *The Ecologist*, this involves "criss-cross bonds that permit the establishment of a veritable cobweb of associations of one sort or another, all of which transcend each other in a way that each individual is linked to each other member of society in at least one and preferably more ways." With this kind of social elasticity, all parts of society are in contact with each other. Without this, there would be "no bonds, no organization—in fact, no real society" (Goldsmith, Allen, Allaby, Davoll, & Lawrence, 1972).

If human settlements grow too big and too fast, there is a serious threat to how well and how long these bonds can hold together. We need to build small and grow slow in order to optimize the value of the arts of collaborative coping. Can we do it globally?

References

Adams, J. E., & Lindemann, E. Coping with long-term disability. In G. V. Coelho, D. A. Hamburg, & J. E. Adams (Eds.), *Coping and adaptation*. New York: Basic Books, 1974.

Brooke, E. M. (Ed.). *Suicide and attempted suicide*. Public Health Papers, No. 58. Geneva: World Health Organization, 1974.

Bruner, J. S. The uses of immaturity. In G. V. Coelho & E. A. Rubinstein (Eds.). *Social change and human behavior: mental health challenges of the seventies*. Rockville, Md.: National Institute of Mental Health, 1972.

Cassel, J. Health consequences of population density and crowding. In National Academy of Sciences, *Rapid population growth: Consequences and policy implications* (Vol. II). Baltimore: The Johns Hopkins Press, 1971.

Clayton, P., Desmarais, L., & Winokur, G. A study of normal bereavement. *American Journal of Psychiatry*, 1968, *125*, 168–178.

Cohen, A. Effects of noise on psychological state. In W. Ward & J. E. Fricke (Eds.), *Noise as a public health hazard*. Washington, D. C.: American Speech and Hearing Association, ASHA Reports, 4, 1969.

Esser, A. H. A behavioral perspective on crowding. In J. F. Wohlwill & D. H. Carson (Eds.), *Environment and the social sciences: Perspectives and applications*. Washington, D. C.: American Psychological Association, 1972.

Flax, M. *A study in comparative urban indicators: Conditions in 18 large metropolitan areas*. Washington, D. C.: Urban Institute, 1972.

Fried, M. Grieving for a lost home. In L. Duhl (Ed.), *The urban condition: People and policy in the metropolis*. New York: Basic Books, 1963.

Fried, M., & Gleicher, P. Some sources of residential satisfaction in an urban slum. In J. F. Wohlwill & D. H. Carson (Eds.), *Environment and the social sciences, Perspectives and applications*. Washington D. C.: American Psychological Association, 1972.

Glass, D. C., & Singer, J. E. *Urban stress: Experiments on noise and social stressors*. New York: Academic, 1972.

Goldschmidt, W. Ethology, ecology, and ethnological realities. In G. V. Coelho, D. A. Hamburg, & J. E. Adams (Eds.), *Coping and adaptation*. New York: Basic Books, 1974.

Goldsmith, E., Allen, R., Allaby, M., Davoll, J., & Lawrence, S. *Blueprint for survival*. Boston: Houghton Mifflin, 1972.

Hamburg, D. A., & Adams, J. E. A perspective on coping behavior: Seeking and utilizing information in major transitions. *Archives of General Psychiatry*, 1967, *17*, 277–284.

Hamburg, D. A., Coelho, G. V., & Adams, J. E. Coping and adaptation: Steps toward a synthesis of biological and social perspectives. In G. V. Coelho, D. A. Hamburg, & J. E. Adams (Eds.), *Coping and adaptation*. New York Basic Books, 1974.

Health Aspects of Urban Development, The. World Health Organization report on seminar, Stuttgart, December 1973. Copenhagen: WHO Regional Office for Europe, 1974.

Holmes, T. H., & Masuda, M. Life change and illness susceptibility. In B. P. Dohrenwend (Ed.), *Stressful life events: Their nature and effect*. New York: Wiley, 1974.

Inkeles, A. Social structure and the socialization of competence. *Harvard Educational Review*, 1966, *36*, 265–283.

Janis, I. L. Vigilance and decision making in personal crises. In G. V. Coelho, D. A. Hamburg, & J. E. Adams (Eds.), *Coping and adaptation*. New York: Basic Books, 1974.

Lazarus, R. S. A cognitively oriented psychologist looks at biofeedback. *American Psychologist*, 1975, *2*, 553–561.

Levi, L. Psychosocial stress and disease: A conceptual model. In E. K. Gunderson & R. H. Rahe (Eds.), *Life, stress and illness*. Springfield, Ill: Charles C Thomas, 1974.

McGee, T. G. *The Southeast Asian city*. London: G. Bell, 1967.

Mechanic, D. Discussion of research programs on relations between stressful life events and episodes of physical illness. In B. P. Dohrenwend (Ed.) *Stressful life events: Their nature and effect*. New York: Wiley, 1974. (a)

Mechanic, D. Social structure and personal adaptation: Some neglected dimensions. In G. V. Coelho, D. A. Hamburg, & J. E. Adams (Eds.), *Coping and adaptation*. New York: Basic Books, 1974. (b)

Michelson, W. M. *Man and his urban environment*. Reading, Mass.: Addison-Wesley, 1970.

Milgram, S. The experience of living in cities. *Science*, 1970, *167*, 1461–1468.

Mitchell, R. E. Some social implications of high density housing. *American Sociological Review*, 1971, *36*, 18–29.

Organization of Mental Health Services in Developing Countries. Sixteenth Report of the WHO Expert Committee on Mental Health, WHO Technical Report (Series No. 564) Geneva: World Health Organization, 1975.

Platt, J. R. The world transformation and what must be done. In G. V. Coelho & E. A. Rubinstein (Eds.), *Social change and human behavior: Mental health challenges of the seventies*. Rockville, Md.: National Institute of Mental Health, 1972.

Psychosocial Aspects of Health and Health Care. Report of a consultation, World Health Organization, Geneva, December 1974.

Schneider, M. The quality of life in large American cities: Objective and subjective social indicators. *Social Indicators Research*, 1975, *1*, 495–509

Stokols, D. Some implications for future research. *Psychological Review*, 1972, *79*, 275–277.

Urbanization. Sector working paper, World Bank, Washington, D. C., June 1972.

Washburn, S. L. Aggressive behavior and human evolution. In G. V. Coelho & E. A. Rubinstein (Eds.), *Social change and human behavior: Mental health challenges of the seventies*. Rockville, Md.: National Institute of Mental Health, 1972.

3

Identity, Politics, Planning: On Some Uses of Knowledge in Coping with Social Change

Richard Harvey Brown

> Monday
> Cloudy today, wind in the east, think we shall have rain. . . . We? Where did I get that word? . . . I remember now—the new creature uses it.
>
> Mark Twain, (1904, p. 3)

> We felt that the right to say "we" required so much more than the simple "revolution" that was to resolve everything.
>
> Richard Zorza, (1970, p. 21)

Reprise

This essay explores relations between moral agency on the personal level and collective self-direction on the societal level. We first discuss what may be called "the politics of identity"—the constraints and resources that individuals discover in their efforts to create a viable self. Then we suggest ways in which planned social change may be humanized—i.e., made more accessible to the control of societal members. The self-conception of persons as agents, we argue, is dialectically interdependent with a societal image of self-renewing institutions designed for and through citizens' collaboration.

Richard Harvey Brown • Department of Sociology, University of Maryland, College Park, Maryland 20742.

Identity is not given, either institutionally or biologically. It is achieved through continuities in the ordering of one's conception of oneself. This evolving self-image is supported and shaped through meanings that are derived through interactions and infused with affect. To the extent that the polity is expressed through the everyday contexts of such meanings, that polity may be said ideally to embody not only the formal organization of social relations, but also the personal participation in and collective celebration of moral community. One context that is critical to such shared meanings is that of intimate interactions between persons of different social positions and generations, for it is here that an integral significance can be conferred on otherwise separate stages in the life cycle and places in the social system.

The wholeness and integrity of the person depends in great part on his possessing a coherent vision of his own essential continuity through life stages, social roles, uprootings, and bereavements. This personal integrity is logically and pragmatically bound to the wholeness and integrity of the polity. The essence of citizenship is the participation of the individual in the polity *as a whole person*. But when the self is fragmented into its various personae and roles, the citizen is replaced by the functionary, with a concomitant fragmentation of that society's notion of political obligation.

In advanced industrial societies, the codes for social relations and self-concepts are no longer transmitted through continuing relationships of accountability with real persons. Instead, these codes are translated by the communications media into impersonal informational bits. Rather than preserving the restorative potentials of stress, such technicist translations define our predicaments through a series of dislocating and isolating messages that, as such, cannot serve as foci of collective commitment. We thus have public moralism and personal activism, but little collective moral action. A humanistic method of social planning would engage citizens collaboratively in constructing meanings that strengthen both their personal identities as agents and their collective self-concept as a political community.

Identity and Culture

To cope means to act as a moral agent, to determine one's existence rather than merely suffer (or enjoy) it. Moral agency in this sense is a definition of freedom, but it is not equivalent to happiness or adjustment. It rather means the affirmation of choice, and hence the possibility of ethical accountability. This view is different from the usual notion of freedom as the ability to fulfill one's desires, which has led to an equation of freedom with indicators of "satisfaction" or with certain civic institutional arrangements. We favor happiness and formal democracy; but our focus is on coping with uprootedness, and though happiness or voting may reflect an ability to cope, moral agency is a precondition of the possibility of any coping at all.

By these assertions we place our essay squarely in the tradition of Western humanism. Marsilio Ficino drew on this heritage when he defined the person as both finite and free, as a rational soul participating in the intellect of God, but operating in an intemperate, destructible body (Ficino, 1948, p. 211). That to be human is to be both subject and object also is recognized in Pico della Mirandola's essay "On the Dignity of Man." Pico does not say that man *is* the center of the world, but only that God placed man in the center of the universe so that he may be conscious of where he stands, and so become "with freedom of choice and with honor, . . . the maker and molder" of himself (Pico, 1948, p. 225). It was from this ambivalent conception of *humanitas* that humanism was born. It has been defined as

> the conviction of the dignity of man based on both the insistence of human values (rationality and freedom) and the acceptance of human limitations (fallibility and frailty); from these two postulates result responsibility and tolerance. (Panofsky, 1955, p. 2)

This Renaissance image of man—with his powers and liabilities—was overshadowed by the atomism of Newtonian science, Protestant psychology, utilitarian individualism, and commodity consciousness. Yet in the age of Ficino, Kepler, and Machiavelli, human life was seen as an attempt to create and impose form, to master *fortuna* through *virtus* (Harré, 1978; Yates, 1966).

Whatever luggage the term "humanistic" may have acquired since the Renaissance, its core referent remains the person as a conscious and intentional actor, capable of exercizing *virtus*, the quality of courage in choosing one's conduct and standing responsible for its consquences. In the human sciences a counterpart of this concern for moral agency is a theoretical approach that conceives actors to be organizers of meaningful acts, and social events as patterned according to reasons, intentions, and imagined consequences. With moral agency for its starting point, a humanistic theory of coping would investigate the person in his social and moral settings, his construction of such settings, the role of power in the imposition and resistance to such

[1] This is not to imply that the Western humanistic tradition is necessarily superior to others, but only that it is *our* tradition, and that our attempts at transcultural generalizations must begin from there. Our examples from the Renaissance are chosen to show the early modern character of this tradition, the roots of which are of course much older.

constructions, and the ways in which cultures, societies, and political economies serve as resources and constraints.[2]

In such a theory, coping would be understood as the capacity for culture creation. This capacity is central to being human, and, as ordinary language philosophers have shown us, every person has acquired this ability by the time he has learned to speak. Indeed, to be a *person*, as opposed to an object, means to be able symbolically to construct "reality." *Human* realities are not external to human consciousness, out there waiting to be recorded. Instead, the world as human beings know it is constituted intersubjectively. The facts (*facta*) of this world are things made. They are neither subjective nor objective in the usual sense. Instead, they are construed through a process of symbolic interaction. A revision of our symbolic expressions, of our shared forms of perception and expression, is thus a revisioning of the world. To cope is to have some mastery of this process.

Thinkers from Vico to Dilthey to Mead have told us that man is the symbol-making animal. Unlike animals who merely live, we have lived experience. Our worlds are apprehended and organized through the mediation of structures of thought and sociation. To say this is to say that all knowledge, including self-knowledge, is perspectival: anything that we know is known *as* something, it is apprehended from some conceptual viewpoint and some social location. A "library," for example, becomes a different object of experience for the accountant, the scholar, and the custodian. Likewise, the rules of baseball define what will be seen as a "ball" or a "strike," much as the "rules" of psychopathology or of sociology respectively define what is to be apprehended as "schizophrenia" or "role conflict." We cannot know what reality is in any absolute or objectivist fashion; instead, all we can know is our symbolic constructions, the *symbolic* realities that are defined by our frames of vision and our paradigms for practice.[3]

If reality is accessible to us only through such mediations, it follows that mastery of reality requires a mastery of our symbol systems. Coping involves competence in symbolic construction. In this sense, coping is not merely a matter of job satisfaction or personal adjustment. Instead, it is a question of potency in creating meaning and form (Brown, 1977).

[2] There is growing evidence from developmental biology and psychology that awareness and exercise of moral agency arises through personal actions that are organized by the child himself. Instead of being stamped from without as on a *tabula rasa* and eliciting reactive responses, such actions are characterized by proactive initiatives that may begin in fantasy or play. See, for example, the essay by Inge Bretherton in this volume. This literature suggests that agency is not only a conceptual or ethical commitment, but also a basic human need.

In this view, the artist or therapist or scientist, as well as the politician or citizen who is seeking to create a new mode of public action, all are seen as having a basic affinity—they are creating paradigms through which experience acquires significance. By stressing the world-creating aspects of conceptual innovation, this perspective also provides a bridge between theoretical and political practices, as well as between what experts do and what all of us do in our everyday lives. We all create worlds. The more we are able to create worlds that are internally cogent and politically viable, the better we are able to cope. Such an assertion is consistent with our earlier statement of humanistic intent. Our perception of the symbolic nature of reality involves a recognition of the fragility of institutions that once were thought to be obdurate and concrete, and this parallels our prior recognition of the frailty of persons; yet the admission that social order is a construction also invites us actively to reconstruct our worlds.

An example may illustrate our powers—and depowerment—symbolically to construct identity and culture. Today the spokesmen for cybernetic systems theory argue that society is (or is like) a great computer, with its input and output, its feedback loops, its programs.

[3] The philosophic schools that tend to support the symbolic realist view include pragmatism, ordinary language philosophy, and existential phenomenology (and even parts of *neo*positivism, with its concensualist theory of truth). See Richard H. Brown, "Symbolic Realism and Sociological Theory," in Richard H. Brown and Stanford M. Lyman, Eds., *Structure, Consciousness, and History,* London and New York: Cambridge University Press, 1978. See also Laurence Foss, "Art as Cognitive: Beyond Scientific Realism," *Philosophy of Science,* 1971, *38,* 234–250; Nelson Goodman, "The Way the World Is," *Review of Metaphysics,* 1960, *14,* 48–65; Roman Jakobson, "On Realism in Art," in Ladislav Matejka and Krystyna Pomorska, Eds., *Readings in Russian Poetics: Formalist and Structuralist Views,* Cambridge, M.I.T. Press, 1971 esp. pp. 41–42; Richard J. Bernstein, *Praxis and Action. Contemporary Philosophies of Human Activity,* Philadelphia, University of Pennsylvania Press, 1971; Karl Popper, *The Logic of Scientific Discovery,* New York, Basic Books, 1959, p. 11; Willard Van Orman Quine, *Word and Object,* Cambridge, M.I.T. Press, 1960, p. 1; and Gilbert Ryle, *Dilemmas,* London and New York, Cambridge University Press, 1954, pp. 75–77.

[4] That thought is a practical social activity, and that everyday life is a process of world creation, is adumbrated by Marx in his *Notes to the Doctoral Dissertation* (1839–1841), *Economic and Philosophic Manuscripts* (1844), and *Theses on Feuerbach* (1845). Relevant passages are reprinted in *Writings of the Young Marx on Philosophy and Society,* edited and translated by Loyd D. Easton and Kurt H. Guddat, Garden City, N.Y., Doubleday, 1967, especially pp. 61–62, 307–308, and 400–402. See also Joachim Israel, "Alienation and Reification," in *Theories of Alienation,* R. Felix Geyer and David R. Schweitzer, eds., The Hague: Martinus Nijhoff, 1978. The term "alienation" has endured a long history and much abuse. For an account of these travails, see Richard Schacht, *Alienation,* Garden City, N.Y., Doubleday, 1970.

This machine—society—is in turn guided by a servomechanism—the technoadministrative elite. To see this imagery as a thing made, as a cultural artifact rather than as the fact, is to reject it as a literal description of how society "really is," and to unmask it as legitimatizing rhetoric. By making a close textual analysis, it becomes clear that in the language of social cybernetics there is an atrophy of the very vocabularies of citizenship, moral responsibility, and political community. In place of these, the means of governance, initially conceived as serving human values, become a closed system generating its own self-maintaining ends. The polity—the arena for the institutional enactment of moral choices—dissolves upward into the cybernetic state, or downward into the alienated "individual," whose intentionality is now wholly privatized, and whose actions, uprooted from their institutional context, are bereft of social consequence, and hence of moral meaning (Stanley, 1973).

The corruption of language and the privatization of morality also are seen in contemporary approaches to the problem of coping itself. With the idea of the citizen displaced by that of the consumer, with civic discourse reduced to opinion polls in which political judgment is transformed into a "bit" (or a wee bit) of computerized information, it is small wonder that popular discourse on coping is infected by technocratic thought.[5] We hear, not of creating a life, but of getting a life-style, not of building one's character, but of improving one's personality. Adult education, which was conceived as a movement to bring higher learning to the masses, or at least the middle classes, often becomes a place where "coping skills" are learned—that is, a place for consumerism of new personae. The adult-education branch of the New School for Social Research, for example, which once sponsored the first courses on race and on psychoanalysis, taught respectively by W. E. B. DuBois and Sandor Ferenczi, is today offering classes on "How to Educate Your Dreams to Work for You," "Body Awareness," and "Personal Growth." Hofstra University, a relative newcomer to the coping market, will teach us about "Masculinity" as

[5] Popular versions of "coping therapy" should be distinguished from more scholarly treatments of the subject. Yet the "serious" or scientific literature is itself flawed by a tendency to employ positivistic or behavioristic assumptions and models, even when speaking about moral agency. Moreover, on the ideological level, much of the literature expresses a liberal reformist bias, generally unacknowledged, that often contradicts the actual findings of research. Important efforts to overcome these difficulties can be found in George V. Coelho, David A. Hamburg, and John E. Adams, eds., *Coping and Adaptation*, New York, Basic Books, 1974. For a critique of popular approaches to coping, see Edwin Schur, *The Awareness Trip: Self Absorption Instead of Social Change*, New York, Quadrangle, 1976.

well as "Pitfalls and Possibilities: A Couples Workshop." If we are to believe the announcements in the catalogs, these courses essentially have no content. Instead, their topic is "life" and how to cope with it. But coping, as we are using the term, means the acquisition of competence in jointly shaping one's world. Instead, like vulgar forms of the human-development movement in general, what such courses provide is an "experience of participation" and a "sharing of insights." Such ersatz activities are offered to persons whose needs are very real. But instead of elevating their personal troubles to the level of political and moral issues that might be addressed through collective action, these activities privatize discourse and feeling even further.

The privatization of values has been accompanied by a public justification of corporate and governmental actions as morally neutral. The substantive rationality of values, purposes, or ends is separated from the functional rationality of means. Since there is no longer a vocabulary in which to consider the morality of collective purposes (e.g., a fair price, a just war), the State instead explains its actions in terms of efficiency. The question, "Efficient for what?" is rarely asked and often subversive. In such a context, commitment to values becomes largely a matter of taste: "I like ice cream, you like freedom." Both become emotive expressions. Having lost the valence of truth and the power to elicit public credence, discourse is reduced to exhortation. "Out of Rhodesia Now!" "See your Druggist Today!" In structure and style, it becomes difficult to distinguish political discourse from ads for Preparation H.

Excessive rationalization produces absurdity. The more that social practice is dominated by the instrumental calculation of experts and managers, the less scope is left for average people to exercise practical insight and moral judgment. Unable to participate in formulating the purposes of their institutions, or even to know what such purposes are, the experience of citizens or workers is one of alienation. The common soldier, for example, can carry out an endless series of actions that, though rationally functional to command objectives, are meaningless to him. (Having recruits dig trenches and then fill them up again serves precisely as training in the efficient performance of meaningless tasks.)[6]

[6] Although the Army must coerce and exhort the unwilling consent of recruits, Madison Avenue has been able to persuade us that we really are happy in our work. The enfeeblement of our critical capacities is illustrated by a case cited in another connection by Kurt Lewin, that of a feeble-minded child who, though failing in his efforts to throw a ball, finds a substitute happiness in the vigorous motion he has undertaken (Lewin, 1935, p. 205). The subject is referred to as a "gesture child," because he is pleased with gestures where others are happy only with results.

It is not only the means of economic production that become concentrated into fewer hands, but also the means of theoretical reflection. As peak organizations emerge, elites exercise their power more broadly, controlling complex interconnections over an ever widening field. But at the same time there are far fewer positions from which the major structural connections between different activities can be perceived, and far fewer men and women who hold such positions (Mannheim, 1940, p. 59). As Seeman (1967) puts it:

> the structural conditions of mass society (e.g. high mobility, rationalization of industrial processes, bureaucratization) encourage a sense of powerlessness which leads the individual to be insensitive to, and uninformed about, an environment over which he believes he has little influence. (p. 106)

Independence and personal autonomy, if removed from their political context, imply social isolation, lack of support from others, vulnerability, and ultimately a mindless conformism. But if we recognize the political culture as central to both automony *and* cooperation, then these ideas come to have a different relation. It is only through a humane polity that collective goals can be established in a noncoercive, nondogmatic way. Such goals have meaning and efficacy, in turn, to the extent that individuals, through personal decisions, become committed to them. By *committing* oneself to the jointly established goals, one affirms one's autonomy; and because the goals are *collective,* such a commitment brings group support. As Joachim Israel (1978) puts it, "whereas extreme individualism in a dialectical way breeds extreme conformism, the existence of collective goals in a similar dialectical way breeds independence and personal responsibility" (21–22). True emancipation of individuals is not a liberation from society, but the redemption of society from atomization.

The great transformation from traditional to industrial society has also involved a transformation of the concept and experience of ourselves as human. Modernization, urbanization, industrialization, secularization, and other such processes not only involve social-structural changes; they also are lived experiences of changes in the meaning of one's self. What results is a steady contraction away from ones history, ones ancestors, and other persons, leaving as residue a Hobbesian calculative ego seeking to optimize its "utils" in a world of pointless motion. In ancient Greek tragedies, the horizon of moral awareness was fate and the gods—their powers set the ultimate limits for human will and reason. Today fate still holds sway, but the gods have been replaced by the economy and social institutions. "As Hellenic man at times sacrificed to Aphrodite and at other times to Apollo, and, above all, as everybody sacrificed to the gods of his city, so do we still nowadays, only the bearing of man has been disenchanted and denuded" (Weber, 1946, p. 148). The gods for modern persons are not only the competing ideologies of political groups; they are also the various compartments into which modern life has been divided—the worlds of work, of family, of schooling, of friends, and so on endlessly, each with its own rituals and deities, each seemingly unrelated to the totems and rites of the others. Everyday life becomes a kind of manic sacrificing, propitiating now one god, now its rivals, with no interconnections that might give a transcendent unity to these acts. As authentic individuality loses its practical base, the person depends less on his own agency and more on the gods of market forces or bureaucratic norms.

The possibility of choosing a career appeared in the nineteenth century and, even then, entailed both the evaluation of various professions and political regimes as alternative forms of life and value, and dissatisfaction with any ultimate option. What once was available only to such heroes as Eugene de Rastignac or Julien Sorel is today available to Everyman. When Balzac and Stendhal were writing, few imagined that society could be restructured. Yet today most persons can envision alternative futures. The relativization of time, space, and status through heightened communication and geographic and social mobility has had its counterpart in a *psychic* mobility, the capacity to imagine oneself into different identities and situations. The view that "anything is possible," held in the nineteenth century by social philosophers and madmen, today almost universally seems both an article of faith and a cause for dread. Modern man has a "choice" of spiritual visions. "The paradox," as Herbert Fingarette (1963) says, "is that although each requires complete commitment for complete validity, we can today generate a context in which we see that no one of them is the sole vision" (p. 236). In this sense we are all uprooted.

This process began in England about two centuries ago, then spread to Europe and America, and is now pervading the entire world. The quotation below resonates with the experiences of Latins or Asians today, but it was written 150 years ago by Thomas Carlyle in England.

> Not the external and physical alone is now managed by machinery, but the internal and spiritual also . . . The same habit regulates not our modes of action alone, but our modes of thought and feeling. Men are grown mechanical in head and in heart, as well as in hand. . . . Mechanism has struck its roots down into man's most intimate, primary sources of conviction; and is thence sending up, over his whole life and activity, innumerable items—fruit-bearing and poison-bearing. (quoted by Williams, 1960, p. 79)

That this phenomenon is not unique to Europe or America, or to capitalist societies as such, is suggested by the following excerpt from a Soviet biography:

> The peculiar feature of this [Soviet] society—after it had been gripped in an iron vise and reduced at breakneck speed to a state of what is called here "unanimity"—was the fact that it proved to consist of individuals working for their own self-advancement either singly or in groups. . . . Such cliques are not proof of the existence of a sense of fellowship, since they consist of individualists who are out to achieve only their own aims. They refer to themselves as "we," but in this context the pronoun indicates only a plurality devoid of any deeper sense or significance and always ready to fall apart the moment a more enticing aim catches the eye.
>
> We witnessed the disintegration of a [pre-Revolutionary] society which was as imperfect as any other, but which concealed and curbed its wickedness and harbored small groups of people who were truly entitled to refer to themselves as "we." I am quite convinced that without such a "we," there can be no proper fulfillment of even the most ordinary "I," that is, of the personality. (Mandelstam, 1970)

A similar point is made in innumerable non-Western writings, of which the passage below from a Sierra Leonian novel is but one example.[7]

> [In the old days] your brother was anyone who originated from the same village or town, and to refuse to help in time of need was unthinkable. But all this was changing fast now. The sense of family interdependence, cement of our society, seemed to be going out of the country with the diamonds; and the European's exaggerated individualism, his constant exaltation of the single human being, at the polls, in the classroom, and in the sight of God was sweeping in. Social cohesion was pawned for the material well-being of the individual, and our mental hospitals were beginning to fill up as a result. Birthrights were being sold daily in our markets for a mess of pottage. I cannot describe myself as being so much disillusioned as unhinged.

The application of rational calculation not only affects the external environment; it also effects transformations of consciousness. To be made an object of technical manipulation is not merely to "feel"

[7] For similar examples drawn from cross-national studies of social change, see S. Milgram, "The Experience of Living in Cities," *Science*, 1970, *167*, p. 1461; M. Fried and P. Gleicher, "Some Sources of Residential Satisfaction in an Urban Slum," in *Environment and the Social Sciences: Perspectives and Applications*, J. F. Wohlwill and D. H. Carson, eds. American Psychological Association, Washington D. C., 1972; T. G. McGee, *The Southeast Asian City*, G. Bell and Sons, London, 1967; D. Stea and E. Soja, "Environment and Spatial Cognition in African Societies," ITEMS, *29*, p. 3, Social Science Research Council, New York, 1975; L. Yap, *Internal Migration in Less Developed Countries*, Report to the World Bank Urban Poverty Task Force, World Bank, Washington, D. C., 1975; and G. V. Coelho and Jane J. Stein, "Coping with Stresses of an Urban Planet: Impacts of Uprooting and Overcrowding," *Habitat*, 1977, *2*, pp. 1–12.

manipulated. It is to *be* manipulated. The human result is not merely a personal opinion or attitude, it is an objective existential state with its concomitant political correlates. We all may become free to do our own thing, but freedom to choose an ideology may remain freedom to choose what is always the same. Our sense of self is also made uniform. Freedom comes to mean freedom from body odour, bad teeth, and passionate emotions. The triumph of manipulation in mass industrial culture is that we come to derive our identities from its products, even though we no longer believe in them (Horkheimer & Adorno, 1973, p. 166).

We lose our ability to cope when we are unhinged from our roots, when our sentiments are manipulated, when we are transformed into objects. "The oppressed have been destroyed precisely because their situation has reduced them to things" (Freire, 1970, p. 55). Such a condition is illustrated by Dostoevski's character Mitya Karamazov (see Schachtel, 1961). On trial for the murder of his father, Mitya feels most vulnerable when the prosecutor asks him to take off his socks:

> They were very dirty . . . and now everyone could see it. All his life he had thought both his big toes hideous. He particularly loathed the coarse, flat, crooked nail on the right one and now they would all see it. Feeling intolerably ashamed . . . (Dostoevski, 1950, p. 587)

Mitya has reified his entire self into the appearance of the nail on his right big toe; it has become the focal point of his identity. He is like the woman who feels safe because her breath smells clean, or the man who can cope because he knows that Jesus loves him. In all such cases the concept of the self is reified, made into an object rather than a subject. The self is not conceived in terms of choices manifest in one's conduct, one's relations with people, one's labor, one's practice. Rather than my feeling that *I* am *doing* this or that, there is a *something* in me or about me that *makes* me this or that, that fixes my identity. The self becomes a commodity. When we say that something is "good for my ego," we do not necessarily mean that it is good for me. To have a good personality is like having a good car or a good job. Our ego or personality is seen as capable of attracting attention, even love. It is like a bond or stock endowed with self-awareness: if demand rises, it is blown up, feels important; if demand falls, it shrinks, feels it is nothing (Schachtel, 1961, p. 122). In such a conception, the self is not the agent of its own actions, it does not create its essence out of contingent existence. Instead, such a self is wholly a victim, or beneficiary, of forces beyond its control. Like Mitya Karamazov, it may be punished for having ugly feet, but not for murdering its father. Because such a self is not *doing* anything, it need feel responsible for nothing.

The denial by individuals of their own agency has a counterpart in institutional bad faith—the lies that organizations tell to their members, and themselves believe in. What on the psychological level is an illusion of constraint masking actual choice, on the social level becomes the appearance of participation covering real exclusion. In the early days of capitalism, workers were explicitly forbidden to unionize. Church and reform groups did make some efforts at "status acceptance," but there was relatively little ideology and practice aimed at cooling out subordinates and co-opting their means of protest. Today such official propagation of inauthenticity has become part of the accepted structure.[8] As power comes to be more concentrated and everyday life becomes less subject to direction by citizens, more and more investment goes into front activities by official agencies aiming to engineer consent. Public-relations firms promote soaps and politicians, the assumption being not only that stance is more important than substance, but also that differences between various brands is so small that "image differentiation" is crucial.

The pseudo-Gemeinschaft generated by such activities results in political schizophrenia. Studies of schizophrenic children provide striking parallels of surface responsiveness and underlying exclusion between the familial and the political spheres. To illustrate these parallels we first report a conversation that reveals schizophrenic family conditions (Clausen & Kohn, 1960, p. 305):

> Son: Well, when my mother sometimes makes me a big meal and I won't eat it if I don't feel like it.
> Father: But he wasn't always like that, you know. He's always been a good boy.
> Mother: That's his illness, isn't it doctor? He was always most polite and well brought up. We've done our best by him.
> S: No, I've always been selfish and ungrateful, I've no self-respect.
> F: But you have.
> S: I could have, if you respected me. No one respects me. Everyone laughs at me. I'm the joke of the world. I'm the joker all right.
> F: But, son, I respect you, because I respect a man who respects himself.

Given the father's totally unresponsive response, the son neither receives support nor can he easily blame his father for not supporting him, because the exclusion is presented in a supportive fashion. The

[8] For examples of pseudoparticipation in industry see Bendix, 1956, pp. 301–318, and Baritz, 1960; in education, Friedenberg, 1963, and Clark, 1960; in politics, Levin, 1960. Rates of mental and psychomatic illness tend to be higher for groups—American Negroes, for example—who are officially invited to participate in the American Dream but practically excluded from it (Broyard, 1950, pp. 56–64, and Seeman, 1956, pp. 142–153; and see also Sartre, 1960, and Watzlawick, *et al.*, 1967).

son is treated as a feebleminded "gesture-child" who is expected to be gratified by the motion even though nothing has been delivered.

Speaking in a similar connection, R. D. Laing (1961) notes that the characteristic schizophrenic situation

> does not so much involve a child [a citizenry] who is subject to outright neglect or even to obvious trauma, but a child [citizenry] whose authenticity has been subjected to subtle, but persistent, mutilation, quite often unwittingly. . . . No matter what meaning he gives his situation, his feelings are denuded of validity, his acts are stripped of their motives, intentions, and consequences. The situation is robbed of its meaning for him, so that he is totally mystified and alienated. (pp. 135–136)

Most research on "political schizophrenia" tries to provide a psychological explanation for alienation that is socially structured. But it is much more important to account *sociologically* for these existential states. Psychological talk, whatever its adequacy by logical and therapeutic criteria, functions on the ideological level as a mask and symptom of the very problem it discusses. That is, by "psychologizing" issues that are rooted in *social and political* conditions, such discourse atomizes consciousness even further.[9]

In his *Science Nuova* of 1725, which laid a foundation for modern human studies, Giambattista Vico gives a political interpretation of Solon's famous saying, Know Thyself. Solon meant to instruct the plebeians, who believed themselves to be of animal origin, unlike the nobles whom they thought to be descended from gods. "Solon . . . admonished the plebians to reflect upon themselves and to realize that they were of like human nature with the nobles and should therefore be made equal with them in civil rights" (Vico, 1970, pp. 92–93). Vico then adds that Solon might have been a poeticized historical creation of the Athenian plebes themselves, who used their own history as a means of liberation. Vico thus transcends the dichotomy between self-knowledge and political emancipation. Understanding for Vico is neither an accumulation of facts and definitions nor a solitary rumi-

[9] Etzioni (1968, p. 882), citing Kornhauser, Sheppard, and Mayer (1956, p. 194) notes that "large segments of the citizens of contemporary industrial societies feel powerless and excluded, and are uninformed about societal and political processes which govern their lives." In the United States only 44% feel that they can influence local and national political events (Almond & Verba, 1963, p. 226). According to Milbrath (1965, p. 21), "About one-third of the American adult population can be characterized as politically apathetic or passive; in most cases, they are unaware, literally, of the political part of the world around them. Another 60% play largely spectator roles." The figures, though quite incomplete, tend to be even more depressing for countries that are experiencing the atomization of industrial social processes but that lack historical traditions of democratic participation.

nation. It is rather the mastery of one's inner life, the consciousness of one's own historical value, the affirmation of one's political rights and duties. Only when we have trained ourselves in grasping the congruencies between thought and action, between personal values and political commitment, can we say we are truly human, for only then can we foresee the probable outcomes of our acts and so be responsible for them. The dictum Know Thyself does not invite introspective brooding. Instead it is founded on the belief in the power of ideas, and the corollary refusal to permit fear in any way to restrict our capacity to think. By understanding ourselves through critical reflection on our social and economic milieux, we can transform conditions of servitude into symbols of rebellion (Gramsci, 1975, pp. 21, 29; Horkheimer, 1974, p. 162).

Humanism and Planning

Even if we were to envision an emancipated future, the question remains as to how such a future may be achieved. On one hand, the manipulativeness of conventional means for effecting social change tends to negate our humanistic ends. On the other hand, in accepting the criticisms of these instrumental techniques we appear to consign ourselves to either mindless activism or contemplative passivity. In this context it becomes vital not only to imagine alternative social orders, but also to invent new methods for achieving them. The problem is not only the purposes of social policy, but also the means of social planning. In our times, even the advocates of human emancipation have employed the most dehumanizing techniques. Yet what methods would represent a liberating *process,* as well as yielding presumably liberated products? Though the exact nature of an emancipatory approach to social change has yet to be defined, we may speculate on some of its likely features. These features may be illuminated by contrast with conventional approaches to intentional social change: elitest planning versus mutual learning, deterministic versus humanistic design, and ultimate utopia versus excellence by stages. In addition, we also will discuss a reformulated relation between fantasy and rationality.

Elitest Planning versus Mutual Learning

In industrial society knowledge has been conceived largely as a stock that social-change agents draw upon as needed. When a situation emerges that calls for strategic interventions, ad hoc models are put together from this stock by special interest groups seeking to define options that will enhance or consolidate their positions. This is true despite large government expenditures for research, because such research is chiefly directed at identifying or evaluating substitutes, and not at producing new alternatives based on new definitions of the problem. Some additional options are generated through research sponsored by public-service organizations that ostensibly impart minimal "value loading." But in practice most research produces options that reinforce the predilections of its establishment sponsors.

If various groups are to be helped in shaping their own destinies, new sets of optional futures and new methods of generating them are needed. Hence the capacity for rapid learning on the part of social-change agents will not be enough. Unless potential client groups can be taken along on his learning trip, the expert's models are unlikely to be enacted. Rather than serving as handmaid to elites, the planner instead must become expert at helping client groups redefine their own situations. He or she must act as a catalytic personage rather than a script writer, structuring the learning experience so that he and his clients can jointly reconceptualize problems and discover new possibilities for concerted action (Friedman, 1973).

This emphasis on learning, on exploration and discovery, also throws light on the traditional dichotomy of ends and means. In the pure instrumentalism of functional rationality, the choice of ends and means divides into two stages involving an appropriate division of labor between political actors and technical analysts. The former are seen as setting the ends, and the latter are thought of as preparing cost-effective means for achieving them. Such an approach is highly unsuited to policy analysis as mutual learning. An unwillingness to consider a change in goals and objectives renders most learning experiences irrelevant, in that the distinction between the ends and means of action, though conceptually arbitrary, functions pragmatically as a means of control.

A morally reflective, citizen-controlled approach to social change would recognize that means, no less than ends, have normative implications. Mutual learning involves a reorientation of action and a reshaping of organizational goals. Given such a process, the planner, executive, or community leader ceases to be seen as a bias-free analyst of optimal courses of action. Instead, he becomes a key actor in an interorganizational process that is essentially political.

Deterministic versus Humanistic Approaches

Deterministic social design proceeds as if all factors are susceptible to advanced prediction by experts whose roles in the process are restricted to stereotyped tasks under the control of a centralized executive. Most of our familiar physical and social structures derive from this deterministic model. Such products as industrial plants, apartment buildings, legal contracts, and systems of administration usually are built according to predetermined forms and procedures. The assumption is that the desired outcome requires relatively little fundamental conceptualization by the planners, and only marginal participation by the users.[10]

A deterministic approach is appropriate where the environment is predictable, where the context is static, where agreement is well established, and where the planning authority controls all the resources; in other words, where people are to be treated as reactors instead of as proactors. Normally, however, such authoritarian control is neither feasible nor desirable. Of course, it is possible to "decree" that communities will conform to predetermined patterns. But external conformity will not by itself assure the desired results. Deterministic certainty is achieved at the expense of local initiative, openness to opportunity, and appreciation of wholeness. By seeking to reduce error to the degree that is possible in the design of roads and bridges, such a method can brutalize everyone involved.

In contrast, the value of humanistic design can be measured by the extent to which it provides more encompassing reasons for being on the scene. In such an approach, ordinary aspects of everyday life also can be part of some larger quest. For example, through one's job one can plan one's life as a career. Goods and services not only can attend to functional needs; they also may serve as a basis for communication, a source of learning, and a model of resource use. Good

[10] A related point concerning deterministic planning is made by Randall Collins and Michael Makowsky, *The Discovery of Society*, New York, Random House, 1972, pp. 200–201: "What about Mannheim's solution to the crises of democracy—planning? Mannheim was rather vague about exactly what to do, and since Mannheim few thinkers of independent stature have addressed the question. It is true that a group of social scientists, jealous of the success of economists in achieving a policy voice through the Council of Economic Advisers, have been clamoring for a Council of Social Science Advisers to offer policies on the overall state of society. Although the general theme is within the compass of Mannheim's hopes, these men have little of Mannheim's substantive insights into modern society or indeed of the knowledge accumulated by the major thinkers of the last century. Their proposals consist of little more than the old social-problems philosophy that has guided American sociology, without striking success, since its inception: Keep a survey team trained on ghetto 'hot spots,' and pour in a few more welfare dollars when the riot temperature is rising. This philosophy sounds much more like a well-known political strategy for domestic counterinsurgency than anything based on sociological knowledge. The mentality of the would-be planners at this point resembles that of the bureaucrat, who, as Mannheim said, reduces all policy questions to questions of technique and administration and blindly accepts and maintains the implicit values of the status quo."

community designs are those that allow individuals and groups to discover and value more of themselves, to recognize their abilities and take responsibility for their lives.

The humanistic approach to social change is dedicated to future opportunities as well as instructed by past lessons. It aims at enhancing people's acumen and talent for using uncertainty creatively, and turning obstacles into opportunities. Since the future is unknowable, taking risks is viewed as endemic. Rather than waiting for complete knowledge to be available to serve as a basis for action, the humanistic change agent begins by initiating a self-developmental process. Doing is viewed as an exercise of practical intelligence rather than an execution divorced from reasoning. The conceptual process of planning is integrated with practical experiments of action.

In the ideal world, programs for social or economic development are carried out with vigor and on an optimum schedule. In the real world, slippages and shortfalls abound. It is impossible to identify with precision the extent to which one or another element of the program may lag. Perhaps the completion of the steel mill will be delayed, or perhaps the staff-training effort will fall behind. Rationalistic systems and purely deductive models fail to provide guidance when shortfalls occur. Attempts to suppress unforeseen events through deterministic planning tend to increase the perception of chaos, rather than insure order.

By contrast, a key point in the humanistic approach is that "things," such as cultural centers, roads, work places, or public spaces, will help create and express the meaning of any given area. However, if such physical entities are simply transferred, without transformation, to a new place, they can become sterile appendages, anomalous presences that are unresponsive to the needs that are immanent in the particular setting. Instead, ordinary features of the community should be used to encourage a pattern of life that elicits the participation of residents in making their actions significant and productive. In this way, also, a reservoir of talent will be created for coping with unforseen events.

Ultimate Utopia versus Excellence by Stages

The utopian approach expresses an inability to conceive of actors as capable of shaping their own history; it reflects the assumption that the future is already possessed by the planners. Most programs for social change take this view. They depict ideal conditions at the end of the horizon. Although lip service is given to the implementation process, the technical effort and the allocation of resources generally are based on the target date of the plan. Politicians and planners tend to promote the ultimate product while giving little attention to the intervening stages of development or to the patterns of life and values they imply.

The ultimate utopia approach emphasizes the correction of deficiencies. It begins with an assessment of the initial situation, comparing what exists with an ideal community. The ideal is generated by cataloging all of the desirable facilities and institutions that have been found in communities or imagined by philosophers. This becomes the basis for a long list of things to be taken care of. The target for social change is conceived of as a "finished" community that will be achieved in, say, ten years, and then remain little changed for some period. But this image is difficult to maintain. Typically, the initial enthusiasm is soon depleted, and it becomes necessary to restimulate interest in continued development. This in turn results in higher costs, while yielding less mastery and fulfillment.

One example of the ultimate utopia approach is the model town. In such places, social and recreational facilities typically are built in scale and kind far in excess of those available in a town of comparable size. Building these ideal facilities is part of a strategy for attracting and retaining workers or home buyers for these sites. But generally it turns out that the investment in facilities alone was not sufficient to animate community life. In fact, the existence of the preplanned facilities may have acted as a disincentive to individual and group initiative. Social change based on an ultimate utopia leads to the building of programs or facilities for abstract general uses, and not for alternative specific activities. The former approach is plausible when the goal is to satisfy everyone in general; but what is good for everybody is usually not compelling to anyone. As a former resident of Columbia, Maryland, put it, "Around here James Rouse [the model-town designer] is the only person who had any fun" (Wasilewski, 1980).

By contrast, an approach of excellence by stages suggests that many projects can be initiated with the various members of the polity being involved in their completion, including members who may come later. This permits each person to invest some of himself in successively refined efforts. Individuals get behind things, influence projects, try to make a difference. Each improvement is taken as evidence of progress, and every initiative tends to beget additional development concepts. Small achievements increase the motivation of residents to engage in larger and more involved actions to evidence their power and to shape a more distinctive identity. In this view, a "community" exists at all stages of development. The community has to function in a way that is appropriate and adequate for each particular stage. There is thus maintained a perception both of continuing progress and of continuing commitment to the values that informed the project at its inception. Moreover, opportunities remain available for ordinary people to contribute to that progress. The community is never a utopia; it is always being enhanced.

Fantasy and Rationality

In the views of both positivists *and* romantics, reality was defined by reason, and symbols were seen as creations of the imagination or fancy. Yet through all this, thinkers such as Durkheim, Boas, and others recognized that the most fundamental cultural forms, the collective representations, are not the product of isolated reflective intelligence, but are born out of the intense atmosphere of collective effervescence. Collective representations derive their compelling power from the sense of the sacred that they elicit, and it is only through their discipline that rational thought becomes possible. Thus today we can speak meaningfully of "symbolic realism" and "rational fantasies." Symbols are not mere frills or imperfections in our scientific understanding. Instead, they are the very stuff of human consciousness and be-ing. As constitutive of human experience, symbols are *real* in the fullest sense of the word.

From this viewpoint, reason and fantasy, rather than being opposites, are seen to have features in common: both are real in the symbolic sense; both are social in that they are expressed and validated intersubjectively; both, we suggest, can properly be modes for social adaptation or revolution.

As a mode of societal change, fantasy operates in at least three ways: as a form of meaning creation that is directly appreciated or preferred in its own right; as a kind of magical trial through which new social arrangements must pass before becoming widely adopted; and as a means of ordering and legitimizing a society's ultimate concerns.

1. Fantasy and myth, and their expression in ritual and cult, are the social equivalent of Freud's alienated inner kingdom: to repossess it is to come home to oneself. Eric Hobsbawm (1959) in his study of *Primitive Rebels*, and Scott and Lyman (1970) in their essay *The Revolt of the Students*, have pointed to the ecstasy experienced in such revolutionary homecomings. They are a destruction of barriers as well as a participatory creation of collective meaning through fantasy and myth. This value of fantasy to an organization or movement can be seen through the "negative" example of members who are indifferent

to having their goals realized. Apparently the achievement of the "goals" would destroy the reason for the members to continue an organizing process that is itself their source of orientation. Thus rehabilitated drug users may "mess up" just before being released, in order, apparently, to return to the halfway house. Workers also may mess up at the moment their ostensive goals are being realized. They appear to do something similar to avoid filling management's quotas; even though messing up may deprive the laborers of bonuses, meeting quotas would destroy the "game" that may be their chief source of meaning at the work site. Similarly, suffering can function to cement group solidarity, even to a point that members may regret the loss of their oppression. Indeed, this group feeling may be preferable to the state of despair—or of no group feeling at all—that might be left in the wake of achieving "realistic" bonuses, trivial social reforms, or other minor gains.

In the industrial world, fantasy generally has a bad press, perhaps because it is one means of resisting the totalizing potential of instrumental reason. But is it possible that the very form of thought characteristic of an alienated situation could be transformed into a mode of change and liberation? Put another way, can a mythic structure of affectivity and community be combined with a rational structure of moral responsibility, and, secondly, can these together be made to replace the amoral and unloving isolation of instrumental reason? Such questions summon forth a *terra incognita* to social thinkers. Yet some preliminary explorations suggest that "irrationality" (or preterrationality) at the level of affectivity and community, "instead of undermining the rationalized structures of responsibility, may in fact preserve and expand our capacity to sustain them" (Kavolis, 1974, p. 118).[11]

Even if such a mode of being in the world is possible, what chances has it of gaining currency? In the early industrial era, fantasy was associated with artists, women, and the lower orders, and rationality was the preserve of the bourgeoisie. But if fantasy is characteristic of people who are not interested in or have little to gain from bourgeois rationality, such a lack of interest is likely to grow. Processes of organizing, formerly thought of as instrumental, increasingly may be engaged in for their affective and communal rewards. Planning and

[11] Irrationality at the level of affectivity is, of course, characteristic of Japanese institutions, both traditional and modern. The feelings of cohesion, as well as the possibilities for manipulation, in such an approach to industrial organization is noted in "*Seishin Kyoiku* in a Japanese Bank," in George Spindler, ed., *Educational and Cultural Processes*, New York, Holt, Rinehart, & Winston, 1974. See also Peacock, 1969, and Tiryakian, 1972.

action to change things may be engaged in, not to get things changed, but mainly to experience a sense of group cohesion.

2. Also, in a pragmatic sense, cultic fantasy represents a kind of magical trial in which new roles and arrangements may become validated before gaining wider acceptance. The secrecy surrounding group marriages and ritual orgies of contemporary "swingers" is used to recruit as well as to exclude, and may be a first step on the way to the normalizing of new conjugal practices. In a Sorelian sense, occult beliefs may be necessary for engendering and renewing the dissonance from everyday reality upon which the drama of conversion depends. And such dissident organizations, in the very process of creating their innovations, often can gather sufficient membership and media exposure to demonstrate and legitimize their alternative arrangements.

Fantasy can be the source of new identities as well as new social practices, and it is a form of "planning" in which anyone can engage. Even in ordinary daydreaming, though the specific vision is experienced as an individual affair, the dreamer still finds himself immersed in a commonly shared culture. Fantasies generally are more institutionalized, however, as for example in American males' dreams of flying off with the girl in the airline ads, or, more broadly, the American Dream itself. In the United States these myths are fairly public, though their mythic quality is rarely acknowledged. In other tribes the dreamer must keep the contents of his vision secret; in still others he may seek advice on how to interpret it.[12] As Anselm Strauss (1969) tells us,

> The interpretation of the vision, in some measure at least, affects the future action of the man as a member of the tribe. It may yield him a mandate or a command to act in a generalized way, and it may give or confirm a sense of identity. . . . Viewed in this way, the range of covert processes—variously denoted by the terms "reverie," "fancy," "daydreaming," and "fantasy"—are important for the conservation and change of identities. (pp. 68–69)

The point, of course, is not only that a mythic underlife exists in industrial societies, but that in accepting the positivistic denial of its relevance we abandon care and control of it, and leave the way open to the manipulation of our fantasy lives by persons and groups whose intentions are alien to our own. The humanistic change agent wishes

[12] For excellent studies of how various societies have dealt with their dreams, and how they have sought convincing answers to the puzzles of that unknown world, see G. E. von Grunebaum and Roger Caillois, *The Dream and Human Societies*, Berkeley, University of California Press, 1966.

not to "capture the imaginations" of clients, but to help them recapture their imaginations for themselves.

3. At a fundamental level, fantasy and myth are the symbolization of a society's ultimate concerns. Man's response to values in prehistory was a form of idolatry, his appreciation of the goodness of a thing involved that thing's being sacred. This is no less true of such ultimate concepts as dignity or freedom today. These ideas take their power not from reason or utility, or merely as the negation of oppression. Unless they preserve the significance rooted in their mythic origins, they become not only trivial, but untrue (Horkheimer, 1974, p. 36).

Foundational myths cannot be validated by science, if only because they refer to the very nature of Being upon which any science might be built. Thus, faith in the myths is not an act of knowing that has a low degree of evidence. If this is meant, one is speaking of belief rather than faith. Faith is the state of being ultimately concerned, of trusting in one's myths "as if" they were true. For these reasons, though they can never be justified or calculated into a rational social plan, such foundational myths to some degree are presupposed in the expectation of efficacy of *any* effort to cope. By explicitly acknowledging this, we begin to elevate our naive beliefs to the level of faith. To do this fully requires us to bring two features explicitly to consciousness: the "as if" character of our myths, and the existential character of our relation to them. The first feature involves a tolerance of others who do not share our own foundational assumptions; the second calls for an adherence to the ethical requirements of our own commitments.

An example of planning as a process of myth creation is provided by Albert Hirschman (1967; cf. Goulet, pp. 31–37 and Mechanic, 1974, pp. 37–38). In response to the criticisms by rationalistic economists of Third World development plans, Hirschman notes that these plans do not need to be accurate to be effective. We systematically underestimate our own creativity, he says, but such plans tend correspondingly to underestimate the difficulties to be faced. Thus their explicit errors tend to cancel out their implicit ones. The "overstatement" of the goals, and the process of national planning itself, functions as a collective mythmaking through which that society can create an ideal image of itself and mobilize its energies accordingly. A purely rationalistic plan would not have this power. In a similar fashion, more general models of social development (Marxism, for example), even in contexts where they are no longer compelling as hypotheses for the science of history, are still important as moral constructs, blueprints for the intelligent application of public choice. By establishing alternative contexts of mythical interpretation, they infuse our present situations with an eschatological light that reveals their moral implications. Such myths about the nature of history or society serve a "planning" function as scenarios of hope and articulations of value, as means for judging, and then accepting or transforming, the intolerable scarcities in the actual world of the present.

Extending this point, we may say that social planning represents two kinds of truth—the Aristotelian concept of "speaking the truth," or the truth of propositions, as contrasted to truth as an authentic response to reality, a "being in truth" (Wolff, 1970, pp. 45–46). The first refers to hypotheses advanced within some disciplinary framework; it is a truth of discourse. The second is mainly an act, a "mythic" feature of existence; it involves "witnessing," not in the sense of being an uninvolved observer, but in the sense of bearing witness, of being a representative *for* truth, rather than merely re-presenting it.

In this existential or mythic sense of truth, social planning (like sociology itself) can be on the side either of piety or of profanation, of order or of renovation. It can sanctify the conventional by formalizing it into occult scientific language, or it can demystify the sacrosanct by dialectically exposing its contradictions. In opening the sacred to scrutiny, social analysis at once profanes and purifies it. Its being in truth is less a destruction of idols than a revelation of the sacred, of the locus of that society's ultimate concerns. Without such revelation and renewal, traditional symbolizations simply lose their hold, ceasing to convey a sense of continuous connection between individual, community, and nature; "reality" becomes senseless, and truth claims *existentially* invalid. Such scarcity of coherence can delegitimize whatever logico-meaningfulness that culture may possess for its members (Duncan, 1978; Stanley, 1978).

To cope with the falling apart of the modern world we do not need better techniques or more rational plans. The fragmentation of our worlds will not be rectified by calculated policies, no matter how authoritative or energetic their source. Part of the story is always the intrinsic logic and existential adequacy of our foundational myths, as revealed in the consequences of the plans and policies to which they lead. I am doubtful that science and technology, though they function as myths, are adequate in this existential sense.

All this suggests that one criterion for a liberating method of social change is that it see both rationality *and* fantasy as mutually supportive modes of creating a viable future. Such a future will require far-reaching changes in the structure and processes of institutions that, for the most part, continue to behave as inflexible automata programmed to only a thin repertoire of actions.

Acknowledgment

I wish to thank George Coelho, Kurt Finsterbusch, and Jacqueline Wasilewski for their critical and substantive comments on drafts of this essay, and Stanford M. Lyman, D. Sam Scheele, and Manfred Stanley for ideas and insights I have incorporated without specific textual acknowledgment.

References

Almond, G. A., & Verba, S. *The civic culture.* Princeton: Princeton University Press, 1963.

Baritz, L. *The servants of power.* Middletown, Conn.: Wesleyan University Press, 1961.

Bendix, R. *Work and authority in industry.* New York: Wiley, 1956.

Brown, R. H. *A Poetic for sociology: Toward a logic of discovery for the human sciences.* London & New York: Cambridge University Press, 1977.

Broyard, A. Portrait of the inauthentic Negro. *Commentary,* 1950, *10,* 56–64.

Clark, B. The Cooling-out functions in higher education. *American Journal of Sociology,* 1960, *65,* 569–576.

Clausen, J. A., & Kohn, M. L. Social relations and schizophrenia: A research report and a perspective. In D. D. Jackson (Ed.), *The etiology of schizophrenia.* New York: Basic Books, 1960.

Dostoevski, F. *The Brothers Karamazov.* (C. Garnett, tr.) New York: Modern Library, 1950.

Duncan, E. A sense of hope. *Human Behavior,* 1978, *7,* 43–48.

Etzioni, A. Basic human needs, alienation and inauthenticity. *American Sociological Review,* 1968, *33,* 870–885.

Ficino, Marsilio. Five questions concerning the mind. In E. Cassirer, P. O. Kristeller, & J. H. Randall, (Eds.), *The Renaissance philosophy of man.* Chicago: University of Chicago Press, 1948, pp. 193–212.

Fingarette, H. *The self in transformation.* New York: Basic Books, 1963.

Freire, P. *Pedagogy of the oppressed.* New York: Herder & Herder, 1970.

Friedenberg, E. Z. *Coming of age in America.* New York: Random House, 1963.

Friedman, J. *Retracking America.* Garden City, N.Y.: Doubleday, 1973.

Goulet, Denis. *The cruel choice. A new concept in the theory of development.* New York: Atheneum, 1971.

Gramsci, A. *History, philosophy and culture in the Young Gramsci.* St. Louis: Telos Press, 1975.

Harré, R. Architectonic man: On the structuring of lived experience. In R. H. Brown & S. M. Lyman (Eds.), *Structure, consciousness, and history.* London & New York: Cambridge University Press, 1978.

Hirschman, A. O. *Development projects observed.* Washington D. C.: The Brookings Institution, 1967.

Hobsbawm, E. *Social bandits and primitive rebels.* Glencoe, Ill.: Free Press, 1959.

Horkheimer, M. *The eclipse of reason.* New York: Seabury Press, 1974.

Horkheimer, M., & Adorno, T. W. *Dialectics of enlightenment.* London: Allen Lane, 1973.

Israel, J. From level of aspiration to dissonance, or What the middle class worries about. In A. R. Buss (Ed.), *The Social Context of Psychological Theory: Toward a Sociology of Psychological Knowledge*. New York: Halsted Press, 1979.

Kavolis, V. On the structure of consciousness. *Sociological Analysis* 1974, *35*, 115–118.

Kornhauser, A., Sheppard, & Mayet, *When Labor Votes: A Study of Auto Workers*. New York: University Books, 1956.

Laing, R. D. *The Self and Others: Further Studies in Sanity and Madness*. London: Tavistock, 1961.

Levin, M. B. *The Alienated Voter*. New York: Holt, Rinehart, & Winston, 1960.

Lewin, K. A dynamic theory of the feeble minded. In his *A Dynamic Theory of Personality*. New York and London: McGraw-Hill, 1935.

Mandelstam, N. *Hope Against Hope: A Memoir*. New York: Atheneum, 1970.

Mannheim, K. *Man and Society in an Age of Reconstruction*. (E. Shils, trans.) New York: Harcourt, Brace, & World, 1940.

Mechanic, D. Social structure and personal adaptation: Some neglected dimensions. In G. V. Coelho, D. H. Hamburg, & J. E. Adams (Eds.), *Coping and Adaptation*. New York: Basic Books, 1974.

Milbrath, L. W. *Political Participation*. Chicago: Rand McNally, 1965.

Panofsky, E. The history of art as a humanistic discipline. In his *Meaning in the Visual Arts*. Garden City, N.Y.: Doubleday, 1955.

Peacock, J. L. Mystics and merchants in fourteenth-century Germany: A speculative reconstruction of their psychological bond and its implications for social change. *Journal for the Scientific Study of Religion*, 1969, *8*, 47–59.

Pico, Giovanni. Oration on the dignity of man. In E. Cassirer, P. O. Kristeller, & J. H. Randall, (Eds.), *The Renaissance Philosophy of Man*. Chicago: University of Chicago Press, 1948, pp. 223–254.

Sartre, J.-P. *Anti-Semite and Jew*. New York: Grove Press, 1960.

Schachtel, E. G. On alienated concepts of identity. *American Journal of Psychoanalysis*, 1961.

Scott, M. B., & Lyman, S. M. *The Revolt of the Students*. Columbus, Ohio: Charles E. Merrill, 1970.

Seeman, M. Intellectual perspective and adjustment to minority status. *Social Problems*, 1956, *3*, 142–153.

Seeman, M. Powerlessness and knowledge: A comparative study of alienation and learning. *Sociometry*, 1967, *30*(2), 105–123.

Stanley, M. Prometheus and the policy sciences. In F. Johnson (Ed.), *Alienation: Concept, Term and Meanings of Alienation*. New York: Seminar Press, 1973.

Stanley, M. *The Technocratic Conscience: Technology and Conscience in Twentieth Century Social Thought*. New York: Free Press, 1978.

Strauss, A. L. *Mirrors and Masks. The Search for Identity*. The Sociology Press, 1969.

Tiryakian, E. A. Toward the sociology of esoteric culture. *American Journal of Sociology*, 1972, *78*, 491–512.

Twain, M. (S. Clemens). *Extracts from Adam's Diary*. New York: Harper, 1904.

Vico, G. *The New Science of Giambattista Vico*. (T. G. Bergin & M. H. Frisch, trans.). Ithaca N.Y.: Cornell University Press, 1970.

Wasilewski, Jacqueline. Personal correspondence, 1980.

Watzlawick, P., Beaven, J. H., & Jackson, D. D. *Pragmatics of Human Communication: A Study of Interactional Patterns, Pathologies, and Paradoxes*. New York: Norton, 1967.

Weber, M. Science as a vocation. In M. Gerth & C. W. Mills (Eds.), *From Max Weber*. New York: Oxford University Press, 1946.

Williams, R. *Culture and Society: 1750–1950*. Garden City: Doubleday, 1960.

Wolff, K. H. The sociology of knowledge and sociological theory. In *The Sociology of Sociology*. L. T. Reynolds & J. M. Reynolds, (Eds.), New York: McKay, 1970, 31–67.
Yates, F. *The Art of Memory*. London: Routledge and Kegan Paul, 1966.
Zorza, R. *The Right to Say We*. New York: Praeger, 1970.

4

Stress, Strain, and Role Adaptation: Conceptual Issues

Marc Fried

Argument and Objectives

Despite the enormous technological progress of the past century, a high degree of stress accompanies modern urban life. Whether the stresses have intensified or diminished since the rise of industrial society is a moot point, but, clearly enough, there is a large gap between technological potentials for improving the human condition, and the social reality of widespread stress, pain, and deprivation. The manifest stresses are quite concrete and diverse: competitive work, family conflict, economic difficulties, physical illness, and daily frustrations that are minor but frequently cumulate to produce severe stress. Some of these stresses are quite evident and disruptive; others are almost imperceptible at first, and their impact is felt only after a delay. But all of these stresses can and often do affect the central activities and relationships of daily life.

One of the greatest difficulties in clarifying these patterns and their origins stems from the complexity of the issues. By virtue of the circuitous pathways through which psychosocial processes operate, most people do not trace the causal sequences beyond the concrete situations and persons associated with the immediate experience of stress. The foreman, spouse, neighbor, teacher, or one's own personality or behavior is likely to be perceived as the source of stress. Or the immediate situation or event which precipitates or symbolizes the

Marc Fried • Department of Psychology, Boston College, Boston, Massachusetts

stressful experience may bear an excessive burden of attributed blame. Nor have the social sciences been very helpful in providing realistic models and empirical analyses that comprehend the intricate sequences and time lags involved.

The larger proposition which I shall pursue is that our society has generated and systematized a set of macrolevel forces which have both overt and subtle ramifications that invade every sphere of life. In one sense, of course, this is simply the "nature of a society." But it is particularly striking because, in many respects, it contradicts the ideals and even the realistic evaluations of mass democratic society. That these influences are also pervasive reinforces the problem they present.

For individuals within the system, these macrolevel forces operate to increase both the large and the small stresses of daily life. The subjective sense of strain or distress is further exacerbated in that the same macrolevel forces provide limited options and narrow constraints within which responses are possible. Moreover, the affects generated by stressful conditions are readily displaced or transmitted to other people, producing a proliferation of stresses. An event so remote as a tight money supply generated by the Federal Reserve Bank produces many repercussions on individuals through its immediate impact on banking and industry. Only after a sequence of intervening and largely invisible processes do we confront greater difficulty in obtaining mortgages, higher interest rates on loans, a proportionately smaller increase in interest in savings (which often appears to be a gain), the bankruptcy or takeover of smaller industries, and the loss of jobs by low-skilled workers. At each successively lower level, there are fewer and fewer means available for coping effectively with these events, since, unlike the industry that can raise prices or lay off workers, individual strains can only be transmitted to family, friends, and co-workers, and result in reverberations throughout small social systems.

Nonetheless, the substantial costs of functioning within these conditions of stress, limited options, and narrow constraints are largely hidden beneath the more familiar benefits of life in modern, democratic societies. Human beings, moreover, are extraordinarily adaptable, and often manage to find tolerable the most intolerable situations. Once an intolerable situation becomes routine, most people learn to make the best of it, unless (and often if) it reaches the level of overt symptomatology. This capacity for adaptation enables us to submerge conflicts, anger, unhappiness, and despair in the very process of actively coping with stress, and thus diffuses and defuses the sequelae of stress.

The widespread evidence of expressed satisfaction with diverse domains of life (Andrews & Withey, 1976; Campbell, Converse, & Rodgers, 1976) reflects the process of adaptation and the mechanism of denial and restriction that help to sustain it. But to conceive of such adaptations to difficulty as manifestations of well-being may be to underestimate the hidden costs of social adaptation. It might be more neutral to limit the concept of well-being to those instances in which there is evidence of true fulfillment, or, at least, in which the benefits are substantially greater than the costs.

Rather than concentrating on these propositions and their extension, we will concern ourselves with the analytic framework through which a closer, empirical examination of those processes may become feasible. The philosophic position implicit in these views is hardly unique. Freud (1927, 1930) and, following on this intellectual groundwork, Marcuse (1964, 1966) and others (e.g., Moore, 1970), have set out the large tableau. However, only the most rudimentary fragments of theory and models are available to guide more systematic inquiry.

The particular focus of this formulation will be on microlevel phenomena. Individuals and small social systems eventually experience the impact of stresses generated by macrolevel forces. It is at this level that further clarification is most essential. Until we can trace the immediate stresses through the several intervening and mediating processes that operate at the microlevel, a problem of enormous complexity in its own right, we cannot hope to examine systematically the significance of larger forces on daily experience and functioning.

Development and operational formulation of the concepts of stress, role, and psychosocial adaptation form the core of the analysis of the microsphere. These are fundamental concepts in behavioral and social science. But they have remained dimensionless and without established, uniform definitions. The first task, then involves development of conceptual and operational definitions. From this it is possible to move on to the groundwork for models that may facilitate research and analysis of the human costs and benefits of social processes and social policies.

Stress, Strain, and Coping

After a period of relative neglect, research on stress has shown a sharp upward curve during recent years. The proliferation of research and, along with it, increased concern about the concept of stress has focused particularly on the stresses associated with changes in life events.[1] Within this framework there is a trend toward convergence in definitions, in methods, and in directions for future work. Since the concept of stress is important for understanding the ways in which macrolevel forces affect individual lives, I shall devote some attention

.o this concept and its measurement. My discussion is developed in analogy with the use of the concept in physical mechanics.

In its most generic sense, stress is omnipresent in life. The critical problem is not whether stress occurs but rather how much stress, how many stresses, and what kinds of stress are exerted on different organisms with different degrees and kinds of capacity to tolerate and respond under different conditions. Whether in physics or behavioral analysis, the immediate response to stress (or impact of stress) can be conceptualized as strain. Taking a leaf from applied physics, behavioral analysis can approach the relationship between stress and strain as the stress:strain ratio. This ratio is designed to indicate, for any individual or category of individuals, the level of strain generated by the same number of units of stress. It uses the simplifying but useful assumption (which can be challenged subsequently without altering the basic formulation) that all stresses of the same force are equivalent. However, how can we designate units of stress? And how do we measure strain?

A number of methods of measuring stress are possible. Some time ago, Holmes and Rahe (1967) used a panel of expert judges to estimate the different degrees of stressfulness of an array of life-change events.[2] Thus, a stress score could be assigned to any event among this array. The events experienced by an individual could be represented as a summated stress score. The theoretical position behind life-change-event measurements is that changes in life situations have a high probability of inducing organismic strains of a pathological nature, and, in this sense, are stressful.

To say that change is stressful, potentially a cause of strain, and, in this sense, a risk factor for many forms of pathology, should not obscure several other considerations.

[1] The research on life-event stress and its implications for physical and mental illness is not only extensive, but appears to be expanding at an ever more rapid rate. Two major, relatively recent volumes provide an overview of the issues, problems, and results: Dohrenwend and Dohrenwend (1974), and Gunderson and Rahe (1974). Several more recent articles with conflicting findings serve to update the perspective: Dohrenwend and Dohrenwend (1978); Dohrenwend, Krasnoff, Askenasy, and Dohrenwend (1978); Gersten, Langner, Eisenberg, and Simcha-Fagan (1977); Hurst, Jenkins, and Rose (1978); and Myers, Lindenthal, and Pepper (1975). Although this represents only one approach to the study of stress, another volume provides an overview of alternative types of studies of stress in relation to adaptation: Coelho, Hamburg, and Adams (1974).

[2] The procedure has a long history, and is an extension of Weber and Fechner's method of objectifying psychophysical judgments (Thurstone & Chave, 1929). Stevens (1966) developed the more formal properties of such scaling methods applied to social phenomena. More recently the procedure has been refined and applied to a broad range of life-change events by Dohrenwend *et al.*, 1978.

1. Different types of change have different stress values, a consideration to which much attention has been devoted in the literature. In linking the concepts of role and adaptation to the impact of stress, I will indicate that the different stress values are likely to be a function of the extensiveness of subsequent role alterations required by an initial change event.

2. It is widely recognized that life-change stress may induce new levels of achievement as well as new levels of pathology. The processes have to be formulated on the basis of general principles which can be represented, concretely, by such different outcomes.

3. A point that is often overlooked is that there are *conditions*, often persisting conditions, as well as events and changes, that are stressful. Indeed, some conditions become increasingly stressful when they fail to change (in spite of accommodations that may reduce the subjective strain). Severe poverty, chronic illness, racial discrimination, and long-term restrictions of functioning are all such stressful conditions. Not only are these conditions stressful, but they may alter the stress values of discrete events, exacerbating or mitigating them in a fashion as yet unknown.

Another method of measuring stress follows similar psychophysical principles, but does not depend on the use of judges. While the use of judges' ratings seems to follow the legal principle of "arm's-length" judgements, it provides no more scientific basis for objectified scaling of subjective experiences than average scores obtained from large, representative samples of populations. Indeed, the approach used by Robert Kahn and his associates and by my own research group assumes that the average strain scores of individuals from representative samples can provide *objectified* stress scores (Kahn, Wolfe, Quinn, & Sroek, 1964 [in collaboration with R. A. Rosenthal]). The analysis of individual deviations from these averaged scores is independent of the assigned stress score, even though both stress and strain measures have a common origin.

The objective of these formulations is to develop a systematic basis for studying stress:strain ratios. We have several bases for generating weighted stress scores for individual events. These can be summated for individuals who have experienced different numbers and kinds of events and will, thus, receive different total *stress* scores. The strain scores represent the subjective, individual variations in response to a life-change (or stress) event. With these measures, we can generate graphs to represent the distribution of stresses in any sampled population, or the distribution of stress:strain ratios.

A hypothetical, generalized stress curve for a hypothetical population at a hypothetical point in time might look like the following graph (Figure 1). The hypothetical stress curve is a skewed normal distribution. It indicates that a large number of people experience moderate levels of stress. The curve falls off slowly as the number of persons who experience more severe levels of stress declines. This stress curve is merely a hypothetical, descriptive curve suggesting the possible distribution of stress levels in a population.

A second hypothetical curve in Figure 2 shows changes in the level of strain with increments of stress, and is the graphic depiction of the stress:strain ratio. I have drawn it as an exponential curve. Although this is hypothetical, and actual empirical evidence might reveal a very different pattern, it fits a widespread view in both physics and psychology. The significance of the exponential curve is that (a) over the course of increasing increments of stress there is a gradual but slowly accelerating increase in strain, (b) at a certain (unknown) point, the point colloquially referred to as the straw that broke the camel's back, there is a sharp upward turn of strain in response to increments of stress, (c) this point we shall refer to, as in physics, as the elastic limit of the organism, defined by the fact that the effects of stress beyond this point are irreversible, that the organism (or metal) can no longer entirely recover its former elastic or flexible properties, and (d) from this point on, the curve goes rapidly to infinity, which is to say that the relationships between increasing units of stress and strain responses become indeterminate.

In generating such curves around average strain scores for each level of stress, we naturally assume that the discrete scores of individ-

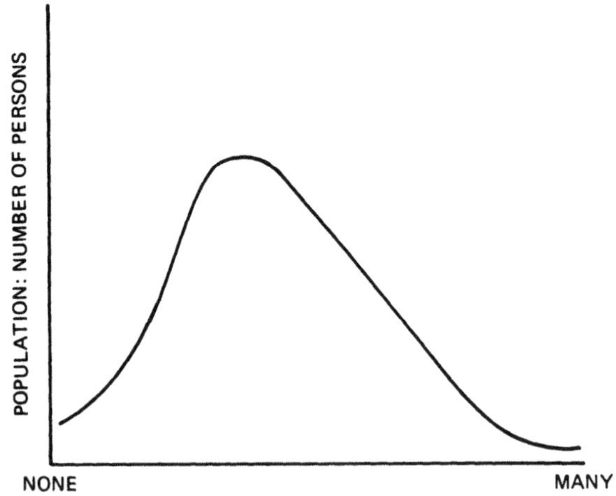

Figure 1. Hypothetical population distribution of stress units.

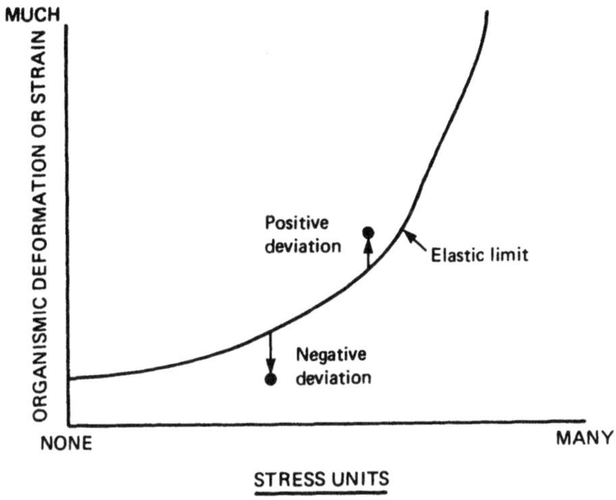

Figure 2. Hypothetical stress:strain ratios.

uals will be dispersed around all of these means. Thus, for any individual, it is possible to calculate the negative or positive deviation from the mean at any given level of stress. In an experimental situation, we might actually chart the discrete level of strain in response to unit increments of stress. In this way we could generate individual curves, since we cannot assume that the curve based on mean scores is individually applicable. Indeed, in clinical medicine an approximation to such an experimental procedure is used to evaluate individual levels of physical tolerance for different levels of physical stress. With psychosocial phenomena such a procedure might prove quite hazardous and, in any case, it would not be particularly useful unless we had reason to anticipate the specific future stress levels to which an individual might be exposed.

In trying to understand the consequences of stress, however, it is useful to differentiate individuals on the basis of different levels of strain at a given stress level. If we want to account for the effects of stress, it is quite relevant to distinguish those people who show a higher than average level of strain at a given stress level from those who are at or near the mean or below the mean. Following a suggestion by Michael Hurst (personal communication, 1979), I should like to designate these deviations from the mean as manifestations of individual differences in the immediate *ability to cope with stress.* I emphasize the immediacy of this measurement mainly because many supplementary processes and conditions begin to operate after the initial impact of stress. Although the immediate response is not entirely free of these factors, and is certainly not a pure measure of individual vulnerability, the early recording of strain in response to stress avoids some of the more complex interactions and adaptations that develop subsequently.

In using aggregate scores derived from subjective strain measures as indicators of the stress of a given event, situation, or condition, it is important to note that there can be slippage between "objective" impacts of stress and subjective representations of these impacts or of conscious feelings of distress. Not only are there differences among individuals in awareness and in the forms of defensive responses to stress, but endogenous system forces may come into play quite rapidly, and modify either the stressfulness of the event itself or its objective or subjective consequences. Thus, at best we can use subjective experiences of strain only as a provisional basis for measuring responses to stress. Oddly enough, the problem is less serious in using *average* strain scores as measures of stress, since, within large samples, those who overestimate are likely to compensate for those who underestimate.

There appear to be large individual differences in the forms of organismic strains that arise in response to stress and in the subjective recognition of these responses (as well as in the more specific awareness of feelings of distress). Broadly speaking, the strains may objectively affect primarily physiological, psychological, or social functioning, or some combination of these. Physiological strains are deformations in physiological or biochemical functioning, or, beyond the elastic limit, anatomical changes. Psychological strains are deformations in cognitive, motivational, or affective functioning, and, beyond the elastic limit, changes in ego structure. Social strains are reflected in deformations in role functioning at the level of activities, responsibilities, or relationships, and, beyond the elastic limit, result in changes in small-social-system structure.

I shall devote most of my attention to social stress for several reasons. Although I respect the evidence of physiological and psychological changes in response to stress, I suspect that some of the most massive consequences of macrolevel stressors are to be found in alterations in role functioning. Moreover, on the basis of a principle of the increasing autonomy of successively lower levels of integrative functioning, I would suggest that psychological and physical deformations have a greater effect on role behavior than the reverse.[3] Indeed, to some extent, bodily and mental changes become most significant when they begin to influence social role functioning.

[3] This principle parallels the phenomena in nervous system functioning (Sherrington, 1920).

In any effort to apply the concept of stress to one of these systems of functioning, a fairly differentiated set of concepts must be employed for system components and forces. Although these have been extensively developed for physical systems and, with less consensus, for psychological systems, role concepts and role terminology remain extremely crude. Thus, as a basis for considering the implications of stress for role behavior, it becomes necessary to extend the formulations about role functioning beyond current usage.

Structure and Dimensions of Role Analysis

Role can be defined as the set of behaviors and associated norms that subserve specific institutional or societal functions and through which individuals fulfill physiological, psychological, and social strivings.[4]

Roles are socially defined through the normative expectations of society about the responsibilities, prerogatives, and boundaries of a set of functions performed by individuals in social systems. These expectations provide some guarantee that socially necessary or desirable functions will be fulfilled, and will mesh with the roles and functional activities of other individuals. The degree of crystallization of these normative expectations, and the rigor of sanctions, are among the routes by which higher levels of the social system have an impact on human functioning, and control role definitions, role conceptions, and role boundaries. However, a number of influences can modify the effects of societal norms on role behavior.

Subcultural variations among different social classes, ethnic groups, age groups, and regions of the country often reveal such normative modification of role expectations. Moreover, in the course of carrying out the activities and functions associated with a role, the mutual influences of role partners upon one another can markedly

[4] In this discussion of role, no reference will be made to the concept of status to which it is ordinarily linked in structural analysis (e.g. Linton, 1936; Parsons, 1951).This approach conceives of status as a unit of the structure of institutions, and role as its dynamic embodiment. In this presentation, role is conceived as a structural unit which has meaning only in view of its functional properties within social systems. Thus it would be superfluous to have two separate terms for different reference points of the same phenomenon. Its dynamic properties, which are variants and extensions of the primary function, are referred to by the broad term "role behaviors" as well as the various concrete aspects of role behavior to be discussed. However, there are a number of points at which the tension between status and role may prove to be a useful conceptual formulation, so the present rejection of status terminology is viewed as provisional.

affect role conceptions and role behaviors. Thus, though the definition of roles must be traced to social norms, behavioral performance of roles reflects the actual social situation and, for many descriptive and analytic purposes, is a more accurate indicator of social structure and process.

It has often been noted that the role concept represents the link between the individual and society. Most of the socially necessary or desirable functions designated by role definitions also represent individual human needs. Indeed, an inevitable consequence of social organization is that the accomplishment of more personal physiological, psychological, or social objectives must largely be contained within the opportunities and constraints of socially defined roles. Social definitions of roles have some flexibility which allows individuals to utilize these roles for more idiosyncratic personal fulfillments as well. The extent to which this is possible is determined, on the one hand, by the options and constraints that are socially imposed and, on the other, by individual motivation and adaptational styles and skills.

With respect to the institutional structure of a society, roles can be categorized by the small social systems, collectivities, or environments in which concrete functions are carried out: work/occupation, family/household, extranuclear interpersonal relations, cultural/recreational participation, and social/personal maintenance. We shall refer to these five clusters of role patterns as *role complexes,* since these systems may (and usually do) include a complex of component roles. For aggregate analytical purposes it is often sufficient to view mainly the different role complexes in a society. For closer analysis, however, and most particularly in considering the significance for individuals who are geared to but never entirely defined by their roles, it is essential to analyze the more discrete roles and the more elementary components of these roles that comprise role complexes.[5]

The component roles of these role complexes are named for the core (instrumental) functions they fulfill. Thus, distinctive roles within the family or household can be categorized functionally as: household maintenance, sexuality, sociability, childrearing, companionship. It is

[5] In the discussion of role theory, we shall limit references to those that have an immediate bearing on the concepts and propositions formulated. The literature is large and diverse, but often does not go beyond the preliminary formulation of the role concept. I am indebted to Anthony Buono for carefully reviewing this literature. The major sources relevant to this discussion are: Biddle and Thomas (1966); Cottrell (1942); Goode (1960); Gross, Masno, and McEachern (1958); Jackson (1972); Kahn, Wolfe, Quinn, and Snoek (in collaboration with R. A. Rosenthal, 1964); Katz and Kahn (1966); Linton (1945); Merton (1957); Nadel (1957); Nye (1976); Parsons (1951); Sarbin (1970); Sarbin and Allen (1968); Thibaut and Kelley (1959); Turner (1962).

important to retain the distinctiveness of these roles even though, in our society, they are generally carried out by the same individuals and are most in evidence in a singular-role complex. Work/occupation roles are less numerous, and can be divided into productivity roles (serving social/organizational and individual needs for accomplishment of production goals) and income-gaining roles (serving individual and/or family support functions). Roles in other role complexes are generally less clearly named, or can be less clearly categorized.

For present purposes, it makes sense to distinguish roles only if (a) different instrumental or core-role functions are involved (e.g., household maintenance versus companionship), (b) the different functions can vary independently of one another along several dimensions of performance (e.g., high productivity–low income, sexuality without companionship), and (c) any one of the core (instrumental) role functions may be carried out, at least in principle, in different role complexes (e.g., companionship at work, at home, or in the neighborhood).

Roles and role behavior are internally structured in hierarchical fashion according to the contribution of each component of role behavior to overall social functioning. We shall refer to these as different *levels of integration of role behavior.* The different levels of integration refer to the fact that each successive level involves increasing ties to other individuals, networks, groups, and organizations. By virtue of this, lower levels of integration permit greater individual variability without provoking disjunctions with or deviations from social norms. At the same time, reactions to stress generally occur initially at the lower levels of role integration. However, they may successively invade higher levels, generating both individual and social problems of a more serious nature. Although there are levels of role integration beyond those we discuss, these are relevant primarily for macrosocial analysis. For present purposes, we shall restrict the levels of role integration to four major forms of coordination.

1. *Role activities,* the very concrete and discrete decisions and tasks that people perform, are the elementary units of roles and of role analysis. Household chores, consumer behavior, the specific actions around work, and the interactions with family or friends can all be understood as regular or transient behaviors associated with diverse roles. Much of the individual and social meaning of the activity derives not so much from the nature of the activity itself as from the role functions these activities subserve.

Any activity may serve several different roles although it is usually guided by a primary role purpose (function). Individual differences, personal preferences, cultural styles, and subjective fulfillments can often influence the choice of role activities or the embellishment of role activities to meet personal desires. Provided that the role activities subserve central or peripheral role functions, however, they meet the fundamental criteria for role fulfillment. All behaviors can be assimilated to roles because, no matter how private or personal, they are influenced by role expectations and role commitments. Only in viewing them as role-related behaviors can one understand the sociocultural or politicoeconomic influences that determine why a given behavior is carried out in a particular way, at a particular time, or in a particular place. Even highly individualistic behaviors, including efforts to escape from norm-dominated patterns, can best by appreciated as variant patterns of role functioning. Some activities retain a considerable degree of autonomy from incorporation within roles (e.g., autoerotic activities, private hobbies, daydreaming). Nonetheless, since sociocultural patterns may influence these private activities, and the activities may influence role behavior, role analysis is an essential component for a full appreciation of the individual and social significance of these activities.

2. *Role function,* and particularly the core or instrumental role functions which are the basis for classifying roles, represent the next higher level of integration. The core or instrumental role function simultaneously serves several objectives: (a) it insures the fulfillment of basic physiological, psychological, and social needs in a normatively acceptable form, regardless of the flux of motivation; (b) it involves the individual in at least minimal forms of social participation in developing interactions with others while fulfilling commitments within the small social systems and collectivities of daily life; and (c) the personal variations in defining role functions, as well as the individual patterns of role activities that fulfill these functions, allow the possibility of achieving more personal, individualized objectives.

Although a role is defined by its core or instrumental function, roles may serve several functions in addition to the core-role function. Indeed, some of the confusion in role terminology stems from the failure to distinguish between core and peripheral role functions. Such functions as expressiveness, communication, personal gratification, nurturance, or control are peripheral role terms. These can be fulfilled as separate activities or in the very process of carrying out core-role functions such as earning a living, maintaining friendships, or child rearing. Although secondary to the main *social* objective of the role, they may be critical for *individual* objectives, or for the *role relationships* within which the core-role function is carried out.

3. A higher level of role integration appears in the *role relationships*

among individuals who are role partners in carrying out instrumental role functions: husband and wife, work colleagues, foreman and worker, consumer and salesperson.[6] The social norms that influence role behavior are partly conveyed through these relationships and, in a very central sense, these role relationships are the societal microcosm. However, concrete role relationships involve transactions, negotiations, alliances, and reciprocal adjustments between role partners. Thus, they also establish the likelihood of fulfillment or failure of fulfillment of psychological needs and desires for both partners. It is this aspect of role relationships to which the symbolic interactionist approach has been most sensitive, and it is one which has been largely ignored or inadequately expostulated by the structural approach. Indeed, the process of developing and sustaining a role relationship in the course of fulfilling role functions helps to give specific form to the role behaviors of all role partners. Thus, for most purposes of analysis of ongoing social behavior, concrete role functions are specified by the social norms and sanctions operating within the role relationships in which they are embedded. In conditions of stress or temporary (sometimes long-range) deficit, role partners can be major sources of social support or of concurrent stress, extending to helping individuals or hindering them in performing their role activities.[7] Sets of role relationships comprise the small social systems within which people function in daily life.

4. Yet another type of role integration, the *role array* links the different roles and activities of an individual. It is based on the different roles in which a person engages, and the degree of investment in each role. On the continuum of levels of role integration, the role array is higher than the role relationship primarily in two respects: (a) each role in the array has a bearing on all other roles and their implied role relationships, and (b) the role array forms a transition between

[6] It is an open question whether the concept of role set (Merton, 1957) or its discrete components that Gross *et al*, (1958) describe as role sectors should be assimilated into this conception of role relationships or viewed as a separate level of role integration. The latter, in fact, appears more useful, particularly for broader social analyses. For present purposes, within the narrower framework we have imposed for studying relationships between society and the individual, we shall distinguish the role set as the encompassing term for all relationships entailed by any role.

[7] Interactions among role partners and the extent of personal fulfillment in role relationships must not be conceived as a static feature of a role relationship. As Zilbach (1968;) has pointed out with respect to the family, personal development, life-cycle changes, and situational factors may necessitate marked changes in family-role relationships, and the necessity for such changes may itself become a critical point of conflict or pathology.

the small social systems and the larger social systems in which some of these roles are embedded.

The role array involves individuals in a variety of role functions and in a diversity of social units. It is a reflection of the extent and type of social participation. However, the role array is influenced by a number of sociocultural and situational forces. Similarities in the range and the priority positions of different roles can be anticipated among people in similar socioeconomic, sociodemographic, and subcultural positions. From an individual viewpoint, the array of different role opportunities enables people to achieve alternative and even compensatory satisfactions in different roles. Indeed, a critical dimension of role analysis involves the congruence or discrepancy between the behavioral role array and the subjective importance of the different roles in which the person engages.

It would carry us too far afield to discuss issues of measurement in any detail. However, the subsequent discussion of adaptation and the process by which stress and, especially, strain "invade" the different levels of role integration requires at least a preliminary conception of the measurement possibilities. Quantitative treatment is possible for the four levels of role integration on the basis of a set of dimensions, each referring to the most critical feature of each given level.

1. Role activities are measured as *desirable* versus *onerous*. Although onerous activities may have to be carried out, the flexibility inherent in the decisions about role activities to fulfill core-role functions allows for numerous choices based on the desirability of the activity in its own right.

2. Role functions contribute, in varying degrees, to the accomplishment of the core-role function that defines and gives social meaning to the role. Thus, any role activity can be measured on a *central-peripheral* scale with respect to the extent of fulfillment of core-role functions inherent in the activity.

3. Role relationships are among the most difficult to measure, since so little is known about the conditions for or consequences of different types of role-relationship patterns. Clearly there is considerable cultural, social, and individual variability. However, a dimension of great generality which can be assessed from the role activity is the extent of *sharing* versus *separation* between role partners in carrying out role activities.

4. The role array, as we have indicated, depends on the hierarchy of behavioral investment in different roles. Thus, we can classify the different role activities according to the different functions they fulfill, and the amount of time devoted to each (as a proxy for personal

investment). However, this behavioral measurement must be compared with a more direct assessment of the subjective importance of different roles in the role array.

These dimensions provide the basis for measurement of role activities referring to the different levels of role integration. *The basis for analysis lies in the relative weight of desirable* (versus onerous), *central* (versus peripheral), *shared* (versus separate), *and high-priority* (versus low-priority) *role activities.* Several alternative types of measurement are possible. In cross-sectional analysis, it is possible to compare the scores of individuals along all these dimensions relative to a similarly defined subpopulation. It is also possible to compare subpopulations (on the basis of aggregate scores) with one another. However, a more dynamic analysis requires the comparison of scores of individuals (or aggregates) at different points in time. *The expansion or contraction of desirable, central, shared, and high-priority role activities is a primary criterion of the process of adaptation.* Although these dimensions of role activities form the primary units of measurement, they lend themselves to numerous composite measures which may eventually prove more useful in empirical analysis.[8]

Despite the many possibilities for changes in role behavior in response to internal and external forces, the role concept is particularly useful in designating the forces of stability and conservatism in the social order. Social change, of course, entails changes in societal role definitions. Moreover, as we have indicated, several structural and dynamic forces produce modifications in role definitions and behavior, either for aggregate subpopulations or for individuals. Yet the host of physical, social, cultural, and economic realities which define the options and constraints of daily life, superimposed on a socialization process that prepares the way for conventional role conceptions, confine behaviors within a relatively narrow range. Social conformity is generally maintained in role functioning despite widespread discrepancies between desires and opportunities or widespread role dissatisfaction. Social conformity within roles is only one aspect of the potential for conflict built into the social order, but it is a very important component of societal constraints.

Some individuals, through the very force of their independent achievements, may introduce new social perspectives on role functioning. And occasionally, the sweep of collective change in role behavior

[8] This formulation permits us to incorporate the contextual and intervening variables that mitigate (e.g., social support) or exacerbate (e.g.,concurrent stresses) the impact of stress, using the same conceptual framework. A given stress may induce role contractions which are "compensated" by role expansion in other spheres (e.g.,increased contact with friends).

begins to alter traditional norms about central roles. More generally, no matter how rigid or flexible the social norms regarding role performance, processes of social change entail individual and aggregate changes in role behavior over time. These behavioral changes will be referred to as psychosocial adaptations manifest in changes in role functioning in response to changing relationships between individuals and their physical and social environments.

Psychosocial Adaptation, Role Behavior, and Personal Fulfillment

Most definitions of adaptation are extremely broad and general. A more restrictive definition has theoretical and operational advantages. Psychosocial adaptation can, thus, be defined as *the process of modification of role behavior in response to changes in psychological or physiological functioning, or to changes in sociocultural, politicoeconomic, or environmental processes.* The definition includes changes within the role system itself which may result from the impact of events outside that role system on any role partner.

Several assumptions lie behind this definition. (1) All behaviors can be relegated to different social roles whether they are central or peripheral to these roles. (2) Change may occur in psychological, physiological, or social processes without requiring psychosocial adaptation, but most such changes impinge on individual functioning and entail minor or major adaptations. (3) Through prior adaptations, individuals develop relatively stable patterns of role behavior, which tend to resist adaptational change. (4) Stable role behavior represents the establishment of some degree of concordance between individual and environment, and only those internal or external changes that produce discordance beyond an individually variable tolerance threshold entail adaptational effort.

For evaluative purposes, it is necessary to designate the success or failure of adaptational efforts in achieving concordance. Although the term adaptation is often used to indicate successful change, such usage confounds the process and its consequences. In retaining the neutrality of the concept of psychosocial adaptation, we can avail ourselves of the term *maladaptation* to indicate adaptational failure. We introduce its obverse, *bonadaptation,* to signify effective adaptational efforts. Bonadaptation and maladaptation, I suggest, can be measured as the degree of *role satisfaction or dissatisfaction weighted by the subjective importance of that role.* This can be extended to include situational satisfactions that affect role behavior (e.g. housing, community) or generic satisfactions (e.g. satisfaction with political process, life satisfactions).

Adaptation itself is conceived as a "satisficing" rather than an "optimizing" process. When the individually variable threshold of concordance is reached, most people are willing to accept the suboptimal conditions. Naturally, it may be less than tolerable; or the consequences of that adaptation may lead to maladaptation. Most people, however, develop a sense of satisfaction in achieving a quasi-equilibrium, even if it is far from initial expectations or ideals. Difficulties in effecting inner or outer change, normative pressures toward conformity, the attention demanded by the many stresses of daily life, all require economizing of role effort, and limit the attainment of optimal psychosocial adaptations. Moreover, numerous conditions limit the likelihood of achieving optimal adaptations. Many role-behavior changes involve unanticipated consequences that may be less desirable than the initial consideration of decisions or actions could predict. And if the change in role behavior is of any import, it is likely to involve supplementary changes in role functioning to subserve the major change, or may have consequences and reciprocal effects on the partners in the role relationship. However, some people pursue different paths: a persisting struggle to optimize adaptational efforts is a modern, Western ideal.

In describing adaptation as a process of "satisficing" rather than "optimizing," I have extended the idea that there are costs as well as benefits entailed by adaptation processes. Bonadaptation signifies that, given the available external options and constraints, as well as the internal resources and expectations that guide decisions, the net benefit:cost ratio has proved favorable. Total satisfaction with an outcome of choice behavior is infrequent under the best of conditions. Freud noted that, in social beings as we are, conflict and the necessity for reciprocity in interpersonal relationships involves the surrender of some desires or aspirations in pursuit of other, presumably prepotent, objectives. The costs of psychosocial adaptation may be of varying types. Frequently, in the face of limited options and narrow constraints, the choices adopted are unrealistic, and may result in dissatisfaction in spite of objective achievements. In many instances, role behavior and the processes of negotiation among role partners necessitate the submergence of conflict in the pursuit of a more "satisficing" set of role relationships.

To pursue the analysis of adaptation it is important to develop methods of studying the specific costs and benefits that lead to a net result of bonadaptation or maladaptation. Concepts of bonadaptation and maladaptation and the use of operational corollaries such as satisfaction and dissatisfaction provide a reasonable basis for social indicators. But a more refined analysis is possible on the basis of component costs and benefits implied by the contractions and expansions of role functioning.

Previously I said that the expansion or contraction of desirable, central, shared, and high-priority role activities is a primary measure of the process of adaptation. The expansion of these role attributes is the benefit component of adaptation, and the cost is represented by the contraction of these role attributes. Naturally, both may occur simultaneously, so that the contractions or the costs must be weighed against the expansions to produce an individual cost:benefit ratio.

It is evident from these observations that the costs and benefits of adaptation must be estimated for the entire array of roles. Although a given stress is likely to affect one role initially and directly, its repercussions may lead to role alterations of diverse types. This is due both to the endogenous relationships among roles (e.g. income-gaining at work subserves household maintenance), and to the fact that a single individual whose psychological and physiological attributes are affected (directly or indirectly) by stress is the link between different concrete roles in the role array. Adaptational failure in one role may be simultaneous with success in another role, and the two must be weighed in a full analysis. Similarly, a process of invasion in response to stress may affect only one role, or may result in deterioration of role functioning across the entire array of roles. The critical dimension, therefore, of actual or potential malfunctioning is the level of resultant discrepancy between the total behavioral array of roles and the subjective significance of the set of roles. Stated differently, the results of psychosocial adaptation can be evaluated on the basis of the costs entailed by adaptation (including forgone opportunities) and the residual strain after role alteration has occurred.

One of the most evident bases for evaluating the hypothesis that links role contractions to costs and role expansions to benefits involves an examination of the relationship between stress and changes in role behavior. On the more general proposition that increments of stress lead to increased probability of malfunctioning we should expect, on the average, that increased stress would result in role contractions. One would not anticipate, of course, that this would be universal. In a fashion similar to the development of a stress:strain ratio and its application to individuals on the basis of deviations from mean scores, some people would show marked contractions and others marked expansions of role behavior as a consequence of stress. The direction and degree of deviation would be a relative measure of bonadaptation or maladaptation.

From another vantage point, it is possible to develop another theoretical framework concerning the process by which stress impinges on an organismic system and produces different degrees of bonadaptation or maladaptation. For the sake of simplicity, I shall limit myself to the case of maladaptation. If the formulation of the different levels of role integration is a reasonable reflection of human and social processes and structures, then we can say that the different role-integration levels represent a progressive series of opportunities for personal and social fulfillment. Whether from the point of view of the individual or from the vantage point of the small social systems in which role functioning occurs, each successively higher level is increasingly important for objective stability or subjective satisfaction.

Since each successively higher level of role integration is more closely coordinated as an endogenous system and bolstered by constraints and supports within the system, higher levels of role integration are less responsive to the immediate impact of stresses that might affect the individual. Thus, a stress such as unemployment would first affect the level of role activities and interfere with daily, role-dominated behaviors. As the stress became more prolonged or more severe for other reasons, it would begin to make incursions on higher levels of role integration by impeding core role functions (e.g. job-seeking), role relationships (e.g. family interaction), and the role array (e.g. other roles that are not directly affected by the work role or unemployment). Similarly, the stresses leading to out-migration precipitate the decision to migrate only after the limitations on role activities and role functions result in changes in role relationships (e.g. those which result from frustration of occupational-mobility strivings) and in the role array (e.g. due to limited opportunities).

I have already referred to this process as *invasion*. It implies that there are successive incursions of stress over time on different levels of role integration. As the forces which denote stress manifest themselves with greater severity or regularity, and the level of strain increases, there is a gradual modification of higher levels of role functioning. This spread from lower levels to higher levels of role integration is the invasion process itself. At each successive level, the problems of maladaptation become more serious because a wider set of activities, more central functions, and a larger number of people become involved. However, at any point in the sequence other factors may enter to modify the invasion process and to impede or accelerate its progress.

A discrete role activity may be impaired (that is, carried out without the requisite or expectable regularity, efficiency, or supplementary functions) without being a major impediment to the social functioning of the individual or the social unit. Of course, the more central the role activity is to the core-role function, the more ramified its effects. More generally, however, a worker may perform one particular task badly without generating sanctions either by co-workers or by the foreman. A wife does not ordinarily evoke marital conflict if she performs some of her marital-role activities inadequately but only as these begin to proliferate and move toward more central or core-role functions. The process of invasion first spreads to a wider array of role activities or behaviors which are of greater importance to the role function. As the role function begins to undergo attrition, it places an increasing burden on the role relationships. There are compensatory mechanisms available. Other people may supplement the halting performance of a role function. A wife may go to work when her husband becomes unemployed, or she can otherwise diminish the significance of the impact on herself or on the family. Friends and neighbors may draw closer to a person who wants to relocate in the hope of counteracting the strains due to overt losses or the lack of opportunity. But to the extent that the role relationship represents an integral component of a structured social unit, such measures are mainly transitory. Unless there is a reequilibration, such as might occur with a better job, most social units in our society do not have the flexibility or resources to sustain a member who fails to fulfill central role functions, or is manifestly seeking new roles or role relationships. Under the impact of continued role failure, role relationships themselves are likely to become disrupted.

There is yet another dimension to the process of invasion of role impairments or achievements. In addition to moving through the hierarchy of integrative levels of role behavior, the invasion process may move through diverse role spheres. For our purposes, we can limit these role spheres to the major broad social activities in daily life. There is no evident hierarchy among these, since different people, in different life situations and in different sociocultural contexts, may give quite different priorities to these role spheres. Thus, it is the number of spheres invaded, along with the hierarchical level of invasion, that is an indicator of the degree of impairment or achievement in social behavior.

These concepts and propositions need further development. The measurement procedures need to be further concretized. The concepts can be defined more precisely. And the propositions (if not the assumptions) need to be expanded, formalized, and validated. Nonetheless, even in their current form, they offer some promise of allowing more accurate and richer formulations and measurements of the impact of stress and of adaptational responses to strain. Certainly it is possible to link these formulations about processes at the small-social-system level to macrosocial phenomena; but this would carry us too far afield.[9] I will limit the subsequent discussion to applying the conceptualization presented to the phenomena of residential movement in forced relocation.

Implications for Understanding Residential Relocation

Two different purposes may be served in exemplifying this conceptual framework by considering the stresses associated with residential relocation. In the application of these concepts and propositions to the impact of forced residential relocation they can attain greater concreteness and clarity. Moreover, though this brief discussion and conceptual translation of residential-relocation issues cannot do justice either to the phenomena or to the conceptual framework, it provides a short-circuit method of considering the implications of the framework for psychosocial analysis. In striving to attain these objectives, I shall indicate the ways in which this approach facilitates the reduction of diverse manifest patterns of behavior and experience associated with forced residential relocation to a narrower set of analytic categories, and provides a basis for a dynamic, theoretical understanding of the processes involved.

In the United States, forced relocation has been impelled mainly by urban renewal fostered by federal legislation, by the development of a vast interstate highway system, and by dam construction. The devastations of war, invasions, earthquakes, and hurricanes have generally, at least in recent years, spared the highly industrialized nations though they continue to be major tragedies affecting residential relocation and migration throughout most of the world (Coelho & Stein, 1977).

In democratic societies, the initial stress of forced relocations generally occurs at the announcement of plans for urban renewal, a highway corridor, or some other large-scale building project involving mass exodus from an area. Partly because of the intense opposition these programs evoked, and partly because of economic constraints, these forms of redevelopment entailing extensive relocation have decreased. A new form of development, however, has begun to spread and to affect ever larger populations: the conversion of single-family homes and multiple dwellings to condominiums. In some respects, it

[9] A preliminary statement of the macrosocial-microsocial relationship is presented in a prior and more synoptic discussion of these issues (Calhoun, in press).

is a more insidious form of residential extrusion. In the shift from a public to a private enterprise, not only have the legal weapons to challenge specific instances of conversion diminished, but the requirements for substantial preparatory information, availability of relocation housing, and financial compensation for loss have disappeared. The spread of condominium conversions across the metropolitan areas of the nation has been rapid, and a total calculation of the numbers of housing units involved, or of the full psychosocial and politicoeconomic impact, is unknown. However, the mounting evidence suggests that it has begun to pose a major problem of relocation stress. Although the specific issues involved may differ somewhat, the impacts on individuals and, thus, efforts to understand the sequences of events, are similar in principle to those redevelopment and relocation programs in the public domain.

As with most stresses that affect life situations, attention and blame for relocation are largely focused on the *mediate* causes of the problem: the landlord or developer, the local urban-renewal authority, or the highway relocation agency. In the face of immediate threat, there is little recognition that economic forces in the private market or political forces in the public domain, beyond the control of landlord, developer, or relocation official, necessitate or encourage such actions. As a consequence, the possibility of mass action to reduce the scope of the problem or diminish its impact is markedly reduced. However, just as the relocation impact of urban renewal and highway construction gradually came to the attention of political authorities, so is the problem posed by condominium conversion gradually drawing the attention of local municipal governments.

The widespread resistance to forced relocation which had served and continues to serve to politicize the phenomenon also provides a means by which individuals confronted with the threat of relocation can cope with the stress. There is, of course, much variation among individuals in any neighborhood in subjective antagonism to forced relocation and in willingness to participate in resistance activities. In some instances, people who had planned to move find the necessity of moving a welcome opportunity for relinquishing binding ties. In other instances, despite strong antagonism to relocation, resistance activity is impeded by a sense of futility. There are variations, as a consequence, in the stress:strain ratio and, even given similar levels of initial strain, in the methods of adapting to the strain.

Differences in the strain engendered by a common stress like forced relocation are largely due to the extent to which role activities, role functions, role relationships, and the role array have been concentrated within a local area. Although it is possible to describe the difference either in psychological terms, or as a difference in intensity of investment in a local area, and though both of these sets of factors are modestly correlated with localism of role behavior, the variation in role functioning is most specifically and strongly associated with differences in reaction to the threat of relocation.[10] Moreover, even if personality factors or embeddedness in the area are conceived as explanatory variables, the mechanisms by which they affect the level of strain in response to impending relocation involve actual or anticipated alterations in role behavior.

Residential relocation and the role adaptations developed in response to it involve at least four major phases of change: initial awareness of developments that entail residential relocation, a situation of imminent change while planning resettlement, relocation itself, and postrelocation adaptations to a new spatial, social, and psychological environment. Differences in the conclusions from different studies of relocation can be traced to differences in modal patterns of role behavior in the prerelocation situation and, as a consequence, to the pre-post transition in role functioning. Ultimately, the extent to which there is a net bonadaptation or maladaptation to relocation for the relocated population as a whole is a function of the extent and types of role alterations required and of the internal or external resources available for the facilitating of the transition. A weighted analysis of the forgone role opportunities due to the loss of the former place of residence and the achievements or failures in role adaptation in the new environment is, thus, critical. The role-adaptation formulation permits us to consider the transition using parallel terms for the pre- and postrelocation situations, and to develop initial quantitative statements about significant human and social events.

The major prerelocation factors that account for differences in the stress:strain ratio can now be stated synoptically. (a) There may be differences in the extent of anticipated loss due to differences in the concentration of role behavior in the local area. (b) There may be differences in immediate coping behaviors entailing role expansion or role contraction. (c) There may be differences in the economic, psychological, or social "support" resources during the early transition process. Prior to relocation as well as after relocation, a qualitative differ-

[10] To include an extensive bibliography and discussion of the literature on forced residential relocation would carry us far afield. In this discussion, I lean heavily on the former collaborative study of relocation in which I was engaged and which involved the most intensive examination of the process that is available (see particularly Fried, 1963, 1965, 1973, and Gans, 1962). However, without citing full references, I have also taken account of concordant and divergent results from other studies of relocation (see also Burkhardt, Boyd, & Martin, 1973, Key, 1967, and Lipman, 1968).

ence occurs when the level of strain precipitates a process of role invasion so that the changes, particularly the contraction of role activities, lead to contractions in role functions, in role relationships, or in the role array. Under these conditions, the alterations in role behavior develop some degree of functional autonomy from the situation that originally produced them, and begin to involve processes of adaptation within role systems other than those associated with local relationships. The risk of pathological developments increases correspondingly.

The many differences in prerelocation role patterning within local areas, particularly in working-class areas, which are most frequently the targets of redevelopment, can be comprehended within four major types. Each of these types presents a different potentiality for postrelocation adaptational success. Moreover, the relative potency of prerelocation role patterns compared with postrelocation situations as determinants of adaptation is different for each type.

One very interesting type, generally more familiar in middle-class and higher-status neighborhood than in the working class, involves a dispersion of role behaviors in both local and nonlocal areas. Frequently, the people who manifest this pattern are already socially mobile, and their spatial mobility is a reflection of the social transition. The fact that they continue to live in a working-class area despite occupational and economic achievements that place them at a new-status level itself suggests role commitments that they do not want to relinquish. But the fact that they have already broadened the geographic base of their role behaviors also implies that the alterations in role behavior required by relocation affect only a limited set of the total role array. On the surface, they reveal a characteristic pattern of bonadaptation to the new environment, many lingering regrets about the transition, and many efforts to expand their role behaviors in the new situation to correspond as closely as possible to their former, local involvements.

Similar to this type, but different in crucial ways, are those people who have never developed the concentration of role activities in the local area to the same degree, or are quite ambivalent about their local commitments. Like the previous group, they are likely to have more spatially-dispersed role patterns. In some instances the spatial dispersion of role behavior is a desire rather than an accomplishment and, thus, the relocation serves an enabling function. In contrast to the previous type, role adaptations can be achieved fairly readily. Although a short-term contraction of role relationships may be involved, these are generally compensated for by an expansion of role activities and role functions in new spatial and social directions.

A third type poses some very specific problems by virtue of the fact that their total pattern of role behavior is, from the very outset, contracted, and the disruptions of relocation frequently force an even greater contraction. In most instances, these are people who have never been truly invested in the prerelocation residential area. While their role activities may have been concentrated locally, they have not maintained this localism because of any intrinsic subjective desirability of local interaction, but rather as a matter of convenience. Thus, though the difficulties they often experience after relocation appear to classify them among those who suffer severely from the process, it tends to be an exacerbation of long-term problems rather than one that can be directly traced to the adaptational demands associated with relocation. In many cases, however, the endemic strain is sufficiently great that the small increment in relocation stress encourages a new process of role invasion, and serious social or psychological problems.

A fourth type, the one to which I devoted the greatest attention in my analysis of relocation impacts in the West End of Boston, are those for whom the prerelocation neighborhood can truly be described as "home."[11] Predominantly working-class in origins and current status, these were people whose total array of roles was largely bounded by the local area. Although they might work or shop outside the neighborhood, such extralocal activities were viewed as excursions into foreign territory. Not only were their role activities predominantly local, but their role relationships in different role complexes were also influenced by the dominance of the local neighborhood in their lives. Thus, though family ties might be close, most of their time was spent with neighbors and local friends rather than with family members. As a consequence, regardless of the opportunities available in the new residential area, the experience of relocation was an experience of major disruption, dislocation, and strain. A marked improvement in some of the postrelocation conditions, like superior housing or relocation into the same neighborhood of former friends or kin, might compensate for the loss and ameliorate the strain. More often than not, however, the level of bonadaptation or maladaptation could be traced to the possibility of major role alterations within the conjugal unit. To the extent that husband and wife, particularly, could engage in expanded role relationships (e.g. increased sociability, companionship) to compensate for role contractions in other spheres, the likelihood of

[11] Although emphasis on this type led to the view that this was the characteristic response to residential relocation in working-class populations, it represented only a large proportion but less than a majority (approximately one-third). Other studies have found smaller proportions of this type, but, as I have indicated, this is largely a function of differences in the degree of localism of role behavior that obtained in the West End of Boston among many people.

successful adaptation was increased. Conversely, when the altered situation increased the role demands of one spouse upon the other beyond the possibility of ready adaptational response, increased strain and a process of role invasion generally occurred. Sometimes it was limited to family disruption; occasionally it extended to major maladaptations, in physical or mental illness.

Although the forces involved are clearly multidimensional, these can be encapsulated within the conceptual framework of role-adaptation theory. Clearly, a major consideration in developing a model that takes account of the costs and benefits of role alterations necessitated by social change involves the comparison of prior and subsequent role patterns. In the case of relocation, however, the extent to which even major benefits of postrelocation conditions can compensate for prerelocation losses is very much a function of the degree of role investment in the prerelocation residential area. Thus, for the last type described, a severe grief reaction resulted quite typically from the relocation changes. Regardless of the new opportunities offered by a new environment, the losses and the contractions of role behavior were largely noncompensable. Moreover, while psychological factors like a depressive orientation might account for a grief reaction among those whose roles were less concentrated locally, there was no association between the extent of grief and depressive orientations among those with intensive local role commitments.

The theoretical emphasis on role adaptation as a process, as a dynamic entailed in meeting changes in life conditions, lends itself equally well to the analysis of voluntary migration. To the extent that long-distance moves are voluntary, the range of opportunities for postmigration role adaptation is a more prominent determinant of adaptational effectiveness than in the situation of forced relocation.[12] Since these opportunities vary quite systematically with the social class position and race/ethnicity of the migrant, there is a marked status gradient in bonadaptation after migration. However, at all status levels, the availability of economic, psychological, and, most particularly, social resources (e.g., contact, former friends or kin, compatriots) has a powerful effect in moderating the stress of transition.

Although both forced relocation and voluntary migration present situations that involve processes of major role adaptation, this conceptual framework is more broadly applicable to the changes in the conditions of role behavior in the ordinary course of life. Whether we

[12] The literature on migration is vast, but mainly concerns itself with the causes of migration rather than with its consequences or with the adaptational requirements of the transition. For a few of the discussions of the issues involved, see Eisenstadt (1954), Fried (1970), Murphy (1965), and Shannon & Shannon (1973).

are concerned with the many overt and subtle changes in economic, political, and social conditions that influence our daily lives, or with those changes internal to the dynamics of role systems themselves, confrontation with change requiring alterations in role behavior is a pervasive experience. Indeed, one of the most persistent difficulties leading to physical, psychological, or social malfunctioning stems from the many resistances to role alteration. Whether we focus on work/occupation roles, household/family roles, or on roles in other role complexes, the appearance of stability hides a host of changes that require alterations in role behavior in order to maintain effective functioning.

Whether due to the comfortable familiarity of established role behaviors or to the threats implicit in initiating a process of role change, people often fail to make alterations in their role activities, to modify their conceptions of role functions, to negotiate changes in role relationships, or to reconsider their investment in different roles in the role array in response to changing conditions. Thus, even though there may be no overt, focal stress, the accumulation of implicit stresses resulting from changing circumstances without corresponding alterations in role behavior produces increased discordance between the individual and the small social system. Such discordance is, indeed, an alternative way of defining strain. But much as these changes may occur within the small social systems of daily life, they are often exacerbated by supplementary stresses stemming from the larger environment, with relatively few widespread resources available to moderate the impact of change.

To return to the theme with which this inquiry was undertaken, the confluence of stresses, major and minor, that stem from structural forces in the economy, in the political processes, and in larger patterns of social organization (often with inegalitarian impact) and the stresses of daily life creates widespread conditions of strain. The fact that macrolevel stresses are only minimally affected by individual efforts in itself implies a constraint on the possible forms of role adaptation. Individuals may participate in movements to resist forced relocation, but the upward impact on bureaucratic processes or legislation is, at best, slow. Their efforts, thus, may be valuable, but can only produce effects from which they are themselves unlikely to benefit. A major impetus to voluntary migration is the disjunction between individual strivings for improved life conditions and the availability of local opportunities for fulfilling these strivings. If potential migrants are to pursue their strivings, they have little choice but to *escape* into new opportunities. Those who cannot confront so major a set of role alterations suffer the restrictions of their residential environments.

Certainly people can and sometimes manage to superimpose their desires on an extant situation, or to rise above the constraints. It is easy to lose sight of the fact, in the midst of our ideology of individualism and the values of personal strength, that this path of deliberate, self-imposed alteration in the total set of role behaviors requires a characteristic that can only be meaningfully called *supermotivation*. And though such accomplishment almost inevitably conjures great admiration in our society, it begs the question of the costs that need to be weighed against the potentially beneficial consequences. That these individualistic achievements cannot solve the adaptational problems of a mass society is clear. The solution must lie in diminished structural stresses, in a more egalitarian distribution of the resources that can mitigate stress, or in expanded options for varied forms of role adaptation.

References

Andrews, F. M., & Withey, S. B. *Social Indicators of Well-Being: Americans' Perceptions of Life Quality.* New York: Plenum Press, 1976.

Biddle, B. J., & Thomas, E. J. *Role Theory: Concepts and Research.* New York: Wiley, 1966.

Burkhardt, J. E., Boyd, N. K., & Martin, T. K. *Residential Dislocation: Consequences and Compensation.* Report Prepared for National Cooperative Highway Research Program (No. 20-90), 1973.

Calhoun, J. B. *Perspectives on Adaptation, Environment and Population.* New York: Praeger, in press.

Campbell, A., Converge, P. E., & Rodgers, W. L. *The Quality of American Life: Perceptions, Evaluations, and Satisfactions.* New York: Russell Sage, 1976.

Coelho, G. V., & Stein, J. J. Coping with stresses of an urban planet: Impacts of uprooting and overcrowding. *Habitat,* 1977, *2,* 379–390.

Coelho, G. V., Hamburg, D. A., & Adams, J. E. (Eds.), *Coping and Adaptation.* New York: Basic Books, 1974.

Cottrell, L. S., Jr., The adjustment of the individual to his age and sex roles. *American Sociological Review,* 1942, *7,* 618–625.

Dohrenwend, B. S., & Dohrenwend, B. P. (Eds.), *Stressful Life Events: Their Nature and Effects.* New York: Wiley, 1974.

Dohrenwend, B. S., & Dohrenwend, B. P. Some issues in research on stressful life events. *Journal of Nervous and Mental Disease,* 1978, *166,* 7–15.

Dohrenwend, B. S., Krasnoff, L., Askenasy, A. R., & Dohrenwend, B. P. Exemplification of a method for scaling life events: The PERI Life Events Scale., *Journal of Health and Social Behavior,* 1978, *19,* 205–229.

Eisenstadt, S. N. *The Absorption of Immigrants* London: Routledge & Kegan Paul, 1954.

Freud, S. *Civilization and Its Discontents.* (Standard Edition, Vol. XXI) London: Hogarth Press, 1961.

Freud, S. *The Future of an Illusion.* (Standard Edition, Vol. XXI) London: Hogarth Press, 1961.

Fried, M. Grieving for a lost home. In L. J. Duhl (Ed), *The Urban Condition*. New York: Basic Books, 1963.
Fried, M. Psychosocial adaptation in modern societies: Costs and benefits. In J. B. Calhoun (Ed.), *Perspectives on Adaptation, Environment and Population*. New York: Praeger, in press.
Fried, M. Transitional functions of working-class communities: Implications for forced relocation. In M. B. Kantor (Ed.), *Mobility and Mental Health*. Springfield, Ill.: C. C. Thomas, 1965.
Fried, M. Deprivation and migration: Dilemmas of causal interpretation. In E. G. Brody (Ed.), *Behavior in New Environments*. Beverly Hills, Calif.: Sage Publications, 1970.
Fried, M. *The World of the Urban Working Class*. Cambridge, Mass.: Harvard University Press, 1973.
Gans, H. *The Urban Villagers*. Glencoe, Ill.: The Free Press, 1962.
Gersten, J., Langner, T. C., Eisenberg, G., & Simcha-Fagan, O. An evaluation of the etiological role of stressful life-change events in psychological disorders. *Journal of Health and Social Behavior*, 1977, *18* 228–243.
Goode, W. J. A theory of role strain. *American Sociological Review*, 1960, *25*, 483–496.
Gross, N. A., Mason, W. S., & McEachern, A. W. *Explorations in Role Analysis*. New York: Wiley, 1958.
Gunderson, E. K., & Rahe, R. H. (Eds.), *Life Stress and Illness*. Springfield, Ill.: C. C. Thomas, 1974.
Holmes, T. H., & Rahe, R. H. The social readjustment rating scale, *Journal of Psychosomatic Medicine*, 1967, *11*, 213–218.
Hurst, M. Personal communication, April, 1979.
Hurst, M. W., Jenkins, C. D. & Rose, M. The assessment of life change stress: A comparative and methodological inquiry. *Psychosomatic Medicine*, 1978, *40*, 126–141.
Jackson, J. A conceptual and measurement model for norms and roles. *Pacific Sociological Review*, 1966, *9*, 35–47.
Jackson, J. A. (Ed.). *Role*. Cambridge (England): Cambridge University Press, 1972.
Kahn, R. L., Wolfe, D. M., Quinn, R. P., & Snoek, J. D. (in collaboration with R. A. Rosenthal). *Organizational Stress: Studies in Role Conflict and Ambiguity*. New York: Wiley, 1964.
Katz, D., & Kahn, R. L. *The Social Psychology of Organizations*. New York: Wiley, 1966.
Key, W. H. *When People Are Forced to Move*. Final Report of a Study of Forced Relocation. May, 1967, unpublished.
Linton, R. *The Study of Man*. New York: Appleton-Century, 1936.
Linton, R. *The Cultural Background of Personality*. New York: Appleton-Century-Crofts, 1945.
Lipman, M. H. *Relocation and Family Life*. D. S. W. Thesis, University of Toronto (School of Social Work), 1968.
Marcuse, H. *Eros and Civilization*. Boston: Beacon Press, 1966.
Marcuse, H. *One-Dimensional Man*. Boston: Beacon Press, 1964.
Merton, R. The role set: Problems in sociological theory. *British Journal of Sociology*, 1957, *8*, 108.
Moore, B. *Reflections on the Causes of Human Misery*. Boston: Beacon Press, 1970.
Murphy. H. B. M. Migration and the major mental disorders: A reappraisal. In M. Kantor (Ed.), *Mobility and Mental Health*. Chicago: C. C. Thomas, 1965.
Myers, J. K., Lindenthal, J. J., & Pepper, M. P. Life events, social integration and psychiatric symptomatology. *Journal of Health and Social Behavior*, 1975, *16*, 421–427.
Nadel, S. F. *The Theory of Social Structure*. Glencoe, Ill.: Free Press, 1957.

Nye, F. I. *Role Structure and the Analysis of the Family.* Beverly Hills, Calif.: Sage Publications, 1976.

Parsons, T. *The Social System.* Glencoe, Ill.: Free Press, 1951.

Sarbin, T. R. A role theory perspective for community psychology: The structure of social identity. In D. Adelson & B. L. Kalis (Eds.), *Community Psychology and Mental Health: Perspectives and Challenges.* Scranton, Pa.: Chandler, 1970.

Sarbin, T. R., & Allen, V. L. Role Theory. In G. Lindzey & E. Aronson (Eds.), *The Handbook of Social Psychology.* Reading, Mass.: Addison-Wesley Press, 1968.

Shannon, L. W., & Shannon, M. *Minority Migrants in the Urban Community.* Beverly Hills, Calif.: Sage Publications, 1973.

Sherrington, C. S. *The Integrative Action of the Nervous System.* New Haven: Yale University Press, 1920.

Stevens, S. S. A metric for the social consensus, *Science,* 1966, *151,* 530–541.

Thibaut, J. W., & Kelley, H. H. *The Social Psychology of Groups.* New York: Wiley, 1959.

Thurstone, L. L., & Chave, E. J. *The Measurement of Attitude.* Chicago: University of Chicago Press, 1929.

Turner, R. H. Role taking: Process versus conformity. In A. M. Rose (Eds.), *Human Behavior and Social Processes: An Interactionist Approach.* Boston: Houghton Mifflin, 1962.

Zilbach, J. J. Family development. In J. Marmor (Eds.), *Modern Psychoanalysis.* New York: Basic Books, 1968.

II

Meanings and Impacts of Uprooting

Introduction

In this section, the authors argue that uprooting cannot be understood merely as a phenomenon of physical relocation. Uprooting disrupts, however temporarily, the sense of security and self-continuity of an individual moving through a changed physical and social environment. In Chapter 5, Marris draws from his experience in urban planning to emphasize the importance of "the structure of meaning" in socialization. The structure of meaning is defined as the conceptual organization of understanding of one's physical and social surroundings. It consists, according to the author, of the unique understandings which the individual forms out of his experiences, as well as the shared knowledge which individuals learn about a culture that provides them with ways of solving the problems of crises and change. Structures of meaning may be highly specific, as they are embodied in unique emotional attachments and commitments to persons, places, and political entities.

Marris argues that uprooting produces emotional stress by disrupting this structure of meaning. Change means loss and bereavement of intimate bonds. The death of a loved one, the loss of a job, the loss of a neighborhood or of a way of life are interruptions of the continuity of structure of meaning. Marris reiterates a key health issue which has been raised by development planners in urban renewal, namely, that the loss of neighborhood and community can lead to grief and mourning syndromes. Coping strategies, therefore, are to be viewed not only as instrumental means for dealing with the immediate psychological distress, but also as efforts to recover meanings that may be irretrievably lost when one is exposed to rapid environmental change.

In Chapter 6, Back refines the concept of uprooting. Not all changes are equally disturbing. Back distinguishes continuous changes, which are predictable developments in the course of the life cycle, from catastrophic and disastrous crises. Using as his model the *chreod,* Back defines uprooting as a change taking place in an inappropriate time span. The author suggests that uprooting disrupts the individual's efforts to maintain predictability and to preserve a positive self-image. Echoing Marris' concern for the capacity to reintegrate meaningful attachments, Back argues that, in order to be able to achieve a certain continuity of self-image (which is a major developmental task in all transitions), the individual must recognize the essentials of the previous environment and use them to adapt to the changed circumstances.

In Chapter 7, Tiryakian reviews four historical accounts of uprooting, and places the phenomenon in the context of large-scale processes of institutional change which are central to modern society. These processes include industrialization, urbanization, political identification, and class-structure formations. In analyzing the study by Maurice Barrès, *The Uprooted,* he observes that secular or nontraditional education results in the creation of intellectual elites who, even as they assume positions of leadership or creativity, are removed both physically and psychologically from their people and their traditions. In studying Simone Weil's *Need for Roots,* Tiryakian presents the spiritual malady of uprootedness which a class of people experience when they are unable to participate in the mainstream of society. Weil focused on the working class, but Tiryakian suggests that an analogy can be drawn with the poor and the aged in modern societies. In a discussion of Oscar Handlin's *The Uprooted,* and K. B. Pakrasi's *The Uprooted,* the author views economic conditions and political changes as the causes of the uprooting of large groups of people in Europe and in India respectively. Thus the uprooting phenomenon concerns entire populations as well as individuals. Furthermore, although uprooting is predictable when social change is analyzed by detached observers at the societal level, it is still disruptive to the individual who may not be adequately prepared for the changes. Thus it can be argued that societal events as discussed by Fried in Chapter 4 are an important context that shapes coping-behavior strategies. Tiryakian also considers the costs and benefits of uprooting. Although uprooting may have pathological consequences for individuals and for social units, it can also result in greater individual self-actualization and creativity. Uprooting may enable individuals to escape the perceived provincial stagnation or economic backwardness of their rural situation.

Whereas Tiryakian emphasizes the sociological dimension of exile and uprootedness in contemporary societies, Pfister-Ammende in

Chapter 8 analyzes *uprooting neurosis* as a pathological condition of the victim of war trauma and dislocations. A clinical case study is presented to sharpen the role of institutional support. Anna Maurer, a refugee who was separated from her family and country upon marriage to a Swiss during World War II and who was later alienated from both her husband and country of residence, suffered from *uprooting neurosis*, a behavioral disorder symptomized by shoplifting, illicit sexual relationships, acute and chronic depression, and attempts at abortion and at suicide. The author gives a moving account of the psychotherapy which the patient underwent as a result of the intervention of authorities, and presents a vivid description of the resulting restoration of her self-image and her capacity to function.

Although Pfister-Ammende focuses mainly on the personal situation of this victim of uprooting, and emphasizes personality and emotional issues, she dramatizes a major theme of the volume, namely, that psychological coping mechanisms must be supported by institutional mechanisms that are mobilized by resettlement agencies in the local area. The author recommends preventive strategies of intervention which host agencies may sensitively use in order to reduce the health risks faced by war refugees and other uprooted populations. She recommends (1) forming liaison teams to work both with migrants and with the indigenous population, (2) conducting epidemiological surveys of the immigrant population with a view toward locating potential sources of stress, and (3) preparing the native population to accept immigrants and to provide social support for them.

5

The Uprooting of Meaning

Peter Marris

When we are asked to describe our roots, we talk most naturally of the past—of parents, teachers, friends, the experiences which influenced us—and perhaps of ancestors whose lives more remotely formed the setting of our childhood. But asked what we mean by uprooting, we think rather of the grief of exiles, of people being torn from the contemporary setting which sustained them. We conceive our roots as both what we have grown from and what we grow into; as both the setting and the attachments which bind us to that setting. Let me try to unravel this evocative but ambiguous metaphor, to understand more exactly what the experience of being uprooted is, why it can be so painful, and why we have condemned ourselves to a form of society which seems constantly to threaten us with it.

The closest human counterpart to the root structure by which a plant nourishes itself is, I suggest, the structure of meaning by which each of us sustains the relationships to people, work, and the physical and social circumstance on which our lives depend. Like roots, these structures of meaning are at once generic and sensitively adapted to the particular setting in which they are embedded; like roots, too, they transplant from this established setting only at the risk of wilting and stunting. But the metaphor simplifies the much more complex way in which these human structures manipulate both form and content, creating the relationships in which they become embedded.

From the beginning of life, children begin to act upon their

Peter Marris • School of Architecture and Urban Planning, University of California, Los Angeles, California 90024.

surroundings, discovering that there are things—things which stand in relationship to their wants and movements. Action explores and confirms the predictability of these relationships: babies soon discover that they can make things happen. As they grow up, they begin to identify, classify, group, and compare, evolving notions of space, speed, time, cause and effect, of reversibility and conservation. These structures of operations by which children interpret the world become progressively generalized and interrelated, until in maturity we may begin to perceive them as theories, ideologies—systems of thought which define reality in terms of its possible transformations.[1] Unless we could learn to structure reality in this way, we could not grasp it at all—nothing exists for us except as it is structured in a set of actual or potential operations. So, by our interaction with our surroundings, we give them form and meaning, creating the intelligible, generalizable regularities which enable us to predict, manipulate, and outwit the threats to our survival. At the same time, as this conceptual grasp develops, we are more and more able to choose and control our relationships: the structures we derive from exploring the possibilities of action become the structures we impose, eliminating possibilities. The social relationships of our lives, especially, are determined by discriminating amongst many possible relationships—and each choice tends to elaborate and further determine the structure as a whole, refining the possibilities of the future.

In part, these structures of meaning can be represented as the common knowledge of a culture—its science, cosmology, ideology—and even more fundamentally in terms of the principles of reflexive transformation which underlie its logic. But they are also in part the elaboration and confirmation of a particular personal history, an understanding of life which is unique because it derives from experiences which no one else has exactly shared. And in so far as they are unique, they can only approximately be communicated. We can never say exactly what life means to us: yet the stability of that sense is crucial to the meaningfulness of our lives.

These structures of meaning have two paradoxical qualities. First, their power to assimilate an enormous variety of new events depends upon their stability. Our versatility, our ability to learn and adapt rests on a fundamental conservatism, since once we lose the thread of continuity in the structure everything threatens to become meaningless. Second, these structures are at once highly generalized, as in maturity they become increasingly abstracted from particular content,

[1] For an account of the correspondence between childhood learning and the formal properties of conceptual structures, see Piaget (1971).

and yet highly specific, as they become embodied in unique attachments.

From the first, what we make of experience derives from our intentions. All meaning is purposeful, however abstracted from actual operations its formal structure may become. A hammer, a chair are only such in the context of their uses. That is, there is no given meaning, no necessary structure of thought, which has not been organized and defined in terms of actual or potential purposeful operations. These purposes may be themselves generalizable, in the sense that they can be satisfied by many people or things, even different classes of things, which can be substituted one for the other. If I want, simply, to make money, there is an innumerable variety of potential ways to do so. But I suggest that the purposes which most crucially define the meaning of our personal lives can not characteristically be satisfied except in highly specific relationships, for which there are no substitutes. The needs for emotional companionship, to nurture, be respected, achieve something, come to be associated with particular people and situations from which they can no longer be detached.

This tendency for the emotional structure of our lives to resolve itself into unique attachments follows, in part, from the nature of the relationships where attachment occurs. They are all, I think, relationships which depend upon a degree of selection and commitment. We cannot love impartially, if love means priority of attention to the needs of one person, nor be a friend of all the world, if friendship means a regular exchange of dinners and confidences. If my child is in danger, I will drop everything to come to its aid, with an assurance that I cannot promise to anyone else's child. Even respect, the honoring of achievement, loses its meaning as it becomes more remote from the particular group of people in whom judgment has been invested, because only a few can truly appreciate what you have done. But this does not explain why one person cannot substitute for another, serially, in that mutually exclusive relationship. And, of course, we may change our loves, our colleagues, our friends or country. Yet to do so seems harder than it need be, if all that we sought was an assured commitment. These exclusive relationships become the context in which we continue to learn; we accumulate an insight and competence which we cannot reproduce in any substitute setting except after a long time. Learning becomes invested in the intimate subtleties of the personal history from which it arose, and which it has helped to shape.

But these explanations do not, I think, fully account for the strength of attachment. They assume that attachment evolves instrumentally, as an expression of needs which are still essentially generalizable, even though they can only be satisfied in specific relationships enriched by a complex structure of secondary purposes and understandings. There seems also to be a predisposition toward unique attachment in our biological makeup. A baby's responsiveness to a nurturing figure becomes imprinted with the image of a particular nurturer, for whom it will not readily accept a substitute; and this imprinting seems to set the pattern of future attachments. (See Bowlby, 1969). We do not experience deep friendship or sexual love as an instrumental relationship, though it cannot survive unless it is instrumental: it seems more profound than relationship itself, the inarticulate basis of emotion from which communicable meanings and structures of behavior arise.

The necessity of attachment may seem overstated, so long as the examples are drawn from one culture or life-style, because you can always point to other ways of life where people do not appear to seek that particular form of attachment. But though the structure of emotional commitment varies, I believe it must always find expression somewhere. If, say, in an African society, marriage counts for less, the family counts for more: the deep sense of belonging is invested in a particular lineage. Or people may look for an escape from the isolation of bourgeois households in communal groupings of care and affection, but that group becomes, the more satisfying it is, the more specifically the place where they belong. Those who claim to live without attachments are, I suggest, either very lonely—bereft of any central meaning to their lives—or they disguise their crucial attachments from themselves, as the promiscuous may remain rooted in their attachment to their parents. To lead a life at once deeply meaningful, yet independent of any particular relationships, is conceivable only, I think, as the vocation of a saint—and I question whether there have ever been canonized men or women wholly uncorrupted by lonely indifference or human partiality.

I suggest, then, that the roots of being are the structures which, for each of us, organize meaning in terms of the way we operate. Structure and purpose are therefore inseparable aspects of meaning, because purpose cannot be articulated except in the context of structure, and yet without purpose there is nothing to articulate. Each develops and transforms the other. And because these structures are essentially operational, we make the reality we comprehend conform to them: the structures of thought reproduce themselves in structures of social and physical organization, as these organizations in turn react upon our understanding. Hence, the structure of meaning which enables anyone to make sense of his or her own life is a unique evolution of both abstract and concrete, generalizable and specific organizations of physical, social, and conceptual relationships, embodied in intentional behavior, and attached to particular people and situations. This structure incorporates shared, public knowledge and skills, but there is a continual interplay between the generalizable principles of organization and the way they operate in the context of personal intentions, loves, hopes, fears, and vocations. Any major disappointment or confusion of understanding, therefore, tends to react upon the structure as a whole, threatening a pervasive sense of aimlessness and helplessness. We can no longer make what happens intelligible and cope with it purposefully.

Uprooting represents any severe disruption in this structure. It may come about through the loss of any of the elements on which structure depends—purpose, attachment, regularity in events, or conceptual coherence. The death of the person you most love, the breakdown of a marriage, loss of an apparently stable job, the disappointment of ambition may all undermine the sense of life's meaningfulness by robbing the structure of a crucial integrating relationship. So, too, shared losses—of a neighborhood through slum clearance, of a way of life through colonization or economic change—can overwhelm the continuity of purpose and structure, leaving people bewildered and demoralized. The sense of meaning may also decay without any readily identifiable loss of relationship, through the emergence of contradictions within the structure itself: as a career unfolds, the latent incompatibilities within its purposes become manifestly irreconcilable; or a society develops conflicts to which its institutional and ideological structures cannot accommodate without radical transformation.

I have tried to explain why meaning is so crucial to survival, and to sketch the complex structure of both actual and conceptual relationships on which it depends. It forms reality into something we can grasp: so whenever the structure threatens to disintegrate, we become anxious: and if we cannot readily accommodate to what has happened, the anxiety turns to grief. Those who have lost a crucial relationship characteristically describe their distress in language which evokes this overwhelming annihilation of meaning. "The bottom fell out of my world," "The meaning went out of my life," " I went dead," "I wasn't there anymore," "Nothing matters now." (See Marris, 1958, 1974). But grieving is more than despair. It expresses the painful search for a way to recover meaning, and the working out of grief expresses dramatically the fundamental requirements of meaningful structure.

The bereaved are torn between contradictory impulses, each reacting against the acute anxiety and helplessness of loss. They take refuge in the past, reliving the time when their lives were vital with meaning, or trying, symbolically, to act as if the dead were with them.

They may dwell in memories, cling to associations, repeat as rituals the old routines. But this retrospective meaning is only reassuring so long as the illusion of a past still present can be sustained: and the illusion is continually shattered by abrupt reminders of loss. Unless the bereaved withdraw altogether into a fantasy where time has stopped before their moment of bereavement, they cannot recover the past without constant stabs of recognition that they can no longer live by the reality they have reconstructed. So they are also pulled by the opposite impulse, to escape altogether from the past and begin a new life, where the loss can be forgotten. But this leads to an acute sense of selfbetrayal, because it denies the meaningfulness of the loss, of the attachments which made loss so painful. The new life is hollow, its purposes without roots in a continuous identity. Hence those who seem not to grieve, but briskly go about the organization of their future, are only disguising an inner emptiness too frightening to face.

Each of these impulses, in seeking to escape from the need to confront and come to terms with loss, would by itself, unchecked, be self-destructive. But together they reflect the two crucial aspects of the work that grieving must accomplish. The bereaved have to find a reason to go on living—a meaning which acknowledges their loss; but that meaning can only arise from the self who suffered loss and be continuous with it. The meaning of the past has, therefore, to be at once retrieved from annihilation and consolidated, and yet transformed, so that it can come to inform the present and the future. The working out of grief is the impulsive, ambivalent, at first desperate struggle to make that transformation. The essential nature of the attachment has somehow to be abstracted from the living relationship in which it was incorporated, and reinterpreted so that its purposes may still be relevant. Hesitantly, erratically, sometimes in anger, sometimes in despair, its realization will evolve toward a reformulation that restores the thread of continuity, reintegrating a structure of meaning strong enough to sustain life. But grief can take many months, perhaps years, of intense psychic effort to reach this resolution (See Marris, 1974,Chapter 3).[2]

The experience of grief shows that the loss of a crucial attachment cannot be made good by substitution. If, for instance, a widow remarries, the new relationship cannot have the same meaning; it can only represent a resolution of grief if it is integrated into a structure of purpose which has already been restored. Otherwise it only inhibits the working through of grief, a dangerous illusion of belonging. Unless we understand this, we risk continually trying to solve the disruptions

[2] For a fuller discussion of personal grief, see Murray-Parkes (1972).

of change and loss by substitutions, intolerant of the seeming irrationality with which the bereaved resist them. And the risk is all the greater in those situations where we do not even recognize that people have been bereaved.

We respect the grief of widowhood, but tend to ignore or disparage the emotional attachments which give meaning to a home, a neighborhood, a group of colleagues. We treat these relationships as if the generalizable need for shelter, work, and recognition were the meaningful structures, and their particular expression merely instrumental. Even in our own lives, we assume that actions which are rational in terms of general principles of career advancement will lead to structures which are meaningful. Yet, though it may indeed make sense to move house, take a new position, we will have to undergo the transformation of the emotional attachments that must be given up: and until that has been worked through, we have not truly reintegrated the meaning of our lives. We may not experience this process of accommodation as grief, because the anxiety is contained within an overriding structure of intention. But the psychic disintegration may also turn out to be more profoundly troubling then we had foreseen—a grief all the harder to resolve because we do not acknowledge it. When those for whom changes make sense are other than those who must live with them, the risk that these changes will be grievously disruptive is obviously much higher.

The nature of grieving shows that meaning, as a structure which sustains life, is a whole: the loss of any crucial relationship tends to undermine the sense of all relationships. And this whole is made up of a subtle, complex interplay between generalizable principles of operation and specific attachments, where the vitality of the structure as a whole depends on both abstraction and commitment. The length and pain of grieving, the need at all costs to hold onto the essential meaning of what has been lost, confirm the uniqueness of attachment. Yet the ability to recover from a crucial loss shows, too, that these attachments can be transformed: the principles of abstraction and reformulation operate at many different levels to ensure the continuity of meaning.

What we call wisdom is, I think, essentially an awareness of this interplay of thought and feeling in the structure of our lives—the sense of how generalizable principles of operation become realized in the emotionally charged vitality of particular relationships, and how these relationships in turn inform the purposes of generalization; the sense that each life is at once unique, and organized by universal principles of structure. To be wise is to understand what is substitutable and what is not. The rationalist who treats all relationships as generalizable is as foolish as the diehard who cannot recognize the possibility of alternatives. Like the impulses of grieving, each represents one aspect of the need to conserve the continuity of structure. The diehard clings to the specific circumstances in which structure has become embodied; the rationalist tries to release it from the vulnerability of any mental attachment. And either, unchecked, represents a self-destructive evasion of the necessities of change. Once we see what is being painfully reintegrated through the process of grieving, we can understand better the conditions of an integrated, meaningful life—a life at once open to learning, growing, adventure, and able continuously to comprehend reality and endow it with personal relevance. Any organization of society sensitive to the well-being of its members must enable them to maintain that continuity. The operations through which it assimilates changes must, as far as is humanly possible, respect the need for a structure of emotional attachments in which these actions can alone be meaningfully realized. Not that benign societies are conventionally conservative: but the way they learn to endure must be, in my sense, wise.

The dominant structure of thought, and therefore of organization, in modern industrial society seems to me to frustrate that kind of wisdom. Its extraordinary power to grasp and manipulate generalizable orders of relationship, to abstract form from content in more and more inclusive systems, represents a dangerously one-sided search for stability of meaning. It operates characteristically by reducing content to undifferentiated elements, which can then be recombined to represent any content, according to the formal properties of abstract systems of relationship. Thus, ultimately, any content whatsoever—anything that may happen—may become intelligible in terms of universal principles: the ideal is a structure of meaning so comprehensive that we are never at a loss. Physical science is only the most impressive expression of this search for an invulnerable ordering of reality. In every aspect of life, we have tried to distinguish the elements of a generalizable system—in economic relationships and the psychology of human behavior, in social organization, communication, and planning. The unique event—a group of people producing something, someone learning, a conversation—becomes reduced to a constellation of universal relationships by which its essential nature can be understood and manipulated.

The power of this mode of understanding is unquestionable; and it stems from the most fundamental principles of human learning. But it does not, of itself, create meaning unless it is informed by attachments which cannot be so generalized. Yet whenever this understanding is used in action, it implies substitutability—that an event can be treated as equivalent to any of the events with which it is classed. Hence the more rational or scientific, in this sense, our forms of social organization seek to be, the more they must treat people as interchangeable. So, for instance, the more scientific management becomes, the more it breaks work down into discrete processes which can be accomplished indifferently by a class of workers; the more rational administration, the more it conceptualizes relationships as a formal interaction; the more sophisticated a political campaign, the more it computes the classifiable biases of the electorate; the more comprehensive a policy analysis, the more it reduces actual choices to a universal calculus of cost and benefit. Each of these ways of grasping and manipulating social reality depends on abstracting the relevant properties of events and subsuming them under the most inclusive possible conceptual system. Their operational counterpart is aggregation and imperialism—the incorporation within a single structure of control of elements whose random variations cancel each other out. That is, the more comprehensive the structure, the less the risk that human idiosyncracies will upset the predicted aggregate pattern of relationship.

Rationalization, therefore, progresses by dismantling the unique event, so that its components can be distinguished and discerned as generalizable formal relationships. The ultimate reality is not an atom or particle, nor a human perception, but a mathematical formula. The practical consequence is a society dominated by an operating logic, where, to make use of its potential theoretical grasp, people must be interchangeable. If it makes sense to close a factory here, open another there, to shift resources from one country to another, the necessary human labor must be substitutable from one location to another. The workers must either move or be replaced: either way, the unique attachments which embodied their purposes have been disrupted. Any generalizable operational system for maximizing benefits presupposes that outcomes can be expressed in equivalent forms—whether of profit, or economic growth, or well-being. But these outcomes can never, in themselves, represent for anyone the sense of their lives. So, paradoxically, the very power to understand and manipulate ends by draining everything of meaning.

The founders of modern sociology all, in one form or another, grasped this central paradox. Marx, who could write almost lyrically of the productivity and resourcefulness of capitalism, saw it also as profoundly alienating, reducing the meaning of human labor to an exchangeable commodity. Weber traced the progress of rationality as a liberating understanding which ended by imprisoning society in an iron bureaucratic cage. Durkheim perceived the division of labor as an inevitable response to the growing density of human settlement, which threatened the basis of moral solidarity on which the integrated meaning of society must work.[3] Since they were concerned with the construction and reconstruction of industrial society, rather than the constitution of meaning itself, they did not deal directly with uprooting and bereavement. But as sociology sought to emulate the formal properties of natural science, creating its own generalizable categories and structural relationships, it lost the ability even to articulate the problem. Differentiation, the abstraction of behavior as roles and norms, the structural transformation into more and more inclusive systems—which for Marx represented the alienation, oppression, and ultimately self-destructive imperialism of capitalism, and for Weber at best an ambivalent triumph of bourgeois rationalism—become in the sociology of Talcott Parsons (1967) the normal coping operations of social adjustment. A tradition of thought born out of concern for the meaning of human experience evolved toward its own abstraction from that experience.

I do not mean to imply that sociological understanding has become empty and useless: only that it cannot, on its own terms, resolve the dilemma which seems to underlie our vulnerability to uprooting. How can we reconcile our need for attachment with our need to generalize the structure of relationship? How can we integrate thought and feeling, so that the interplay between them makes, for each of us, a meaningful life; organize the collective mechanisms of adjustment to avoid a continual threat of personal bereavement? Just as in grief, the bereaved seek to escape the painful task of reintegration by escaping into the past or future, so, I think, there are two reactions to the impersonal logic of contemporary organization, each of which, if unchecked by the other, leads to perverse solutions. The first is represented, for instance, in the writings of Ivan Illich (1973). He repudiates modern science and technology, urging us back to a smaller-scale society of more primitive techniques. I doubt whether such a voluntary abnegation of our power of understanding and control is even conceivable. It seems plausible, if at all, only because the advocate is highly sophisticated, and does not share in the ignorance he wishes to impose. The solution is essentially authoritarian. Any meaningful turning away from the possibilities of modern technology has to be more, not less, sophisticated than the techniques it repudiates—that is, it has to include these possibilities within a more comprehensive structure of meaning. The second impulse is, I suggest, more diffusely represented in a culture of nonattachment: a bland receptiveness to

[3] For an excellent comparative discussion of Marx, Weber, and Durkheim, see Giddens (1971).

relationships which makes no commitment to any. Everyone is lovable in general, and no one in particular. Neighborliness, affection, tenderness, community spirit, are treated as feelings without specific content, drifting benignly where chance blows them (See Slater, 1970). But the lack of commitment robs all such relationships of a crucial aspect of their meaning: they become hollow and sentimental. The real attachment becomes introverted in narcissism.

Yet it is not hard to conceive social relationships at once unique and representative of generalizable organization. In the early Middle Ages, vassalage was a crucial principle of feudal order: but it was also a personal bond, sealed with a kiss.[4] And we retain, at least residually, the ideal that marriage, parenthood, friendship should conform to public, shared expectations, while expressing unique loyalties. Their meaning is both commonplace and intimate: and we treat them still as the most meaningful of all relationships—together with the vocational attachments of those fortunate enough to find work with which they can personally identify. But it becomes harder and harder to maintain this constellation of attachments in the face of institutional pressures. Racine, the court dramatist, could more easily refuse to dine with the King of France, because it was his child's birthday, than a contemporary executive could refuse a week-end assignment in order to play with his children, or a posting because his wife did not want to give up her job. To preserve marriage as an endurable bond, the wife is expected to treat all other attachments, to work, friends, community, as expendable—narrowing the meaning of her life to a tight knot of family cares which risk becoming overburdened with a stifling intensity. And besides these personal strains, whole communities may be robbed of their sustaining structure by economic decisions which leave many stranded in a bewildering emptiness. The efficient decision is not perceived in terms of reconciling production with the complex texture of life, but is calculated only by a generalized economic logic.

I do not mean to suggest that everyone should remain rooted in one place—nor even that we should always seek to avoid disruptive changes. But we need to understand and respect bereavement, in all its forms—to recognize that it is both painful and demanding of much emotional energy to repair. We need to beware of imposing on

[4] "Imagine two men face to face; one wishing to serve, the other willing and anxious to be served. The former puts his hands together and places them, thus joined, between the hands of the other man . . . [and] utters a few words—a very short declaration—by which he acknowledges himself to be the 'man' of the person facing him. Then chief and subordinate kiss each other on the mouth, symbolizing accord and friendship. Such were the gestures . . . which served to cement one of the strongest social bonds known in the feudal era" (Bloch, 1965, pp. 145–156).

ourselves more disruptive changes than we are resilient enough to withstand—and to challenge the right of anyone to impose changes which undermine the meaning of other people's lives. When change and loss are inescapable, we need to create the circumstances in which people may have time for grief, and ways of working through it. Contemporary industrial society is, I think, unique in abandoning the customs and rituals of mourning which have always before helped to articulate grief and lead it toward a gradual resolution. So too we have become indifferent to the nature of every kind of loss: and all this, I suggest, arises from a one-sided preoccupation with a single aspect of meaning. The conception of society as organized in large-scale formal relationships, where individuals are substitutable for each other, reflects a conception of the physical world constructed from the formal properties of a few elements. The sense of power to manipulate both the social and the physical reality conforms to the same principles. Even the moral universe is brought within the framework of this logic, as a utilitarian calculus of discrete units of happiness. The values which any human life is thought to pursue are phrased so abstractly that they appear as an accumulation of gratifications, continually presented in the bland images of advertisements. The Puritan sense of being alone with God, of wrestling with the moral meaning of one's private life, has evolved into a public tolerance which treats every creed as a celebration of the same utilitarian conformity. By these conceptions, loss becomes merely replacement, uprooting reorganization, and grief a sickness. This reluctance to recognize the nature of grief seems to reflect a deeper reluctance to face the central question of meaning.

Uprooting, then, can be seen as a crisis of reintegration, where relationships, the way they are conceived, and our emotional attachments, all have to be restructured, in a complex transformation which maintains an underlying continuity. But our personal vulnerability to uprooting is reinforced by a failure of integration in the shared ideologies which interpret and shape the structure of social reality. Not only does this leave the resolution of the personal crises unsupported by a common understanding; it imposes such crises by the insensitivity of its constructs.

If continuity of meaning is crucial to well-being, the wholesale relocation of people or work and the disruption of communities by economic change are hardships—even when substitute shelter or employment is offered. And this hardship cannot be treated simply as a cost, whose claim upon society can be discharged by compensation. Once we recognise this, the ruthlessness and underlying selfishness of actions which dispossess people of the context of their lives can be disguised no longer. The politics of welfare has been preoccupied with the distribution of goods, where I consider that we should be concerned, more fundamentally, with a politics of meaning. The ultimate questions are, not how many people live in substandard houses, or whether welfare roles or gross national product per capita are on the rise—but, for whom is life most meaningful, whose meanings predominate, who has help to learn, and for whom is change a bewildering imposition? Even if some general need can rightfully disrupt the structure of relationships some people have built for themselves, have we a right to ignore the grief we cause, and shrug it off as the price of progress? Unlike welfare, meaning cannot be given or withheld: it must be learned, and no one can learn on behalf of another.

The politics of meaning implies, therefore, a far more complex interaction between people than assimilating individual cases to general principles, or inviting the public to contribute its views to a plan. Consider this instance of apparently responsive social action: communities in the East End of London were invited to take part in planning the redevelopment of the decaying dockside neighborhoods stretching along both sides of the Thames River. It is an area of overcrowding, declining employment, and dilapidation, lacking parks, schools, and buses. The East Enders knew these hardships well, and had been protesting against them for years: yet they retained an affection for the quality of life that still dwelt precariously in the old dockside parishes. But the redevelopment plan was not designed to respond to what Wapping or Silvertown or Rotherhithe meant to those who lived there, nor to help them to master the changes which threatened them. It set out from the most general statement of the issue, drawing a boundary around the largest conceivable area of eventually developable land, and asking how to plan its future for the benefit of the metropolitan region as a whole. The residents could then only legitimately articulate their needs with a framework of analysis and logical sequence of decisions which subsumed these needs under the principles of a far broader and remoter set of questions. Instead of assimilating planning to the problems they perceived, their problems were to be assimilated to the issues as the planning team defined them. Hence their participation might help the team to learn: but they learned only the frustrations of a dialogue in which their language was continually discounted. The relationship rapidly soured, although the planners insisted, I think sincerely, that they wished to put the interests of the residents first (see Marris, 1975).

If, instead of framing the questions in the largest conception of the issues, the plan had set out from the residents' immediate concerns, they could have grasped and shaped it. As they explored solutions, they would still have come to the wider interests, the relationships with surrounding areas with which their needs had to be reconciled. But they would have encountered them on their own terms, within a context of action where they could negotiate their interests responsibly as a political issue. The planners would then have found their role in working out the requirements of this resolution.

Again and again, actions, seemingly designed to help people, frustrate and bewilder them by alienating the problem from the context of their lives as they perceive it. Community planning against poverty in the United States was intended to be controlled at least in part by the poor themselves. Yet those who funded these programs and their expert advisers preempted any real discussion by insisting from the outset on a comprehensive framework of analysis, derived from social-science theory, in whose terms the only valid proposals were experimental demonstrations of generalizable principles of action. A woman on welfare, harassed by rats, debts, and constant threats to the safety of her children, does not see why her evident problems must be translated into the language of theories of deprivation before they can be tackled: why the immediate, obvious hardships whose meaning is so clear to her cannot simply be accepted for what they are. In the same way, a peasant family does not see a problem of world overpopulation, but the needs of its children, and so cannot relate to the family-planning campaigns provoked by the concerns of international agencies. In all these instances, the intention is to help, but in none of them are the issues defined by those to be helped, or stated in their terms. And therefore the intervention only alienates its supposed beneficiaries, making them dependent on an understanding beyond their experience.

This paternalistic preempting of meaning intrudes insidiously even into relationships where those in need of help do, apparently, control the definition of the problem. For instance, a woman in distress after the breakdown of her marriage consults a psychiatrist. However unassuming the professional guidance, it revolves around questions of personal adjustment, whose definition presupposes some conception of normal behavior: and the psychiatrist's conception of normality predominates, because only within that framework can the validity of his "expertise" be sustained. Unless the personal problem can be subsumed under a general theory of problematic behavior, the therapeutic role loses its professional legitimacy. So, the meaning of this woman's difficulties comes to be defined by the psychiatrist's sense of normality, whereas it might have helped more to redefine normality in the light of her difficulties.

Thus, the primacy that we give to abstract, systematic conceptions of meaning tends continually to overwhelm the personal context of understanding: and this makes the expert, rather than the sufferer, the acknowledged authority on a problem. Yet experts rarely control the conditions of a solution. They do not characteristically own the resources, or govern the economic and political relationships, of society. The definition they give to problems, as practical mediators of social action, inevitably rests on the given structure of power. So it becomes impossible to raise, within their formulation of the problem, the issues of power that are crucial to the terms of its solution: and it is this imposition of universalized constructions of reality on the political question of whose meanings are to predominate, which undermines our groping after a more eqalitarian and less uprooting structure of relationships.

We can nearly always make sense of the conflicts which affect our own interests: whether we win or lose, act wisely or foolishly, these conflicts are meaningful, and we can learn from them. And we can make sense of theoretical knowledge, so long as we can assimilate it to our own structure of meaning. But it is profoundly frustrating to deal with questions of power once they have been mediated by theory, because theory presupposes a generalization of the problem only possible when these questions have been settled. Thus, impersonal structures of thinking, of the kind we call theory or science or expertise, cannot be used to define questions of action, without preempting the politics of meaning. The more we try to obviate conflict by transposing questions of power into problem solving, the further we are from reconciling public and private meanings, from translating the language of science into the language of attachment.

Planners, scientists—doctors of any philosophy—whose vocation is to discover and apply some generalizable theory of relationships are useful essentially because they can interpret one person's experience in the light of another's. The final purpose is not to translate the meaning of particular relationships into a generalizable language, but through that generalization to make this meaning accessible in terms of another, equally particular, personal experience. Unless the sequence of transformations is completed, theory becomes a substitute for the resolution which must arise from political interaction. Instead of clarifying the nature of conflicts it seeks to settle them, and, by invalidating these political meanings, becomes more profoundly alienating than the straightforward exercise of power.

I am not arguing against the need to learn from conceptual systems which form reality into constructions of universal relationships. If our lives are to be meaningful, each of us has to take responsibility for his or her own actions, in all the complex interdependence of human affairs: and we cannot understand that complexity without some theory more comprehensive than our own experience. But we cannot take that responsibility, nor use the power of abstract thinking, unless each of us insists on the primacy of our own context of meaning. To protect that reality, we have to struggle to control the circumstances of our lives—to fight, conciliate, bargain with the interests of others. But first we have to control the interpretation of these circumstances, so that, whatever happens, each of us can learn, assimilating events to our personal understanding. Then, whatever the losses and disappointments we suffer, we have the most strength to withstand grief.

So I think that if we are to make our lives less vulnerable to uprooting, we must first control the uses of knowledge, and beware of alienating the sense of life in search of general systems of operation which, in themselves, cannot endow experience with meaning, but only translate the meaning of experience.

References

Bloch, M. *Feudal society* (L. A. Manyon, trans.). London: Routledge & Kegan Paul, 1965.

Bowlby, J. *Attachment and loss* (Vol. 1, *Attachment*). New York: Basic Books, 1969.

Giddens, A. *Capitalism and modern social theory.* Cambridge, Eng.: Cambridge University Press, 1971.

Illich, I. *Tools for conviviality.* New York: Harper & Row, 1973.

Marris, P. *Widows and their families.* London: Routledge & Kegan Paul, 1958.

Marris, P. *Loss and change.* New York: Pantheon, 1974.

Marris, P. Planning for people: The docklands example. *New Society,* 1975, Feb. 20, 447–449.

Murray-Parkes, C. *Bereavement.* New York: International Universities Press, 1972.

Parsons, T. Evolutionary universals in society, in *Sociological theory and modern society.* New York: Free Press, 1967.

Piaget, J. *Genetic epistemology* (E. Duckworth, trans.). New York: Norton, 1971.

Slater, P. *The pursuit of loneliness.* Boston: Beacon Press, 1970.

6

Uprooting and Self-Image: Catastrophe and Continuity

Kurt W. Back

Catastrophe and Continuity

Uprooting and Self-Image

The term "uprooting" brings to mind a definite picture. "Roots" go down into the ground, "up" represents the opposite direction, counteracting the image of rest which the first part represents. The association of vegetable life, permanency, and vertical direction gives a peculiar interpretation of a human being in a certain predicament.

The use of a term with these associations shows an interesting point of view of residential mobility. We may notice this attitude by using the term "transplanting," which keeps the vegetable image, but changes the direction to a horizontal progress. The same change of residency looks here like a predictable, normal event, a necessary temporary shock in the service of future development. The vegetable and passive image is still maintained, but the direction is now changed from the vertical to the horizontal. Horizontal translation means that a new soil will be found, that the action—even if by an outside agency—is a regular event, and that it may be beneficial to the organism. Uprooting, stressing only the vertical dimension, shows no future and no expectation of normal development.

Kurt W. Back • Department of Sociology, Duke University, Durham, North Carolina 27706.

Etymological discussion such as the foregoing may seem to be outlandish or pedantic. However, any exposition of a phenomenon which can be understood by others is a kind of language. Science has developed several kinds of languages for its use; in fact, it can be looked upon as a collection of languages, with mathematics being the principal one. We shall turn to a discussion of the appropriate type of mathematical theory and terminology for the discussion of uprooting and mobility. Before doing so, we must define what the problem is about which we are talking. For this purpose, we can proceed to an analysis of the common everyday language.

The metaphor of uprooting occurs in many Western and non-Western languages. This choice of metaphor leads us to a very important question. Given that change is bound to occur as a part of life, how can we distinguish the continuous change and development which is the essence of life, the changes which are major but can be expected as part of any life, and the unique "uprooting" changes which may disrupt the whole life of an individual?

While we are discussing uprooting, we must also define the complementary term, namely, what it is which is disrupted by uprooting. Tentatively, we can call it the self-image. Our task now is to show the meaning and relationship of these terms in common language and in science.

Subjective Language

We can note first the importance of time and space in the notion of uprooting, the time in a person's life and the direction and distance of change. Time here means even more than a short or medium range; it implies a larger view of the life course by a person, including interpretation of the past, expectations of the future, and fittingness of a change at the present time. The same act of mobility, looked at only from a purely spatial point of view, may be an uprooting experience or not, depending on its place in the projected course of a person's life.

Any move, even to an extremely different environment, can be accepted as a regular part of life. Sometimes a change, though abrupt in the physical environment, is of short duration and repetitive—such as commuting, or the transhumance of pastoral people (Braudel, 1972). Changes are sometimes anticipated as part of the course of life; examples are conditions which are initiatory into adulthood, the journeyman's travels, relocation for education, the Grand Tour, or military service. What would be closer to our present concerns are traditional migrations to cities, centers of trade, or even to distant countries in order to accumulate funds to be able to live in chronically depressed areas, or the tradition for some family members, such as younger sons, to do so. Such migration is built into the projected life course and, if accepted as such by all participants, becomes a step in the life cycle, transplanting rather than uprooting. The crucial point of the subjective experience of uprooting is therefore the combination of change in space and inappropriateness in time. This combination of pattern and change is the problem of the appropriate mathematical language. This problem is not common in mathematical tradition. We shall therefore discuss in detail the development of mathematical language for our purpose.

Mathematical Language

The Problem

Basically, the mathematical representation of time is that of sequence, linear or at least continuous; that is, an arithmetical representation. The representation of spatial arrangement is that of patterns, a geometrical language. The ideal mathematical system for us would take account of the patterns, the constellation, along the time continuum, analogous to ways in which we can describe the life-cycle and migration dimensions. Supplementing the functional analysis which brings individuals from one point in time to the next, we need a geometrical representation which describes the constellation at any one time. These can be composed to see whether continuous development or sudden breaks are occurring.

Description of patterning, of the "gestalt," has been a principal aim of social scientists, though usually they have had to settle for a purely arithmetical scheme. Advances in mathematics and in social research have made it possible to make a new attempt on this problem, and see how it can help us in discussing the different effects of rapid changes within a person's life. We shall sketch the development of the problem in its historical context.

Mathematics of Physics

The first success of a mathematical approach to empirical phenomena, and still its prime showpiece, has been in the physical sciences. This success has been so stupendous that the methods employed for this field have been looked at as the standards for any use of mathematics in science. Especially, theoretical physics has been resolved to a branch of mathematics. This has been accomplished by the work of physicists as well as that of mathematicians. For our present purposes, the latter contributions will receive a closer look.

Originally, say in Galileo's time, there was no mathematics available for the new physical science. Two kinds of mathematics had to be developed to accommodate the science of mechanics, that is, to treat the movement of rigid bodies. They were analytical geometry and differential calculus. Both were necessary for reducing the facts of movement, which occurs in space, to a relation among numbers. Analytic geometry does this by translating position and distance into measurement on coordinates, and calculus by introducing computable relationships into the measurement of continuous change. The physical assumption for which these new departures were so important is the reduction of bodies to a point with specific measurable characteristics (e.g., a body can be represented by its center of gravity with a certain mass, position, and velocity) and an exact measurement: the relations stated presupposed exact numbers, deviations from which were considered errors.

Even with these assumptions, theoretical mechanics is restricted to close limits. Mechanics is best adapted to continuous movement. Where there are too many jumps, as in small particles, other auxiliary methods have to be devised. Even in the movement of rigid bodies, there are difficulties if the shape of the body becomes important, as in air resistance. In ballistics, for instance, the exact movement of a projectile is not determined purely mathematically, but empirically, shape factors being introduced as arbitrary constants.

It can be seen that the mathematical advance was, in effect, a change from geometry to arithmetical representation. The limitations are apparent when geometrical features of the path of the object start to become important, as they will in concrete applications. This will be true in many branches of science, but it becomes all important if we deal with living systems. We shall show how this new requirement has affected biology, and how it can produce a new language for behavioral science.

Mathematics of Biology

The importance of form in biology is evident. An organism develops from two germ cells; the pattern and shape of the organism are determined by the genetic code, and determine its species as well as its survival. Further, more or less the same type of organism will result, even with a number of vicissitudes during the development.

However, two different germ cells will result in a radically different organism. We have here several important differences in the necessary mathematical representation: first, the configuration is at least as important as the number (e.g., the number of cells is less important than their arrangement). Second, the outcomes are discontinuous, and intermediate outcomes are not likely (e.g., between species or between sexes). Finally, there is no single time path to be computed, but a set of acceptable trajectories which will lead to predictable final outcomes.

The construction of mathematical systems which can meet these specifications has been difficult and slow. Only as late as 1917 did D'Arcy Thompson put together the fundamental theoretical outline of the problem, in his work *On Growth and Form* (Thompson, 1961).

This combination of patterns and functional changes over time makes for a very complicated and almost unmanageable procedure. However, some aspects can be neglected by simplifying the ideas. The course of life of any organism is for us only important in its large outlines. In fact, it is just the fact that the organism is essentially stable which distinguishes the study of living systems. The task of finding the unity in the seeming diversity is the task of the biologist as well as of the psychologist and the social scientists. This task is the determination of the enduring characteristics of the organism. Of course, details of characteristics may vary, and one has to define the units which stay constant with a particular purpose in mind. One wants to talk of the same organism from infancy to senescence, or one might acknowledge abrupt changes where "he is not the same person anymore."

Topology and Chreods

Mathematically, one needs to define a configuration which keeps its permanence, disregarding small changes. Modern biologists have designated such a temporally extended configuration a "chreod," a Greek term for "necessary path" (Waddington, 1957). It is a configuration which keeps its essential stability over time, even though it may be temporarily disturbed, such as a developing organism, which will become a normal adult member of the species, learning under a wide range of disturbing conditions which may lead to temporary deformations.

The mathematical point of view taken here is comparatively simple because it disregards small, short time shifts and concentrates on larger configurations which do not change frequently. This point of view is often advantageous even in physical sciences. Thom, a mathematician in this field, gives several instructive examples (Thom, 1970). If we mix different fluids in a container, we can determine accurately the initial and final states of the mixture. It would be practically impossible to derive the differential equations which could describe the path of each particle to its final destination, though we know the eventual final composition. We do not have to believe in teleology to accept the fact that the exact placement during any moment of transition is not important, but that we know the final goal. One does not have to invoke a mystical conception of fate to account for the observer's knowledge of the final outcome.

Another example would be the division of a country into river basins. There is a comparatively small number of basins in any area; if we want to trace the water cycle in the country, we have to know only the basins into which water enters—at whatever point of the basin a raindrop enters, we know at what river mouth it will end up. Again, we do not have to trace the whole paths of raindrops in order to determine which river will carry the water. Compared to the complexity and size of the topography of a country like the United States, the division into river basins is a great advantage. It makes possible further analysis of the development of the whole system over time.

The last example shows a strong analogy to biological and sociological problems. We do not have to trace individual life cycles to identify the major distributions of individuals. The biologist can predict the future species of germ cells and also distinguish major stages in the development of the organism. Mathematical procedures of theoretical biology have been created to deal with these aspects of development. They recognize that they have to consider complex systems, but that they have only rough stages of development to work with. The change from a mechanistic to a biological framework was accomplished without involving any questionable principles such as vital forces or destiny. It involves just a shift in point of view of organizing and analyzing available facts. It is important, therefore, to look more closely at the procedure as a starting point for indigenous social and psychological procedures.

The principle of a developing system is that it has a predetermined course, and that minor perturbations will not affect the course of events over the long run. Given a certain curve of development, one can develop from it some equations which have the requisite stability. These sets of equations are then called dynamic systems, which are stable if their shape is not affected by disturbance (Bruter, 1974, Thom, 1975). The basic structure will depend on a number of parameters and dimensions necessary for describing the system. The system can be considered stable as long as the basic shape remains. A line representing a function can be deformed gradually to one looking quite different. A power function having maximum and minimum points can be transformed gradually into a straight line. Graphically, a string having peaks and valleys can be pulled tight. However, though the change is continuous, we notice intuitively a sudden change when the peaks and valleys vanish. Here we can see a sudden complete transformation of shape in a continuous movement. The points where these qualitative changes occur are called the catastrophe points of the system. These points can be made to represent meaningful changes in a living system. Beyond a catastrophe point, the whole constellation and the relation between its points changes. What we have called the chreod, the configuration persisting through time, is disturbed at that point.

All in all, the mathematics shown to be appropriate for biological development defines the equivalence of similar shapes or patterns, equivalent to the biological interests in species and other types whose numbers can be recognized but not easily defined. The configuration can then be treated as sets, and in the development of these sets, continuity can be treated as a sort of equivalence, usually called topology.

Topological methods have been applied successfully to such fields of theoretical biology as embryonic development and analysis of nerve sets (Zeeman, 1977). It fits the fact of living systems better than the mathematics or movement model describing the continuous curves of functions of several variables. It is likely of use in the behavioral and social sciences, which also deal with organisms. The transitions here may be as imperfect as the one from mechanical to living systems, but it should be a better approximation to human action than a mechanical model.

Mathematics of Social Science

One clear difficulty in applying the topological model to social science is that in biological conformations we can see the patterns, such as cells and nerve sets, but in psychological and social events we cannot do so. A system of attitudes and values can be represented geometrically, but this is in itself an analogy, whereas the development of cells in an embryo is an empirically given datum. This has been true of any mathematics applied to social science. The arithmetical scaling-of-attitudes model was a complex effort to do what a simple ruler could do in measuring length. The concepts of theoretical biology, such as chreod or catastrophe, are still useful for our purpose, coming from a closely connected subject matter; but we must make some inferential leaps in applying them to the problem at hand.

Intuitively, a geometrical representation of psychological conditions can be a natural idea. One can talk of neighboring attitudes, of one idea's leading to another, or of deep and superficial layers of personality. Visually-inclined scientists have drawn diagrams of all kinds of psychological and social processes. Terms like "depth psychology" and "mapping of semantics," or the drawing of sociograms, attest to this fact. Even half a century ago, K. Lewin (1936) tried to translate intuitive visual notions into a topological psychology. Here a spatial metaphor was taken seriously. Advantages and the attending difficulties in defining shapes soon showed themselves.

In representing the topology of a life course, the biologist can start with the development of the visible cells or organ system and describe their course and abrupt changes. In tracing the course of a human life from a human perspective, one has to build constructs from nonvisible patterns, such as background, attitudes, or expectations. An individual topology would have to be constructed from each individual's perceptual field. A radical reorganization of this pattern would then signal the end of the chreod, and a catastrophic condition. If this condition were triggered by a residential change, we could then call it uprooting.

The definitions proposed here can qualify as a mathematical representation of a subjective view of uprooting, and intensive study of individuals could show the individual expectations of the life course, and the times when they change abruptly. However, this mathematical analysis of phenomenological data becomes impractical when a social event is studied. Uprooting is to a great extent a subjective condition, and an analysis of individual life histories is the only way to represent this point of view faithfully. This would mean that any discussion of the uprooting of larger units is practically impossible. We need, therefore, a compromise, a conveniently objective criterion of continuity which also makes subjective meaning.

By considering comparable experiences of many individuals, we can determine the situational conditions of uprooting. Thus, we are using their insight, which we have gained from the use of a new mathematical language based on the principle derived from subjective experience. From this we can turn to a sociological language, combining two fields of sociology, migration and the sociology of life cycles.

One of the functions of a constant pattern is its predictability. Sociologically, we use terms like "roles" and "ideology" to express this fact. As long as the roles and actions of different people with whom one comes into contact and the relations of different facts and ideas in one's mind stay constant, one's actions and reactions fall into an habitual pattern, guaranteed by past experience. If the pattern is upset, then each perception must be analyzed for its meaning, and each action entails a decision. The reaction to an uprooting situation would then be attempts to maintain predictability. Conversely, the stress of uprooting would be seen in the impossibility of maintaining a predictable universe. On the other hand, a geographic change, which maintains predictability, would then not be an uprooting experience. We can define the chreod—i.e., what is preserved under these conditions—as the self-image.

We can look at the mechanisms for preserving predictability in two ways: from the inside, as a preservation of the self-image, and from the outside, as social efforts to preserve continuity. In both uses, attempts are made by the individual or by social groups to preserve the self-image, to avoid a catastrophic situation where completely different rules apply.

Instead of defining exactly the configuration of attitudes and expectations and their changes in uprooting, we can look at mechanisms which have developed to survive uprooting situations. We can assess from the point of view of efficiency whether they intend to avoid catastrophe (maintain the chreod) or provide an adaptation to an unavoidable catastrophe.

The Language of Social Science

Conservation

A scheme which provides a sociological translation of the topological model has been proposed by Peter Marris (1974, 1977). He reviewed several studies which he conducted on uprooting and related topics, and found a common trait which he called "conservation." The inference of this trait could explain behavior in widowhood, and among native entrepreneurs in East Africa, missionaries in West Africa, and working-class students in the English elite universities. In all these cases, successful adjustment included preservation of some essential factors of the previous conditions and change of the new conditions to adapt to the preserved self-image. In this way, new ways of behaving are constructed. Thus, innovative behavior is derived from a desire to maintain the individual patterns, the self-image. Thus, widows realize the essential meaning of their relationship to their dead husbands, missionaries the principal factors of their religious commitment, and the mobile working class the meaning of class identity.

Marris makes the point that successful weathering of a crisis depends on recognizing the essential features of the previous situation and using it for a creative adaptation to the new situation. This is the way of preserving the essential self-image and acting effectively in the new place. Contrasted with this optimal procedure are the twin dangers: keeping the whole self-image in the new situation, i.e., an excessive attachment to the past, and giving up one's old self-image. As an example, one can consider the aging process. The ideal conservation process is to take the essentials of one's youthful personality and carve a place in society appropriate to one's age. The contrast would be, on the one side, the excessive dwelling in one's past, and on the other, the start of a new life defined by retirement villages or golden-age clubs. The chreod, the necessary path, would determine the normal development even under this hardship. The other solutions either destroy the chreod or make its path impossible.

Adaptation and Its Pitfalls

One can look at the several social phenomena as different adaptative or maladaptative measures for the conservation of the self. One phenomenon which has been frequently noted is that the attitude and behavior of rural migrants to cities are not intermediate between those of natives in the two places, but are further removed from the urban residents than even those of the rural-born residents. Thus, fertility has been shown in some conditions to be higher among rural-urban migrants than among rural residents, with urban residents being the lowest (Caldwell & Okanjo, 1968; Morgan & Kannisto, 1973). A current phenomenon is the exaltation of the previous identity in symbolic forms, and aggressive assertion of nostalgia. The proverbial Irish-American nationalism is only one instance of the increase in identification with the group an immigrant has left. An opposite feature is extremely strong anticipatory socialization, rejection of the background, which leads, for instance, to intergenerational conflict among immigrant groups. In this condition, the individual will try to make a new construction from the available cues of the new situation (Brandel-Syrier, 1971). This new self will then depend on the clearly visible features of the new life, which will give an exaggerated picture of the new society which it is supposed to represent.

Groups

Some other recurring events among uprooted groups can be understood as means of avoiding the twin pitfalls. Chief among these procedures are the creation of special groups, and the appearance of rituals or games. Groups become an extension of the person in the specific situation of mobility. The natural group to do this is the family, and the function of the nuclear or extended family in mobility is probably still underestimated. Litwak (1961) has shown how the extended family was influential in U. S. immigration, and immigration laws are witness to its importance by favoring relatives of current residents. Temporary groups can be created for persons who are in similar situations, such as newcomers to a city. Mintz (1976) has shown how a pseudo-family developed among Africans brought as slaves to the Caribbean. Here, companionship on the same ship led to the formation of almost familial ties, and descendants of these groups are still identified in the islands. The temporary attachments to these groups give a way of maintaining the old identity while the status in the new group is worked out. The status in the temporary group can assume importance for a short time and act as an intermediate stopping point in maintaining identity.

Again, we find these groups created frequently to guide in the transition from one stage to another in the life cycle, that is, in temporal uprooting. Groups of this kind are found in initiation ceremonies, creating a cohesive group of initiates whose membership has meaning during the ordeal and later. Few such groups are institutionalized and recognized in spatial uprooting. But it is significant that in the mobile American society surrogates of these groups have sprung up, namely in the encounter movement. These groups, however, do not perform the whole transition function, but provide only temporary attachments without providing a transition to a new chreod (Back, 1972; Fernandez, 1978).

The effect of groups can be classified in the same way as the adjustments in individual development. Under proper guidance they can lead to a creative response which preserves the individual self-image under new conditions.The maladaptive groups create excessive nostalgia, which makes adaptation to new conditions impossible, or provokes a complete break with the past which destroys the chreod and forces a new self-image (Marris, 1978).

Ritual

Another way of protecting the self-image is rehearsing the change, or some other way of immunizing oneself from the change. A general way of describing this technique is the formal ritual. Rituals are designed to guide people with similar experiences. Following a ritual can transform an individual crisis into a socially recognized predicament. Children sometimes use games for the same purpose. They can rehearse by playing a future identity, or discover by replaying the essential features of past identities. The more rigid features of a ritual retain the essential character of play—the limitation of time and place, the separation from the concerns of the everyday world, and the reduction of the situation to a few formal rules (Erikson, 1977).

Group experiences and ritual are combined in ceremonies which have been called *rites de passage* (Van Gennep, 1960). These rites are especially fitted to catastrophic changes. Special rituals assemble groups of candidates for a major change, explain the traditions from which they came, keep the group in an intermediate step, and then give the initiates further guidance in the meaning of the new status. Mainly, this is a combination of group procedures and ritual. One special feature which should be mentioned is the intermediate step where the group is kept together in no status at all. This phase, which is called liminality, serves to stabilize the self-concept, now only enclosed within the transition group (Turner, 1969). The ultimate outcome of preservation of the chreod may well depend on the successful managing of this phase.

Ritual may involve seemingly superficial aspects of life in keeping with the playful character of much ritual. Thus, the individual self-image is frequently preserved by diet patterns—keeping national food habits as far as is compatible with the products of a new location, such as Chop Suey or pizza. A case can be made for the close connection between eating and the self-image; physically food creates the self, and that is true as well in the social sense (Back, 1977). Transformation of food habits can give a graphic picture of the preservation and transformation of the chreod.

One extreme situation where the self-image is in obvious danger is bereavement. Here, the immediate group is destroyed, and ritual has to recreate the new affective relationships on which the self-image rests. The appropriate ritual here is mourning; excessive mourning can lead to complete loss of the self (Lindemann, 1977), while suppression of the ritual and avoidance of any speculation about future direction puts grave stress on the survivors. It is interesting that investigators have found parallel emotions in spatial uprooting (e.g., in the loss of a home through urban renewal). Mourning for a lost home represents rituals similar to those involved in mourning for a lost person, with creative adaptation threatened by reliving one's life in the old home, or by going through the motions in the new home, with the feeling of having lost one's identity (Dohrenwend & Dohrenwend, 1969).

Conclusion

The language of geometry is helpful for understanding the several problems of transition. The task in managing uprooting is to transplant a new configuration into a radical different setting, into which it may not fit at all. Preserving the configuration, assessing its essential stability while adjusting nonessential parts to the new environment, is the mark of handling the crisis successfully. In the analysis of individual handling of stress, it has been shown that the conditions which surround the person can reduce the physiological symptoms of stress. These include the presence of friends, common experiences, ideological committment and even physical envelopment in obesity (Back & Bogdonoff, 1967). In temporary stress situations, a geometrical, topological picture is a powerful theoretical tool, and the same can be said of the study of catastrophic changes in the course of life, such as uprooting.

Conversely, the power of the theory will depend on the ability of scientists to give operational meaning to geometrical constellations. In discussing the application of the theory we have seen that the different aspects of the self-concept are the most useful interpretations of geometrical constellations. Thus, the task of applying theories analogous to those of developmental biology to social events such as human uprooting lies in the definition and use of fruitful measures of the self-concept.

References

Back, K. W. *Beyond words: The story of sensitivity training and the encounter movement.* New York: Russell Sage Foundation, 1972.

Back, K. W. Food, Sex, and Theory. In T. K. Fitzgerald (Ed.), *Nutrition and anthropology in action*. The Netherlands: Van Gorcum & Co., 1977.

Back, K. W., & Bogdonoff, M. D. Buffer conditions in experimental stress. *Behavioral Science*, 1967, *12*, 384–390.

Brandel-Syrier, M. *Reeftown elite*. London: Routledge & Kegan Paul, 1971.

Braudel, F. *The Mediterranean* (Vol. I). New York: Harper & Row, 1972.

Bruter, C. P. *Topologie et perception* (Tome #1). Paris: Doin, 1974.

Caldwell, J. C., & Okanjo, C. (Eds.). *The population of tropical Africa*. London: Longmans, 1968.

Dohrenwend, B. F., & Dohrenwend, B. S. *Social status and psychological disorder: A causal inquiry*. New York: John Wiley & Sons, 1969.

Erikson, E. H. *Choice and reason: Stages in the ritualization of experience*. New York: Norton, 1977.

Fernandez, J. Passage to community: Encounter in evolutionary perspective. In K. W. Back (Ed.), *In search for community*. Boulder: Westview (AAAS), 1977.

Lewin, K. *Principles of topological psychology*. New York: McGraw-Hill, 1936.

Lindemann, E. Symptomatology and management of acute grief. In A. Monat & R. Lazarus (Eds.), *Stress and coping*. New York: Columbia University Press, 1977.

Litwak, E. Geographical mobility and family cohesion. *American Sociological Review*, 1961, *26*, 250–257.

Marris, P. *Loss and change*. London: Routledge & Kegan Paul, 1974.

Marris, P. Conservatism, innovation, and old age. *Aging and Human Development*, 1977, *9*(2), 121–135.

Mintz, S. *An anthropological approach to the Afro-American past: A Carribean perspective*. Philadelphia: Institute for the Study of Human Issues, 1976.

Morgan, R. W., & Kannisto, V. A population dynamics survey in Lagos, Nigeria. *Social Science and Medicine*, 1973, *7*, 1–30.

Thom, R. Topological models in biology, In C. H. Waddington, (Ed.), *Toward a theoretical biology*, (Vol. 3). Chicago: Aldine, 1970.

Thom, R. *Structural stability and morphogenesis*. Reading, Mass: Addison-Wesley, 1975.

Thompson, D. *On growth and form*. Cambridge: Cambridge University Press, 1961. (Original edition published 1917).

Turner, V. *The ritual process*. Chicago: Aldine, 1969.

Van Gennep, A. *The rites of passage*. Chicago: University of Chicago Press, 1960.

Waddington, C. H. *The strategy of the gene*. London: Allen & Unwin, 1957.

Zeeman, E. C. *Catastrophe theory: Selected papers, 1972–1977*. Reading, Mass.: Addison-Wesley, 1977.

7

Sociological Dimensions of Uprootedness

Edward A. Tiryakian

> To be rooted is perhaps the most important and least recognized need of the human soul. It is one of the hardest to define.
>
> —Simone Weil (1952, p. 43)

Introduction

The present era seems to be a transitional period in terms of cultural time orientation. The progressive and dynamic future, which has been so much and for so long the temporal horizon of our civilization of modernity, seems in the 1970s to have lost much of its glitter, glamour, and magnetic appeal as a source of new waves of great expectations. As it becomes emptied of meaning for contemporary society, the future recedes in existential significance for contemporaries, and the historical past is becoming salient in its place.

For most of the world—certainly for Africa, Asia, and Europe—the past has always had great cultural significance: the standards of good taste, the classics, made the past the yardstick for the present; furthermore, the past has been used by ruling elites to legitimate their power and status. Even societies which think of themselves as "progressive"

Edward A. Tiryakian • Department of Sociology, Duke University, Durham, North Carolina 27706.

revere their past historical events and founders as sources of inspiration and purification. The United States is something of an anomaly, for up to now it has resisted the lure of the past; most of American culture and society has had as its temporal grounding a strong orientation toward the future as the basis for ordering the contemporary world. Yet it is in the United States where I (Tiryakian, 1977) see an emergent recession of the future as progress and an emergent view of the future in apocalyptic terms, while the past has an enhanced meaning in the cultural ethos.

It is within this cultural context of a shift in temporal horizon that we may ground the obverse aspect of uprooting, namely the recent fascination with roots, with dis-covering one's identity in the ancestral past. Quite in addition to the rather sedate commemoration of the American Bicentennial, the 1970s is emerging as the decade which turned its back on the future—a future whose image is becoming bleak and austere with stagnant economies, stagnant populations, depleting resources, and so forth. Instead, cultural curiosity now spotlights the past. So, for example, the commercialization of nostalgia finds the interwar period of the 1920s a good locale for depicting carefree times (with many films based on that period being major box-office attractions, such as *The Sting, Bonnie and Clyde, Paper Moon*); alternatively, the 1950s is another featured decade of normalcy and carefreeness (exemplified in the film *American Graffiti* and its television sequel, *Happy Days*). Genealogical services are flourishing, according to a recent issue of *Time* (1977), which estimates that the hunting of ancestors is the third most popular hobby, after numismatics and philately.

As part of this exploration of the past, or, perhaps we should say, this upgrading of the past in terms of its meaning for the present, we are also witnessing the "rediscovery of ethnicity" (Glazer & Moynihan, 1975; TeSelle, 1973). The thrust for this was to a large part political events of the 1960s. The Civil Rights movement among other things provided a new integration of blacks, with the term "Afro-American" providing the common geocultural heritage waiting to be recovered. In the wake of "the movement," the "white backlash" in part took the form of white ethnic assertion. As blacks got a sense of pride in belonging to an ethnic minority group which had had low social prestige in American society, so also did this become manifest in white ethnic minority groups. After the political thrust of ethnicity had spent itself—at least temporarily—the "rediscovery of ethnicity" became the subject of a collective biographical quest on the past of several minority-group representatives. I have in mind the recent works of Alex Haley (1976) and Michael Arlen (1975), structurally similar in describing the writer's odyssey to the previously unknown and unrecognized homeland, Gambia for Haley and Armenia for Arlen. If we add to these two another recently published collective historical biography, that by Dan Rottenberg (1977), we may note that these works represent a successful quest for roots on the part of a spokesman for each of three Diaspora people.

The quest for ethnic roots is not only a new biographical experience; it is also a phenomenon pregnant with cultural and sociopolitical importance. For one thing, I see an implication of this being a shift in the basis of sociopolitical identity, one with a certain narrowing of the circle of political identity. Stefan Kanfer, writing on the phenomenon of ancestral search, aptly notes:

> The recent "white roots" phenomenon is a reversal of U.S. tradition. In other periods, immigration was the sincerest form of flattery. Many of the populations that came to the U.S. were in flight from the past. (TeSelle, 1973, p. 54)

Obviously, a return to one's ancestral homeland suggests that one is no longer fleeing the past; one is accepting the past, and consequently the collective setting of one's identity in the distant past. But this suggests that one's identity with the national-state polity of the present is losing its charm and meaning. Perhaps not in the sense or to the extent of the so-called politically alienated youth of the 1960s, who were more futurists and utopia-seekers than they were past-oriented; still, the return to the homeland does involve a certain rejection of the present nation-state as *the* homeland. I belong to a generation culturally conditioned and socialized by public institutions (notably that of public education) to think of themselves as *individual* members of the American nation state, who to become functioning persons had to shed the "old clothes" of ethnic attachment and old-world traits. Today, minority groups are no longer social cadres to be escaped from as sources of stigmatization or obstacles to social mobility; rather, they are being reappraised as primary and positive vehicles of self-identity. In the process, there is a blurring of "minority" and "majority" group identity, for the pluralization of American society and the legitimation of plural life-styles signifies the effective disappearance of a "majority group" and its normative standards (of dress, speech, religious values, etc.) as the cultural yardstick for first-class citizenship in the national center. If anything, yesterday's majority group, which really lacked ethnic-historical cohesion, may be becoming demographically and politically an endangered species.

So today we see the reciprocal process of uprootedness/rooting as one involving national identity. The search for origins and for roots in the distant homeland of the past involves an uprooting in identifying with the nation-state, which had been the primary process of political and social identity for so many peoples in the 19th and the earlier part of the 20th century. If the implications of this are grasped, it should direct attention to this contemporary phenomenon as being of more than sociological or psychological interest, in an academic sense, but also very much of national political interest.

Let me terminate these prefatory remarks by proposing that "rootedness," and therefore "uprooting" as its opposite pole, have multiple bases in social being. Ethnicity is one dimension of rootedness, but we cannot assume that it is the only source which relates the person to his or her environment in the person's formative period. What I will do in the body of my essay is examine four works which all have "uprootedness" in their title and as their focal theme, yet present different perspectives and dimensions: uprootedness stemming from secular education, uprootedness stemming from European immigration, uprootedness stemming from a sociopolitical catastrophe, and lastly, uprootedness as a spiritual and civilizational malady. Each instance carries a different meaning for the actors experiencing uprootedness, hence our approach is guided by a phenomenological consideration rather than one which would assume that there is *an* objective condition of "uprootedness" differentiated from other sociopsychological conditions, each being measurably "out there." Uprootedness, like the obverse feeling or experience of having roots, is an intersubjective reality, one of feeling of belongingness and attachment with one's geosocial origins. But what one person defines in such terms is not what another will, for different persons—even those objectively designated as belonging to the same ethnic group—may experience different roots, which for them are the source of their being-in-the-world.

Social Conditions Of Uprootedness

Although uprootedness in its various manifestations is not restricted to any particular society or epoch, its relevance for social scientists is that it is intertwined with central features of modernity. The large-scale processes of sociocultural, socioeconomic, and sociopolitical change which typify the modernization process are the very processes which underlie the magnitude of uprootedness and uprooting in modern society. We can see this more specifically by following how "uprootedness" is treated in the works we have selected.

Let me begin with the earliest work I am familiar with that bears this title, a political, autobiographical novel by the French Academician Maurice Barrès, who published *The Uprooted* in 1897. Barrès, coming from the area of France annexed by Germany in 1871, became a French nationalist of pronounced conservative tendencies; he sought the renovation of French national energy through the affirmation of traditional values, particularly patriotism and religion. Alsace and Lorraine, the regions of the East which German annexation had uprooted politically from France, maintained their cultural roots with France, and the attachment of the population that had remained was still deeply French. For Barrès, there was a sacred duty to recall these ties and work for the reunification of the lost territories. In the novel under discussion, however, he felt a more ominous source of uprooting than political annexation, namely *secular education*.

The novel portrays a group of young men who go through the last years of the *lycée* together in a part of Lorraine that was allowed to remain French soil. These students are marked in their last year by their teacher of philosophy, M. Bouteiller, who is a major inculcator of the values of secular education; it is because of his influence that each will leave his native Lorraine to seek career and fortune in glamorous, glittering, and distant Paris. The public-school and secular education are seen by Barrès as the avant-garde of modernity and the uprooter of tradition.[1] The university is a powerful organ of the State in the formation of minds, Barrès remarks, no matter what political regime; the republican regime is no exception in seeking to mold minds to conform to its premises. Bouteiller, the learned civil servant qua public school teacher, is "a pedagogic product, a son of reason, a stranger to our traditional local or family customs, he is totally abstract" (8:24) (Author's translation).

Living in the abstract realm of reason, freed from natural biases which attach men to their native soil, Bouteiller sees himself above community and prejudices. He being cosmopolitan rather than provincial, why should it bother him to uproot these children, to detach them from the soil and social group of their birth, ponders Barrès? His task is to sow ideas in young men's minds, without examining the cultural reality from which they come (in this case, that of Lorraine). Hence, the secular education that he instills in them robs them of their national consciousness, understood by Barrès in the sense of the consciousness of the past of their native land and the desire to relate themselves to this past. This is the uprooting of the young intellect. Instead of activating their identification with national virtues, Bouteiller's education develops in his students an intellectual energy which

[1] For a lucid overview of Barrès, see Ouston (1974).

urges them on "To Paris!", the place marked for them in which to accomplish their destinies (p. 38).

Barrès is dealing, then, with a very common feature of modern society: uprooted youth, whose uprootedness is to be understood in a double sense. They are psychologically uprooted from their local traditions, and they become physically uprooted, choosing the lure of the metropolis to remaining at home in the country. The yearning for Paris, the feeling that it is *there* (rather than *here* on one's native soil) that one's destiny is to be worked out is indicative of the uprooting effect of modern secular education. This education is "liberal" in the sense of liberating the individual from traditional attachments, hence points to uprooting as intrinsically a part of the secularization of education.

Although the class of philosophy has turned students away from local traditions, from regional perspectives, and leads to their migrating to Paris, Barrès feels that the move is perhaps necessary in the life cycle of modern man. The sphere of action is limited for those who remain all their lives in the traditional locale:

> In Lorraine, isolated and without values, plunged into inertia, boredom, death, they aspired for Paris which they considered to be a center where they could collaborate for great things. (p. 179)

Going to the secular Mecca is or may be an act of self-emancipation, perhaps because, hints Barrès, the modern metropolis has a youth area (such as the Latin Quarter) which offers young adults much greater scope for initiative and self-expression than the aged-dominated traditional setting. He observes:

> Liberty! that is what can save them. It is not beyond repair if at twenty they are uprooted . . . vigorous as they are, they can stand a transplantation . . . A series of trials and errors will allow them to find the position suited for the persons they have become in ceasing to be Lorrains. (p. 99)

What makes it difficult for the graduates to take new roots is a general condition of anarchy, the absence of overriding social bonds uniting various fractions of French society; there are no social bonds, no agreed upon rules of conduct, no common goals (pp. 179–184). The young men reunite at the Tomb of Napoleon in Paris to launch a collective project which will help restore the national energy, but the project—a newspaper intended for concerned Frenchmen—is badly financed, and eventually fails. Two of the group, driven to economic despair, will murder a woman, and one of them be sent to the guillotine. For Barrès, this is a sorry example of the extreme end result of uprootedness, and he castigates:

> Lorraine had sent a number of her sons to Paris to elevate them to a superior ideal, the ideal of Reason. . . . In Racadot and Mouchefrin [the two murderers] the effort has completely failed. Did those who organized the emigration [school teachers] feel they had the responsibility of their soul? Did they realize how perilous was their action? They did not offer a good field for "replanting" those uprooted. . . . From their natural order, perhaps humble but still communitarian, they strayed into anarchy, into a moral disorder. (p. 345)

Of course, not all the *lycéens* fare badly, but the structural condition of what Emile Durkheim termed *anomie*,[2] which might well be seen as moral anarchy, makes the problem of finding new roots a chronic one for the educated youth of modern society. Perhaps—and this goes well beyond Barrès' diagnosis—since it is the educated youth which is the most prone of age sets to be uprooted, it is perhaps also from the trials and errors of this age group that new cultural roots will develop as the basis of modern society's renovation. In any case, the seminal and sensitive analysis of Barrès points out an important facet of modernity, for if it is questionable that secular, rational education *causes* modernity, it certainly is an intrinsic component of it (Cf. Wren, 1977).

To extend this further, those who have gone through the stages of higher education and have acquired the knowledge and the skills highly valued by the organizational, technological, economic, and other demands of the modern world tend to become the elites of this type of social system. Yet, it may be added, the very nature of secular, rational higher education is one which also uproots and estranges these persons from local traditions and customs, from the people, leading to an abyss as great as that between nobility and commoners in *ancien-régime* Europe. Hence the educated elite may find themselves both in privileged positions and yet also vulnerable to accusations of being detached, unresponsive, and alien to the world of rank-and-file human actors. The more the knowledge system of the intellectuals becomes global, world-wide, the less they can relate effectively to the everyday world and cares of the people. For the uprooted intellectuals,

[2] See the preface to the second edition of *The division of labor in society* for Durkheim's presentation of this crucial concept. Durkheim and Barrès, it might be pointed out, were contemporaries from Lorraine; both diagnosed the ills of modern society as reflecting the absence of a normative consensus providing social bonds. Yet, whereas Barrès took secular, rational education as a primary source of uprootedness, Durkheim viewed it as the necessary basis for rooting individuals in modern society and its polity. Still, there was a convergence in their perspectives, for both ultimately saw in religion and collective gatherings a renovation of political unity and national energy. For a further discussion of Durkheim and Barrès in this context, see Tiryakian (1978).

distant indeed is "vox populi, vox Dei." These estranged intellectuals or those operating in terms of "world systems" become subject to accusations of being aliens without roots in the nation. Whether in Africa, Asia, or Europe, radical and/or totalitarian regimes sooner or later turn against uprooted intellectuals—or "rootless cosmopolitans," as they were branded during the Stalinist "purges"—albeit these very persons may have been in the avant-garde of the revolutionary or nationalist movement which had previously fought against the status quo. Such, then, is one dimension of uprootedness. As we have seen, for Barrès the uprooting of the mind or the intellect is an important manifestation of the educational system of modern society.

Let me go on to a second work bearing the same title, *The Uprooted,* one however where the phenomenon of uprooting is a more immediate, brutal, physical process, stemming from sociopolitical disruptions. This is the study by the Indian sociologist K. B. Pakrasi, who examines patterns of adjustment of the millions who found themselves uprooted from ancestral lands in 1946–1948 and subsequently as the result of the partition of India (Pakrasi, 1971). Though some of the materials and the tragedy of uprooting reflect particular historical conditions and relations between Hindus and Muslims, we find in this study a more general sociological dimension of uprootedness in the modern world, since the mass exodus of groups fleeing the dangers of wholesale extermination has also become part of the political landscape of modern society.

Perhaps Durkheim's notion of anomie, alluded to in the above discussion of Barrès' study of uprootedness, is also useful for getting a perspective on the study by Pakrasi. The uprootedness stemmed from the political dismantling of India into two political states. British imperialism in India—as imperialism elsewhere—had only superficially brought together heterogeneous peoples; Western education did not seek to cement different ethnic traditions into one cohesive people. It certainly had not attempted their reconciliation, albeit much was made of the "Pax Britannica." British rule had placed political frames on India and had regularized everyday life. However, this social order was flimsy, particularly in the regions which would directly bear the brunt of partition:

> Partition created the notion that the Muslims were to be absolute masters in Pakistan and the Hindus in the country called India . . . the masses, who were then emerging from a long period of subjection, were suddenly inflamed by a feverish desire for power which frequently broke into mob hysteria. Neither moral principles nor religious beliefs, neither individual reflexion nor the most elementary respect for human dignity, could check the explosion of men's instincts.(p. 124)

The disruption of political rule, the breakdown of regulation of conduct toward "alien" minorities (albeit these minorities may have been neighbors for centuries), in brief, the breakdown or destructuration of the polity was a situation of acute anomie. The fate of Muslim minorities in Hindu demographic strongholds and that of Hindu minorities in Muslim demographic strongholds is a tragic instance of violent, bloody uprooting. This physical wrenching of one segment of the population from its traditional mooring, caught in a situation where it must flee on short notice or else face the threat of extermination or virtual enslavement, is a pathetic, dramatic, and unfortunately widely occurring instance of uprooting in the present century; *it is a phenomenon which seems to have a greater likelihood of occurrence with the breakup of imperial and/or colonial systems*. So, for example, the horrible massacres of the Armenians in the 1890s and again in 1915 took place in the fading hours of the Ottoman empire; Jewish pogroms in Russia became more common at the tail end of the Czarist regime; the Watutsi uprooting in Rwanda-Burundi in the 1960s came at the close of Belgian colonial rule in the mandated territories. One is tempted to say that an empire or a colonial system provides shallow roots to different social communities, allowing some to exploit others; with the empire's coming to an end, the political ecology becomes disrupted. What results is nothing short of a social disaster for the uprooted, who more often than not are of higher socioeconomic status and achievement than the demographic and/or armed majority which uproots them.

Pakrasi's study is informative for the data it provides on the trauma of relocation of the uprooted and the characteristics of those Indians who did leave what became West and East Pakistan. Pakrasi notes that there are important social differentials in those who did migrate under the extreme conditions of 1946–1948. Nearly 50% of all migrants were literate, a much higher incidence than in the settled Indian population in states bordering Pakistan; a higher level of educated and a lower percentage of land cultivators characterized the émigrés coming from East Pakistan (12:129–130). Perhaps because they had enjoyed a relatively privileged position in their previous setting, they found the adjustment more difficult, albeit they no longer were in physical danger.

One of the severe adjustment problems of any uprooted group is to feel at ease and welcome in its new environment. The uprooted—who are essentially disaster victims—come to think of themselves, not as a band of individuals, but as a group, a "we-group," set apart as an enclave in a larger community of outsiders (although it is the migrants who for the host society constitute the outsiders):

> The Government and local public appeared as the "they-group" to the refugee–settlers and a covert prejudice overburdened the uprooted mass of humanity in question. . . . A feeling of being unwelcome was (and is) not totally absent in many quarters with most of the uprooted Hindus and other non-Muslims. (12:126)

Pakrasi notes that for many of the uprooted, who had left ancestral homes for the sake of rejoining their brethren in a greater India, the experience of neglect, apathy, and even discrimination on the part of bureaucrats and local non-migrant neighbors was one of degradation (p. 140). One may say that these uprooted may have experienced in short order both relative gratification and relative deprivation. The former because those who do survive the bloodbath or the threat of extermination when others around them have fallen will feel a certain gratification at survival, relating themselves to their unfortunate companions. But the latter will also be experienced because, relative to their previous social standing and self-image, the indifference, apathy, and hostility of the host society make the migrant's lot harder to bear: the "they-group" is seen as insensitive, not really caring or accommodating to the uprooted. This is essentially contained in the following observation of Pakrasi:

> Wherever and whenever the uprooted families and persons found (or find even today) it impossible to realize their desired levels of "status aspiration" and some degree of economic satisfaction, they manifested negative identification, separate cultural dominance, aggressive feelings, parochial activities, or even organized movements for ensuring existence within the new society of the receiving State.[3]

And so, one out of four households surveyed in 1960–1961 in West Bengal still considered themselves as refugees; it was harder for those living in cities to feel a new sense of rootedness than for those living in towns, and harder for these than for those who had made villages their new homes (12:127).

Although Pakrasi's *The Uprooted* highlights the magnitude of the problem involved in the relocation of the Indian refugees—a demographic magnitude amounting in 1957 to a combined total of nine million migrants from West and East Pakistan—he also notes social factors attenuating the trauma of uprootedness. Traditional social groupings of occupational class, caste, and particularly nuclear and extended family units were vital mediators between individuals and the situation of anomie, providing an important stabilization in the adjustment process. Without the maintenance of what I am tempted to call this "social inertial guidance system,"

[3] This statement might be highly applicable to Palestinians in Jordan and Lebanon, to Pieds-Noirs in France, and to Moluccans in the Netherlands.

> We should have experienced a most dismal and shattered organization and behavior of the constituent family units comprising the uprooted population in question. Inter- and intra-familial cohesion did persist in conformity with the prevailing social norm and cultural expectation in such strong manner that every scope of disintegration and derangement was averted to an amazing extent (12:14).[4]

Pakrasi's study is much richer in sociological data than I have discussed, but my purpose in bringing attention to it is its implicit focus on an important dimension of uprootedness in the modern world: the relatively large number of those who belong to an ethnic minority forcibly uprooted by changing political environments. Of course, in the 19th century, political dissidents were forced to flee in exile due to forces of political repression against "aliens" or "extremists"—the Revolution of 1848 and its reaction helped to drive many out of their homelands—but somehow the uprooting was more a matter of individuals than of substantial populations.

One could say that the "modern" period of society began in the latter half of the 18th century (though, to be sure, any beginning of the modern period is highly arbitrary), and that modern sociopolitical uprooting on a substantial basis began with the American and French Revolutions; it was not so much an uprooting of an ethnic minority as that of a class—the "Loyalists" in the American case and the nobility with a sprinkling of the liberal professions in the French case. In the 19th century, sociopolitical uprooting was more of an individual matter, until the latter decades, as noted previously. In the present century, uprooting has become more democratic, since it is not just an upper class which becomes violently dislodged by political upheavals, but also the middle class (as in the case of Cuba), and even broader ethnic groups—for example, the Asiatics in East Africa (particularly, but not exclusively, in Amin's Uganda). If we also treat as part of the same genus of uprooted other sociopolitical refugees of the postwar period—e.g., blacks of the southern Sudan, Palestinians, Angolans, various ethnic groups from the old French Indo-China, Bengalis from Pakistan, Ibos from Nigeria, etc.—we see that not only has political uprootedness become a matter of international concern, a global phenomenon,[5] but also that increasingly it has shifted from being a Western phenomenon, up to and including World War II, to being a Third World phenomenon in the postwar setting.

[4] Again, here we have an observation which is very comparable to other types of catastrophes and disasters, studies of which have highlighted the importance of primary groups in successful surviving disasters. See Tiryakian (1959) and Hill and Hansen (1962).

[5] For recent albeit sparse information on refugees in the world today, see *Yearbook of the United Nations 1973* (Vol. 27). New York: UN, Office of Public Information, 1976, pp. 492–499.

The third study which illuminates the theme of this essay is one having the same title as the previous two, this one being perhaps the most familiar to an American audience, namely Oscar Handlin's *The Uprooted*. This study is by one of our leading social historians coming to grips with a central feature of the modernization of the United States: its intensive and extensive immigration by a heterogeneous group of peoples culturally alien to the Anglo-Saxon heritage of the country. Like the population studied by Pakrasi, Handlin's *The Uprooted* treats the psychosocial process involved in the European uprooting and American rerooting of migrants. The "push" factor behind the waves of emigration from eastern and southern Europe to the United States was as much economic as political. The pauperization of the rural proletariat, the depletion of the soil, created population pressures among the rural population feeling the repercussions of industrialization and the concentration of wealth in cities; capital had become liquid, rather than rooted in land as had been the case before industrialization.

To be sure, the impoverishment of the marginal farmer was not localized in southern and eastern Europe, since it also became a widespread condition in South Africa among the rural white Afrikaners, and in the United States in many states of the "Bible Belt." But in the 30-year-or-so period prior to World War I, it was to a large extent European peasantry from southern and eastern Europe, together with those who, like the Jews, linked the eastern peasantry to the larger economic markets, that constituted the bulk of the newly "uprooted." And, of course, this major wave of migrants from the soil were the successors of a previous wave of those coming from the European soil, the northern European population on the fringe of the British isles: the Irish, the Germans, the Scandinavians.

Handlin devotes the first part of his analysis to factors in the uprooting process in the native setting, but it is the second half of the journey, the transplanting from the Old World to the New, which figures prominently in his discussion. Among many fine points his study raises, I would like in particular to underscore two: (a) The transplanting of a population from one homeland to another leads to modifications in the characteristic of the population, no matter how much it may yearn to maintain its roots intact, and (b) The rooting process which follows in the wake of uprooting requires at least a second generation to be completed. Handlin does not himself state these propositions as such, but I think they can be gleaned from his discussion.

The theme of anomie, of normative and cultural destructuration, which we have brought into our discussion of the first two studies of uprooting, is also germane in the case of European emigration and American immigration, which is the focus of Handlin's *The Uprooted*. Note, for example, the following:

> The peasants had been cut off from homes and villages . . . which were not simply places but communities in which was deeply enmeshed a whole pattern of life. . . . Thus uprooted [the emigrants] found themselves in a prolonged state of crisis—crisis in the sense that they were, and remained, unsettled. (16:62)

The European village community, nurtured by centuries-old traditions which interlaced kinship, political, and religious patterns into a *Gemeinschaft* ethos—could not be transferred to America. "The communal qualities of peasant agriculture never took root in America. Attempts to restore the village or to colonize whole groups together failed miserably" (16:83).

A theme which Handlin accentuates is the painful psychological adjustment the immigrants had to make in a new and strange social setting: they were strangers to the indigenous culture. In his chapter entitled "The Shock of Alienation," Handlin emphasizes that, for the older generation of immigrants, the adjustment to the new world could only be at best partially positive, since, "the demand that they assimilate, that they surrender their separateness, condemned them always to be outsiders" (16:285). And Handlin concludes his study with a personal statement:

> We are come to rest and push our roots more deeply by the year. But we cannot push away the heritage of having been once all strangers in the land; we cannot forget the experience of having been all rootless, adrift. (16:306)

Much as I am impressed with the skill and sympathetic understanding by which the social historian portrays the uprooting process of immigrants as a very trying, bewildering, arduous adjustment, I think that Handlin tends to accentuate the negative, whereas I see in the very materials he presents some positive aspects of uprooting. Thus, Handlin notes that the transatlantic crossing involved the transition between two different social worlds: it carried "a startling reversal of roles, a radical shift in attitudes. The qualities that were desirable in the good peasant were not those conducive to success in the transition" (16:61). These traditional qualities mentioned by Handlin—neighborliness, obedience, respect, and status—reflect both a sedentary society and one with feudal overtones. They are not qualities functional in the dominant American social order, with its stress on individualism, egalitarianism, and both physical and social mobility. In the transplanting experience of European immigrants, it is a striking feature that in the latter half of the 19th century and the beginning of the 20th, the great numbers coming from a peasant/agrarian setting in southern and eastern Europe (as had the Irish earlier) did not settle so much on the land as in the cities. I think that in part this reflects their desire for the greater freedom of expression found in urban areas, their seeking emancipation from the social climate of the closed agrarian community, where landlords and patriarchal elders are more likely to make their authority felt than more impersonal urban authorities. The land is of course a source of deep identity, but however hallowed we imagine it to be, we must also reckon that in Europe it had for many cultivators become equally hollow. So when Handlin remarks that "wherever the immigrants went there was one common experience they shared: nowhere could they transplant the European village" (16:144), I would contend that this is in part because an important component of the uprooted—particularly the younger elements of both sexes—did not want to reestablish the traditional village community with all its values and division of labor. The reversal of roles, noted by Handlin, benefited the groups in the immigrant community which had remained throttled in old world society.

American society, which undoubtedly formed a landscape strange to the uprooted newcomers, a landscape void of European markers and props, furnished a soil where the individual really had a choice as to formulating his own projects, and where he had to choose himself voluntarily. Handlin seems to view this as a painful adjustment, and indeed it is, since in Europe those firmly rooted in traditional society have their choices made for them. "Leaving the old clothes behind and donning the new" is part of the American rebirth experience; implicitly this was the experience more migrants than not were looking for in voluntarily uprooting themselves. And this applies for another ancillary group of migrants, those whom Handlin designates as "dissenters"; in traditional European agrarian society, they were marginal, basically outsiders in the community where they were. Economically significant, yet socially they constituted what Weber termed "pariah capitalists" who were not incorporated and did not seek incorporation in the moral community defined in terms of religious membership. In Europe "they had been the outsiders who did not belong. In America they found their position the only normal one; there was no established church, no solidary community" (16:114f). Although Handlin does not note this, we may point out another aspect of role reversal in this context. Those Catholic migrants who in Europe had grown up experiencing that community membership and church membership were coextensive—as in the case where the Church was established in Italy, the Austro-Hungarian Empire, Poland—suddenly found themselves in

America members of a religious minority where there was no established religion in the European sense. The conforming cultural majority in the uprooting became the minority, marginal to the religious ethos of the United States; on the other hand, the marginal dissenters in Europe became in the United States more readily adapted to secular institutions, such as higher education.

Using Handlin's study as a springboard, American society is thus one which has many interconnections between uprooting, migration, and rootedness. The majority of the American population of today came here because of collective uprooting overseas—whether we are talking about the English religious dissenters of the 17th century, or the African victims of slave raids, or the Irish victims of the potato famine, or peasant victims of rural pauperization first in Europe and more recently in Mexico, or even the Cuban and Vietnamese victims of political upheavals. Such collective immigration provides a certain continuity in the American experience. It has made for a plural society, one in which second- and third-generation individual members have taken on new roots as Americans, yet once these become secure, they also seek to uncover earlier roots more existentially proximate than those of the Anglo-American sociocultural heritage. It is also the case that the very high rate of physical and social mobility, so characteristic of the United States, means that rootedness is more problematic, less likely to be experienced as part of self-hood than in the societies from whence come most of our migrants. And to this problematic feature of rootedness may be added the consideration that to take roots in the deep layers of American culture is a rather elusive matter. The core American culture, at least as I have come to view it (Tiryakian, 1975), is essentially a Protestant Puritan one which is more of a covert presence than a formalized, institutionalized reality. There is no differentiation between religious comportment and social behavior in other spheres of life, although, paradoxically, there has never been an established church or an official religion. Furthermore, the strong worldly activism so characteristic of American society may deceptively appear to be nothing more than an incentive for materialistic acquisition and gratification; however, such activism is suffused with moral injunctions and stern demands for accountability. In brief, the heart of the Puritan culture is highly moralistic, without a religious institution being the guardian of morality. This is a major factor in making American culture highly nebulous to "outsiders," and also a factor in making it so demanding on individuals. One result of this culture is that to take new roots in it is more difficult than one would suspect.

Let me terminate this presentation of studies pertaining directly to the theme of uprooting by mentioning the most global treatment, that contained in Simone Weil's (1952) *The Need for Roots*. I will limit myself to her observations contained in the second part, entitled appropriately enough "Uprootedness." It is obvious that she treats uprootedness as a pathology, and particularly as the malady of modernity. One source of this is provoked by instances of military conquest in which the conqueror "remains a stranger in the land of which he has taken possession" (18:43). Weil, it might be pointed out, wrote this during World War II, with the experience of the German occupation of France, but she also had in mind other settings, as is clear from her discussion at the end of the volume: the conquest of the Romans who "destroyed systematically whatever remained of spiritual life in the countries occupied by them" (18:297) and the conquest of southern French civilization by the North in the 13th century in what became known as the "Albigensian Crusade." Military conquest may not even be necessary to induce uprootedness, she adds, for economic domination in itself can impose a foreign domination which destroys human roots (18:44). Her immediate concern is for the working class, whose alienation from capital and resulting dependency on money and being employed by others is the salient mode of uprooting for that social group:

> Although [our workmen] have been geographically stationary, they have been morally uprooted, banished, and then reinstated . . . in the form of industrial brawn. Unemployment is, of course, an uprootedness raised to the second power. (18:45)

The uprootedness of the proletariat in prewar (i.e., pre-1940) France rendered them in a state of apathetic stupor. Weil argues that, starting from the uprootedness of the working class (a condition which marks the beginning of industrial modernity), there are two possible revolutionary alternatives:

> One consists in transforming society in such a way that the working class may be given roots in it, while the other consists in spreading to the whole of society the disease of uprootedness, which has been inflicted on the working class. (18:48)

France in the late '30s seems to have followed the latter, for the collapse of France in 1940 reflected internal interclass antipathy and suspicions, stemming from uprootedness, with an ever weakening commitment and loyalty to the country itself. Again, then, uprootedness and anomie seem profoundly interrelated. Weil does not stop her diagnosis with the uprooting of the working class, for she mentions the uprooting of the peasantry since World War I, "demoralized by the role of cannon fodder they had played in it" (18:40).

Economic uprooting—of the peasantry and the workingmen—is a

cardinal feature of modernity, and for Weil a major source of uprootedness. But Weil's discussion goes well beyond any Marxist critique of modernity; in fact, she seems to indicate that much of contemporary Marxism itself contributes to uprootedness. Like Barrès, she sees that modern education makes for uprootedness by imparting an intellectual culture which is "deprived both of contact with this world and, at the same time, of any window opening onto the world beyond" (18:45). Modern culture—really, "bookish" culture, we might call it—is faulted for being insensitive to the truth of the spirit, and for making those who labor feel unworthy for not being school teachers (i.e., intellectuals). Both liberal and Marxist educators find a common ground in not being able to relate to the working class, and in formulating outlandish doctrines that can only serve to "bring about the most intense uprootedness among the working class" (18:47).

Weil not only makes uprootedness a more general condition than do the previous studies, but, in raising uprootedness from an ontic to an ontological aspect of modernity, she also adds a dimension of activism to the notion of uprooted. That is, she mentions that for those who are uprooted (which we may also term being estranged from their native soil), there are two modes of behavior. The first is to fall into a spiritual lethargy. This is the most typical mode, and I think it is not far from our image of those who are uprooted: the act of being uprooted has deprived them of the vigor of their natural environment. The Hindu refugees studied by Pakrasi, the European migrants discussed by Handlin, and the young intellectuals described by Barrès may be thought of as common sufferers of a spiritual lethargy: uprooted and transplanted to an alien soil, they suffer from being aliens and lacking nurturing roots. But Weil points out a second mode, namely that those who are uprooted may in turn seek to uproot others: "Whoever is uprooted himself uproots others. Whoever is rooted in himself doesn't uproot others" (18:48).

Uprooting is therefore not just a passive condition which happens to people and to social strata. It is also a condition propagated by groups who are themselves rootless or have been uprooted, who have been detached from firm moorings and who seek—misery loves company!—to extend the condition to as many as possible. We might say that this spiritual malady has carriers of infection, like Typhoid Mary, who may have developed a certain immunity to uprootedness. Although the early Church which found in Gnosticism its most fearful enemy did not label this heresy as the one that brought uprootedness, it is not farfetched to propose that in fact the Gnostic attitude of profound indifference, abhorrence, and hostility toward this world, and the Gnostic's identification with the "unknown God," separated by an unbridgeable chasm from the empirical world, is a direct reflection of the experience of having been uprooted and being rootless in the alien world. The Gnostic will seek either to flee this world, or else to uproot this world from its present moorings.

Weil, then, proposed that uprootedness has to be linked with the degradation of labor, with the despiritualization of work, of both agrarian and industrial work. Conversely, "a civilization based upon the spirituality of work would give to Man the very strongest possible roots in the wide universe" (18:98). Roman civilization, at its heart atheistic and materialistic, had destroyed the spiritual life of countries it occupied, and the adoption of Christianity as the official religion of the Roman Empire had not helped matters, for the official Christianity had been emptied of its spiritual content, argues Weil (18:297). What this convert to Christianity, though not to any of its formal institutions, sought was to reestablish the authentic roots of civilization in reuniting spirituality and labor.

Perhaps, if I may leave her suggestive albeit unsystematic remarks, Simone Weil expressed with great feeling what Max Weber did with greater scholarship and objectivity in his landmark study. Weber (1958) traced back salient institutional features of modern society—its pervasive capitalist-industrial-bureaucratic network—to profound and unsuspected spiritual roots, namely, those of ascetic, this-worldly Protestantism. In at least the early stage of Western capitalism, when the latter was innovative in comparison with earlier forms of capitalism, because its very ascetic grounding meant that capital had to be accumulated rather than enjoyed in consummatory gratification, there was a unity of spirituality and labor for those adhering to the Puritan/Protestant faith: labor was not to be shirked, precisely because in replacing the sacraments it had become the vehicle of salvation. Weber was well aware that the comingling of spirituality and labor was not a permanent union, for he himself lived during the transition from early entrepreneurial capitalism to a more advanced, corporate, impersonal capitalism wherein bureaucratic structures become "iron cages" of modernity rather than vehicles of worldly salvation. Weber's discussion of the historical process of rationalization in all its forms—in economic, political, artistic, and scientific comportment—not only provides a superb analysis of secularization but, in terms of our present discussion, also contains a sociological dimension of the process of uprootedness. That is, the divorce of the economic system from its spiritual matrix, the separation of modern work from its religious inspiration in the Protestant work ethic, may be seen through Weberian eyes as a basic condition of uprootedness.

Weber's perspective, or at least my reading of his perspective, is congruent with Durkheim's perspective in *The Division of Labor.* In this work, Durkheim noted that it is a normal condition of modern society to have a structural integration of its components by means of a complex division of labor, which is more than an additive arrangement of individuals, since the division of labor has normative aspects. He also noted that there are pathological aspects that manifest themselves, and his discussion of *anomie* indicates that the absence or weakening of normative obligations in the sphere of economic conduct was a source of instability or attenuation of the efficacy of the division of labor. Again, I would like to suggest that Durkheim would have no difficulty in seeing that his perspective is one that points to the uprootedness of the socioeconomic sphere from a spiritual, i.e., moral grounding, one which bridles economic wants and appetites. If economic work, that is, what we do for a living, is detached from its spiritual dimension, then it loses an important dimension of meaning. More generally, the uprooting of all spheres of human comportment from their spiritual core becomes a pervasive pathological condition. Durkheim's notion of anomie, as I have had occasion to use it on several occasions in this essay, is today increasingly applicable to sexual comportment. The deregulation of sexual conduct, its separation from spiritual anchorage, is perhaps the last stage of the uprootedness of modernity. The meaning of being a man, a woman, of being married or single, or having children—of all the myriads of comportments and roles which are regulated and institutionalized forms of sexuality—loses its sense as the religious cadres of sexuality are torn asunder by the producers of dehumanizing mass culture and their entourage. As Weil noted more than a quarter of a century ago, a minority of the uprooted are uprooting a majority, with many of these falling "into a spiritual lethargy resembling death"—prophetic words she used well before the ravages of the drug culture, a lethal instrument of modern uprooting!

Conclusion

Here, then we have viewed a brief sociological panorama of uprooting. In discussing four works dealing with the same theme, I have tried to suggest that uprooting must be related to a social context, that there are essential sociological features of the phenomenon. A broad context of uprooting is modernity, that is, uprooting is related to processes of change which are central to modern society: urbanization (since the global exodus from country to city is a very direct form of uprooting), secularization (involving the uprooting of intellectuals and of masses from their traditional cultural matrix), voluntary and involuntary mass migration, etc.

The sociological perspective on uprootedness forces us to consider that it is more a group phenomenon than an individual matter—it is different groups in different contexts that suffer uprootedness, and consequently uprootedness is a collective experience. Perhaps this also leads us to consider yet another form of uprootedness not treated in the works we have discussed: the uprooting of an individual from his community. In ancient Greece, perhaps the most serious form of punishment meted out was ostracism, that is, the banishment of the offender from his community; this act of uprooting is one found in a great many preindustrial societies. For "primitive man" or even "classical man," to be cut off from one's native society, to be uprooted, was the harshest punishment; our jails and penitentiaries continue this punishment, and tend to make for a permanently uprooted segment of the population. A different kind of uprooting of individuals from their social community is the one promulgated by colonial education. The policy of "assimilation" or Westernization called for the training of a small number of indigenous elites who would function as go-betweens in the colonial regime, mediating between alien rulers and native masses; such persons, wearing Western clothes, using Western languages, consuming Western products, and attending Western schools, were termed "évolués." They considered themselves to have made the grade of modernity and left behind their more "backward," tradition-oriented peoples. In effect, they were uprooted, and for the most part they did not find, or were not allowed to find, roots in the community of the alien colonizers. Interestingly enough, it is from the *évolué* ranks that so many nationalist leaders emerged in Africa and Asia in the 1940s and 1950s. We might venture more broadly to say that in terms of modern society it is intellectuals who constitute one of the most pronounced segments of the uprooted, and it is this stratum which periodically seeks, perhaps unconsciously more than consciously, to find new roots in the agrarian or working-class masses. Certainly, one casual observation of the counterculture is the attraction of the countryside and of folk culture for uprooted urban youth; it is more than a symbolic deed when groups of college dropouts from the cities form a communal settlement away from urban areas!

This is one manifestation of the attempt to find roots in contemporary society. There are other manifestations, notably the emergence of ethnic consciousness or the "rediscovery of ethnicity," including ethnic nationalism. Individuals are reintegrating themselves, or at least their social identity, with their traditional ethnic community, or the one that was lost from sight. In many advanced societies—France, Great Britain, Spain, Belgium, for example—the reactivation of ethnicity in the traditional territory or region of the ethnic group has become more than a cultural movement, and also takes on the form of political movements which challenge the integral wholeness of the nation state.[6] We may see this as a sort of poetic justice, since it is the development of the modern nation state and its centralization of power which was a major factor in uprooting individuals from their ethnic/regional ties.

Before drawing to a close this essay, I would like to raise one last point about uprooting, and this in terms of redressing a bit our image of it. No matter what author we consider who has written on "uprootedness"—at least we may take those cited here as fairly representative—the phenomenon is seen to be pathological. It is very easy to share this perspective. But something needs to be said about the positive consequences of uprooting. Uprooting is after all part and parcel of what is involved in social change, and social change is not only a pervasive process of modernity, it is also a central core value of modernity. If all persons were firmly rooted and anchored in their society, we would have stasis and social entropy. This might be the most desirable condition for those in power, but what about the powerless? Handlin's immigrants formed one class of uprooted, but it is by changing ecological settings that the European migrants, or their children, were able to find an improved life, one that did give them greater degrees of material, cultural, and political freedom in which to maneuver than their old habitat. The educated *lycéens* of Barrès were uprooted, but it was in Paris and not provincial Nancy that most of them could carve a career of their own choosing. I suppose that what this suggests is that uprootedness may be an important stage in the life cycle, in the life cycle not only of the individual but also of the group. For certainly one could argue that by being transplanted to a new soil, the Irish, Italians, Jews, Armenians, Germans, etc. who came here modified old-world traits and gained new vigor of collective expressions. Uprooting is a painful phase in the life cycle, of the individual and the collectivity, but it may be a precondition for greater growth and development than if the self or collectivity remained fixated in its native environment. I say this not merely from theoretical considerations, but also from having personally experienced uprootedness, not once but twice.

And perhaps uprootedness, as a final thought, may be seen as a very ontological aspect of the human condition in general, one reflected in the primordial myth of the progenitors of the human race: the expulsion of Adam and Eve from Eden is the first instance of uprooting. Had they not been uprooted, there would be no history of mankind.

[6] For an appropriately titled introduction to some of these movements in Western Europe, see Mayo (1974).

References

Arlen, M. *Passage to Ararat*. New York: Farrar, Straus & Giroux, 1975.

Barrès, M. *Les déracinés (The Uprooted)*. Citations are from *L'Oeuvre de Maurice Barrès* (Vol. III). Paris: Au Club de l'Honnête Homme, 1965.

Durkheim, E. *The division of labor in society* (2nd ed.). (G. Simpson trans.). Nw York: Free Press, 1966.

Glazer, N., & Moynihan, D. P. (Eds.) *Ethnicity: Theory and experience*. Cambridge: Harvard University Press, 1975.

Haley, A. *Roots*. Garden City, N. Y.: Doubleday, 1976.

Handlin, O. *The Uprooted*. Boston: Little Brown, 1951.

Hill, R., & Hansen, D. A. Families in disaster, In G. W. Baker & D. W. Chapman (Eds.), *Man and society in disaster*. New York: Basic Books, 1962.

Kanfer, S. Climbing all over the family trees. *Time*, 28 March 1977, 54.

Mayo, P. E. *The roots of identity*. London: Allen Lane, 1974.

Ouston, P. *The imagination of Barrès*. Toronto: University of Toronto Press, 1974.

Pakrasi, K. B. *The uprooted: A sociological study of the refugees of West Bengal, India*. Calcutta: Editions Indian, 1971.

Rottenberg, D. *Finding our fathers: A guidebook to jewish genealogy*. New York: Random House, 1977.

TeSelle, S. (Ed.) *The rediscovery of ethnicity*. New York: Harper Colophon, 1973.

Tiryakian, E. A. Aftermath of a thermonuclear attack on the United States: Some sociological considerations. *Social Problems*, 1959, *6*, 291–303.

Tiryakian, E. A. Neither Marx nor Durkheim . . . Perhaps Weber. *American Journal of Sociology*, 1975, *81*, 1–33.

Tiryakian, E. A. The time perspectives of modernity. *Loisir et Societé/Society and Leisure*, 1978, *1*, 125–156.

Tiryakian, E. A. Emile Durkheim. In T. Bottomore & R. A. Nisbet (Eds.), *History of sociological analysis*. New York: Basic Books, 1978.

Weber, M. *The Protestant ethic and the spirit of capitalism*. (T. Parsons, trans.). New York: Scribner's, 1958.

Weil, S. *The need for roots*. New York: Putnam, 1952.

White roots: Looking for great-grandpa. *Time*, 28 March 1977, 43–44.

Wren, S. C. Review of *Education and individual modernity in developing countries*, A. Inkeles & D. B. Holsinger (Eds.). *Contemporary Sociology*, 1977, *6*, 314–315.

8

The Long-Term Sequelae of Uprooting: Conceptual and Practical Issues

Maria Pfister-Ammende

Introduction

This chapter begins with the case history of an uprooted person. It shows the disastrous entanglement a human being can experience, and permits us to sense the loneliness of the uprooted. It concerns a woman's fate which came to my medical attention in 1954. The anamnestic data and the other details have been thoroughly checked; all the names are fictitious.

The Case of Anna Maurer

Anna Maurer, born in 1923 in the eastern part of Poland, was brought up in a respectable, Catholic, middle class family. She was a very sensitive girl. Being eager to learn, on her own account she studied and got a job, although that was not customary in her social class.

Then her country was afflicted by the war fury. The marching in of the Germans brought an abrupt end to Anna's peace (which had

Maria Pfister-Ammende • Swiss National Committee for Mental Health, Zurich, Switzerland.

lasted nineteen years), her family security, her job, and her mental development. For two years she witnessed very closely general massacres of thousands of Jews. And during the same period the place was trembling from raids and indiscriminate deportations. It was at this time that the young girl developed a fear of police and uniforms. She also experienced, however, that distress and danger unite human beings; the people in her village grew into a close community. But then she became the victim of one of those notorious "single actions." Indiscriminately she was driven with others into a field where they were forced to dig potatoes with their bare hands from the frozen ground and also to spend the night there. Persons who were not able to work on were beaten to death. On the second day, a man saved her from death, risking his own life for her. On this "field of murder" Anna was stricken physically with pyelitis, which she was to get again and again in her further life, and psychically with a severe shock trauma which deepened her fear of police and uniforms. The family tried to hide in a lonely place; death, however, followed them, and snatched away her father's brother and his entire family from before the eyes of horror-struck Anna, a new shock for the twenty-year-old girl. The disaster went on. She was caught alone, and met her family again in the "camp," except for her sister Maria. In a thirty-hour race against death—the camp commander had allowed her thirty-three hours—Anna fetched her sister from another camp. This shows her deep family tie. It was probably their unshakeable family bond which kept them all alive in the following years of deportation and forced labor; but just this very trait was to become fatal to Anna in the future.

During the weeks of transport to forced labor in eastern Germany these people were locked up in railroad cattle cars, vegetating in their own dirt, and were later left naked before the eyes of the rabble of soldiers, who enjoyed the show. From this time on, each uniform was Enemy No. 1 for Anna. Three years of hard labor in a farm yard followed. Since her mother was breaking down physically and psychically, Anna took her place. A quality of maternal concern became her main function, but all the hope and expectation for life's joy and happiness in her young soul gave way to a dull apathy. "Just food and warmth—nothing else." This short phrase gives evidence of the loving concern, but also of the hopelessness of this young woman.

In these circumstances a Swiss crossed her way. He was working on the same East Prussian farm, as a head milker. An upright man, who loved his animals passionately; he was, however, an outsider, a sort of odd person. Anna's gentleness and kindness, her motherly way of being concerned, attracted him "mightily," as he had spent a joyless youth with his own mother. For Anna, however, so far just one thing in life had been significant and important: her family. The two young people could only see each other on rare occasions; so Anna denied herself marriage for rather a long time. "This is not enough for getting married." Her parents, however, succeeded in persuading her, as they wanted to see their daughter married to an honest man. So they finally got married, but she did not tell him anything about the experienced cruelties; her father had strictly forbidden any talk about them: "Never will anybody come to know, not from me and not from my family, that they have dishonoured one of us."

In 1945, right after the marriage, the young couple were "repatriated" to Switzerland, the man's home country; and this was probably the hardest moment in the life of the young woman. "We still went together for a little while," she told me later, "then my parents and my brothers and sisters had to go in another direction, back to the foreigners' camp. I went with Werner. It was like a dream—I do not know how it went on. It is not his fault. Why was I so much attached to my parents? We stuck together to the last breath because of all that had happened to us—horror and fright had forged us together. Not even for one week had we ever been separated from our parents. And now, my brothers and sisters could go with them and I couldn't."

With her marriage, Anna lost the physical proximity of her family, her mother tongue, her church. She was brought from Germany into Switzerland, a country unknown to her, with a "strange language and high mountains to be afraid of." I observed thousands of Soviet citizens from the Ukraine during the war who also had this feeling of being oppressed by the mountains—a deeply rooted shyness and fear felt by people from the plains (Pfister-Ammende, 1973a).

All these factors rendered it difficult for this woman to take root in her new environment. And yet, she would have been able to adjust herself thanks to her great—in fact, too great—adaptability, if she had found in her marriage a real partner. This man, however, had neither the quality to be a real partner for her (*Du-bezogen*), nor did he allow her really to love him directly—a deficiency which was to depress his soul severely in the future. He found it difficult to make contacts, and had no empathy with others. He had gone through an adolescence without love, and had lost everything in the war. He had to build up a new existence in Switzerland, and he did so with an admirable eagerness and engagement, supported by his diligent and economical wife. The couple were sent to the man's home community, and received 3,000 Swiss francs, from the official Swiss authorities for repatriates. The man could not work—and this was a typical repatriate's fate—in the field of agriculture, because of the completely different farmers' working conditions in eastern Germany and in Switzerland. So he became an unskilled worker in a country factory.

In her sixth month of pregnancy Anna fell ill again with pyelitis; she had, however, a confinement without complications. The marriage continued without any inner relationship and without a real contact in the erotic sphere. The woman had asked her husband for a divorce. "I was so terribly homesick for my parents, who had moved to Belgium. At that time (the postwar period) it was very difficult to have contact with displaced people across the border. I would have taken the child and gone on foot to meet them. I knew, I am his wife and we are a family. If he only would have paid a little more attention to me. I wanted to go away—I felt so helpless."

The man refused the divorce. Haunted by anxiety dreams, tortured by fright, sexually unsatisfied, she was looking for some kind of support; she found it with "things one can hold in one's hand." In her cupboard she piled up things that she had taken from shops, some of these things that she did not even understand the use of—crockery for dogs, brushes for cows, five darners, medicaments with Latin names, Melitta filters, etc.

A second child was born, and they had to move into a new apartment, which was very dark and situated directly at a mountain wall. "I was always alone with the children. On Sundays, he left as early as three o'clock to go into the mountains. Not that he should have stayed out of love, but that I should not have been so afraid." She became pregnant again, and suffered from a heavy headache; she "collected," took things to herself! The inner impulse "to have something in her hand" became irresistible, and the hour came when she was discovered and "cornered." "It was terrible—I was spoken to from behind by policemen. I'm so terribly afraid 'from behind'—that was like those times with the Germans. I said 'yes' and 'no' to everything they asked, and concealed what I could. This had been the rule with the Germans." Three days later she suffered a miscarriage. Unfortunately, the responsible rural court did not know of psychopathologic mechanisms such as uprooting neurosis and instinctual needs which cannot be controlled or stopped by willpower or by a later understanding. Thus the woman was sentenced, conditionally, to ten days of imprisonment for the theft of objects worth 84 francs. The man was just stupefied, and asked her again and again why she had done that, and she answered always the same thing: "I do not know." He told me later: "She was in despair. Even my mother, who did not like 'the foreign woman,' said to me: 'Have pity on her, she will take her own life." The woman was by far more ashamed of this theft than of a later affair which was more severe. "This is the worst of all," she told us, "but why did I steal at that time so senselessly?"

The man, his family, the village withdrew from the "thief." She was so depressed that her husband in the end tried to console her. Finally, in 1949, they moved away, into a city, in order to build up a new existence.

The couple gave their children to foster parents, and started to work, both in the same factory. And now the man showed very strange behavior: he implicitly, in effect, denied the existence of his wife, and tried to go to work alone. The woman, however, watched for him, and once she waited for him after work, trying to take his arm; but he jumped onto a streetcar—later he said that he had been in a hurry, otherwise he would have missed the streetcar. Their colleagues laughed, and the woman was terribly ashamed. "I wanted to show them that I am not a 'miss' and that I go with my husband; but he was gone in no time and the others laughed." When she came home—crying—her husband said that she should be able to find her way home alone, after all. This was just the thing she could *not* do, since there existed no "home" for her.

Here we find an essential coresponsibility on the part of the husband, which he later fully admitted to me. He refused himself to this uprooted human being who was deprived of her family ties, her "*Heimat*"; so his wife was left a stranger in the void, since she had also been deprived of her children. And then, finally, there happened what necessarily had to happen: there were other men who liked this kind woman.

She met a force and an influence which were stronger than her own willpower and moral strength. A man—we shall call him Bruno—offered her what she needed: concern, consideration, kindness, and later also tenderness. "I would have liked so much to free myself. But I had nobody. If I had had a mother or a sister or a girl friend." This man took her out of her state of feeling displaced (*Ver-rückheit*), of being uprooted, into a genuine relationship. She soon loved him with all her being, but the "*Beseelung*" was for her much more important than the sexual satisfaction which followed one year later. "I do not know—am I bad? I was afraid, it is true, and I had a bad conscience, but I never repented that it had happened (the sexual contact). The experience which a woman has with a man, I had with him." He freed her of her frigidity (on which she had consulted a doctor in the village, since she was looking for the cause in herself).

She became completely dependent on him, blind to the double game that this married man was playing with her and with his wife. "I do not know why, but I had no will of my own. When I was alone, everything was there: religion, education, duties, children; but when he was with me, all that was gone. I had no strength." Anna became pregnant. She wanted to separate from the man and have the child "legitimately," but "I could not help believing Bruno that it would have been unbearable for him to see the child grow up with another name." So he procured a criminal abortion. He went on courting her: " 'Not just the good things, but also the hard things forge people together,' he said to me. This was so moving, don't you think so? It made me cry."

Tortured by the conflict between her duties, her motherly love on the one hand and her love as a woman on the other, she wrote to him about her prison sentence. "I never knew whether my husband had forgiven me. His family despised me, 'the person,' and he suffered from it. Bruno took me in his arms: 'You poor little thing—I love you even more now—nothing can separate me from you.' " She was overwhelmed, and initiated a divorce suit. The entirely bewildered husband admitted, it is true, that he was no "ideal husband," but he suffered so deeply that he lost very much weight; he first looked at her silently and imploringly—then he resorted to begging—she did not have the strength to carry through the divorce. Yet she remained dependent on Bruno. At this point, however, Bruno's wife took action—she forced him to confess his double game to Anna. Let us hear Anna herself: "I saw that he had lied to me. I ran away from the apartment, I ran and ran and ran. Found myself in the forest, ran on, into a church, out of it again—it became dusk, it became evening. I was in a cemetery, sitting on a bench—my mother came, took me by the hand and said: 'Come.' I got up, bells were ringing. I stumbled, fell down; my mother was no longer there. I went on—there were the bells again. I sat on a bench until it became pitch-dark. Two men spoke to me—I ran on, heard the rushing of water—all the time I was hearing the water, I wanted to find it. It came nearer and nearer, but there was the town and so many people. I was looking for the water, for the bank. And there was a bridge—I was looking for the bank. Young fellows were whistling, I escaped from them and lost the rushing of the water. There it was again, and then I knew: This finally is it—the water—I ran very fast—and then I fell down." She was found unconscious on a bridge and taken to a hospital. When she woke up there, she only said, "I want to go home, to the children, I must take the children to bed."

Her husband was advised to accompany her next day to the psychiatric outpatient department. There she remained silent, probably because she wanted to save her marriage. And the confusion of the whole situation went on, until Bruno's wife told Anna's husband about it all. Anna confessed everything to him except for the criminal abortion. Her husband could see that he, too, was guilty, because he had neglected his wife. "Let time go by—I will become different." However, everything went on as before. Again Anna became pregnant. Bruno refused to marry her, but suggested another abortion; during the respective preparations, they were surprised by Anna's husband—and here and now the definite separation between Anna and Bruno took place, influenced by pressure from outside. As the husband had called in the police and Anna had admitted everything at once, a judicial inquest against Bruno and Anna was opened by the public prosecutor.

Later on, she became pregnant by her husband, fell ill with a new attack of pyelitis, and was sent again to the psychiatric outpatient department. There the diagnosis was "Changed in her personality by war experiences." After that it was arranged by her family doctor, who knew from his own experience the war and the psychic damages caused by terror, that the court authorities send her to me for a psychiatric opinion.

My Investigations and Observations

In the beginning the exploration went on well. She gave the impression of a "quiet, decent, clean woman and good housewife," and made every effort to help me to find out about her experiences and the psychological connections. She was quite gifted intellectually, and one could feel a real warmth radiating from her when she was talking about her children. At first she was depressive and apathetic, but gradually she showed more and more confidence in me, and finally—in a kind of mother transference projected on me—she opened up entirely.

In the fourth session she did not go into the waiting room, but hid herself in a dark corner of the corridor, trembling and behaving like a sleepwalker. She pressed her bag to herself, and was wringing her hands so that her knuckles cracked. With both hands she was holding her head, gnashing her teeth and turning around several times with her eyes wide open and with all the signs of fear. Finally, she excitedly uttered the word "police." She stared in front of her, opened her handbag several times—as if automatically—, and looked at all the objects it contained. Then she cried, "I did not steal." I found out that she had intended to do some shopping in a department store, and that she suddenly had had the feeling that someone was pursuing her because she had stolen. She had been running "blind, with a feeling of a pursuer behind me, as in a fog," until she had reached me, and all the time she had been "expecting someone's hand upon my shoulder." After this "confession" her excitement grew less, but not her heavy headache and the impulse "to just run away." We succeeded, however, in soothing her so that she no longer felt so desperately lonely and left to her uncanny inner impulses—she was in the fourth month of pregnancy—as she had always been before. She could now speak, for instance, about "us women," and even felt joy about this new child; but in the sixth month a fetal death occurred, after a rupture of the membranes and a prolapse of the cord. The patient mourned for the loss—"I would have given my heart's blood for this child to live. That's what we women are like."

The Rorschach test and the Szondi test showed a considerable lack of psychological and instinctual satisfaction, as well as a strong family consciousness, in addition to ego weakness and extreme suggestibility. Her emotional world was almost amorphous, and dependent on her extrinsic destiny. She was shy to the extent of being "afraid of the evil eye," and suffered from a deep inner embitterment which was, however, of a strangely passive nature. For her, laws had something "absolute"; when she came into conflict with them, she fell into a deep, dull despair.

Discussion

A nineteen-year-old girl, showing no signs of any hereditary taint, intelligent, interested and open-minded, engaged in an occupation of her own free will, with a sensitive mind, had been brought up well-protected in the bosom of a family in eastern Europe. Brutally, she was drawn into the war happenings and the terror of the Nazis. She experienced the violent deaths of members of her own family and of thousands of others. Just by a hair's breadth the young women escaped the same fate. On this occasion she got a severe psychic shock trauma. For years she had to vegetate in Germany as a deportee, a fact to which her family reacted by drawing as closely together as possible. During the years of oppression and terror she developed a fear of police and uniforms, and a diminution of her willpower toward pressure from outside. The only sense that life kept for her was the care for her family.

Paradoxically, she fell into an uprooting neurosis on the occasion of her marriage to a brave but somewhat odd man. She fell into a psychic vacuum and into complete loneliness when her husband, an immigrant himself, was repatriated to his country of origin, Switzerland, and turned out to be incapable of offering her security, warmth, sexual satisfaction, and contact with her new environment. In spite of all this, she tried to arrange her life in a positive way. She was esteemed for her cleanness, her diligence, and her quiet disposition by all the people of her surroundings—with the exception of her German mother-in-law.

Physically damaged by a recidivistic pyelitis, lonely, tortured by homesickness for her family, left alone most of the time by her husband who was literally obsessed by the mountains (*"bergbesessen"*), she suffered from continuous states of anxiety and severe headaches throughout her first pregnancy. During the second pregnancy, in addition to all this she became afflicted with kleptomaniacal impulses which brought her to court, where she was sentenced to ten days' imprisonment, conditionally. From then on she became entirely confused and helpless, and so she fell—body and soul—for a married man whose double game she was not able to see in spite of her intelligence; he gave her sexual satisfaction and psychologically a feeling of home (*Heimatgefühl*). This was an unconscious fixation beyond the influence of her willpower. For years she found herself in a conflict between an insoluble relation and a moral code based on superego; between maternal love and suggestibility.

As she had not had the strength to say "no" to her suitor who was now her husband, so she had not the strength now to resist the seducer—who also procured the criminal abortions. He became a sort of *"Ersatzheimat"* to her, like the "things one can hold in one's hand." His refusal to marry her, and the respective comments of his wife, finally opened her eyes, and she suffered a psychic breakdown with a suicidal tinge. She wanted, in fact, to die through the begetter, by means of a new abortion which he demanded of her. Her willpower was entirely broken and paralysed, as she felt psychically overburdened. At this time she was in a state of precarious and perilous psychic derangement.

In summary, one must speak of a *traumatic neurosis*, of a grown-up person with ego weakness, in two different shapes: as *shock neurosis*, and as *uprooting neurosis* (Pfister-Ammende, 1945), in a *physically damaged* woman who had twice lost her foetus by spontaneous abortion.

Here my psychiatric opinion for the court could be considered terminated. The meeting between Anna Maurer and myself, however, had taken a turn which led from the situation of giving my opinion as an expert to psychotherapy. I wish to describe this second phase of our working together.

Psychotherapy

The woman had been frightened when I wanted to put down the history of her life. She had to make an immense effort to overcome this barrier. "It was a terrible emotional upheaval, that first hour," she said later; and "as after a horrible movie. Everything came back, the screaming of the Jews when they were shot dead, and the people in the train to Germany, on the floor, in their own dirt."

Once we had overcome this obstacle, she was able to speak. She opened her heart to me. After several sessions, she said, "I don't know—it has helped me so much to be able to come to you. Sometimes, I can think quite sensibly." Or, "When I was pregnant again, it was like lying between millstones. Then I had to come to you. Before, it was always running away—sometimes it is still like that. But now I don't *want* to; you have told me I shall not give in. And you also said that I have already big children with whom I can sing and talk. Oh, my goodness, what they can ask!"

In the session when she had hurried to me, feeling a pursuer behind her, we did not talk much—she just relieved her distress by weeping. I felt that the psychiatric investigation had now to give way to psychotherapeutic help. This woman had come to *me*, to *my* door—in the terrible state of her flight impulse, in her danger of becoming a kleptomaniac again during her present pregnancy, she had come to *me*, instead of running away into the fog, the nothingness, to the water!

She came much too early, to the next session, standing before the door, shyly, but smiling. Eagerly and cleverly, she worked with me for the report. She said, for instance, "We'd better leave that out, about Werner's German mother, who still calls me 'the person.' It is not my husband's fault, and he loves his mother."

Again and again she now spoke about "us women," for instance, when the foetus died in the sixth month: "I would have given my heart's blood for the child to live. That's what we women are like."

At the end of this session, as she was already standing by the door, she quickly opened her handbag, took a piece of paper out, and put it into my hand: "No harm meant!—Good bye"—and off she went. It was a poem that she had written. It reflects the change in this woman's emotional frame of mind. It ran:

> Das Antlitz sanft und milde,
> Vom Lampenlicht erhellt,
> Eine Stimme spricht zu mir:
> "Mein liebes Kind!—Erzähl'!"

"Mein liebes Kind!" drei Worte,
Für mich ein Reich, ein Glück!
Wie etwas einst Geraubtes
Bekomme ich zurück.

Wie einst der Heiland sagte:
"Lasst Kindlein zu mir kommen!"
Hast Trost und Mut gespendet
Und Last mir abgenommen.

Ich möcht' mein müdes Haupt
In Deinem Schoss vergraben,
(Nicht weinen und nicht klagen)
Die Augen möchte ich schliessen
Und—*einmal* Mutter! sagen!

Written on December 18, 1954

This poem written by her in German can hardly be translated as a poem; I will confine myself to stating roughly what it says:

The face gentle and mild in the light of the lamp—and a kind voice talking to me: "My dear child—tell me!"

"My dear child!"—three words—For me they mean a kingdom and a happiness—bringing back to me what was taken from me long ago!

As once the Savior said, "Let children come to me!", so you gave consolation and courage back to me, taking the burden from me.

I wish to hide my tired head in your lap—(not crying, not complaining)—closing my eyes and saying—just once—"Mother!"

The next time she came, she looked timidly at me and anxiously—and blossomed (like a flower) when I passed my hand gently over her hair and thanked her for the poem.

We now finished our common work for the opinion, and found it advisable that she should come to see me until the court would give a decision. I tried to explain to her (from a medical point of view) what had happened to her. We thorougly examined her dreams, and made every effort to find out about the psychological meaning of her kleptomania and the fugues during her pregnancies, her pathological dependence, her readiness to make sacrifices, and the ambivalence toward her husband and her friend. Freed from the pressure of overwhelming inner impulses, and trusting me, she quickly and positively comprehended the psychological connections.

A few weeks later she came to me, overjoyed—the court having decided, on the basis of "our report," to drop the charge and quit the prosecution against her! "For me the war has ended today, ten years later than for the others," she said.

We still saw each other, several times, until she felt strong enough to bear life alone. One day she told me that, for the first time since her deportation from home, she had dreamt of the place where she was born and brought up. She said:

> I dreamt I went through a cornfield. It was summer and the wheat was golden ripe, one could actually taste it in the warm air. The wind moved the sheaves. But it did not blow from north to south and not from east to west. It moved the ears to me and they bent down to me as if they wanted to caress me.

She added: "Now, I am my own self again—after fourteen years."

Epilogue

There had been a psychiatric investigation, followed by psychotherapy, first supportive and then analytical. Since the woman was inclined to close attachment to the point of dependence, I had avoided allowing her to bind herself too closely to me, and so the wish expressed in the poem did not come true. Otherwise, it would scarcely have been possible to detach her from her mother-transference to me. This was especially so since she saw in me, in her transference, a *strong* mother, who did not break down like her own mother. As substitute mother I also lifted her father's ban on talk about the terrible happenings. Thus the demands of the patient's stern superego could be overcome in this sphere. At this point the patient's amazing elasticity and flexibility became evident, and this explains why she was able to relax, to free herself psychically in comparatively short psychotherapy, and to give her further life a new shape.

In this psychotherapy the main point was to help a human being who had been uprooted and who suffered from a grave environmental damage to gain back her composure, by means of psychoanalysis: dream interpretation and the transference method.

The psychiatric treatment accompanied the woman in this way, disentangled her confusion and inner chaos, and then let her go her own way again in the world. Some time later she appeared once again for some little help, and then she went out of my sight. Later I learned that she had gone to her family.

Conclusions

This tragic case underlines the necessity of preventive mental-health measures in times of emergency, like flight, uprooting, and displacement. Doctors, nurses, policemen, and judges should be instructed concerning the mechanisms and the effects of the loss of

"Heimat" and of uprooting; they should be well informed, so as to know how to interpret and deal with such reactions. The woman would hardly have given herself to an unscrupulous criminal seducer if in her helplessness she had had the understanding of the village doctor—such an understanding as she found—almost too late—in her family doctor in the city. Likewise, it has to be stated that, had the court authorities comprehended the incompatible in the thefts of this homesick and uprooted woman whom the Great Flood had drifted to a country unknown and strange to her, Switzerland,—much of the harm could have been avoided.

Primary prevention of uprooting is not possible in a case like that of Anna Maurer, as the primary causes of the pathological events were the war and its consequences. Secondary prevention would probably have been possible, however; for instance, help on the part of the village doctor on the occasion of her first consulting him because of her headaches, or psychiatric help on the occasion of her first conflict with the Law. Now, in spite of the positive results of the psychotherapeutic treatment, it remains uncertain whether in the future this patient will not suffer from some of the symptoms of the "survivor syndrome" (Berger, 1977; Eitinger, 1972). Guilt, self-loathing, and masochistic behavior may never be completely resolved, as has been observed in other victims of severe prosecution.

Basic Theoretical Considerations

Turning to basic theoretical reflections on migration and uprooting, I wish to emphasize that migration, transplantation, or relocation *may* lead to uprooting and lasting uprootedness; but not necessarily so. Uprooting is an inner psychic process by which an individual loses his ties with his familiar surroundings, human or environmental. Uprootedness is a term which does not express a judgement of value, but the problem of not being able to enter into relationships. It means a drifting in the void. In the words of a refugee: "We are like plants whose roots have been cut off. We are floating on the surface of a vast, muddy water, not knowing in which direction the wind will blow us."

These are the words of a Jewish man who in 1944 had escaped to Switzerland from the Nazi terror. Yet, this very day, while I am quoting him, there are thousands of "floating refugees" on the South China Sea, mostly Vietnamese fleeing their home country in small boats and on brittle rafts, sometimes only to be denied landing privileges anywhere—the loneliest people in the world. There may have been twice as many, but this we shall never know for certain, for many of them perished in the sea or by famine and thirst.

There are different types of migration, and all of them may lead to uprooting.

At different periods in the *life cycle* mobility takes place. During or shortly after adolescence, most young people in the Western world leave their parents to settle independently somewhere else. At the time of marriage, or during other types of community life, separation and transplantation happen to almost everyone who starts a new relationship, such as founding a family of one's own. This kind of biologically induced transplantation may be called a positive migratory move.

Retirement represents another turning point in life, when a *partial mobility* takes place, caused by sociological circumstances. Often roots which had grown for decades in the wider social environment of the working place must be pulled out to be replanted somewhere else. If the individual is not sufficiently prepared for this change, his occupational or professional ties will continue, and he will feel abruptly cut off from his work. This trauma may lead to uprootedness.

Sociologically caused moves of executives, military officers, and international staff members can become the cause of uprooting. These people are forced to move and to accept this as a condition of their employment. This demands a certain flexibility on their part, and also on the part of their families, and they must be able to adjust themselves fairly easily to new surroundings and to different civilizations.

Voluntary and planned migration takes place in the worldwide process of urbanization. Coelho and Stein (1977) have described the dangers of uprooting and overcrowding in this respect.

The hiring of foreign labor is also based on voluntary migration. It is by no means exclusively a symptom of today's "*Wachstumsgesellschaft*"; for a long time, very ancient tribes in the Middle East along the Red Sea, the Beja tribes, paid foreign laborers, mostly Eritreans and men from other neighboring areas, to help them with their crops. This attitude was usual in that part of the world. At present, hundreds of thousands of Asians work in the Arabic emirates, and many Africans in Saudi Arabia.

In Western Europe, according to UNESCO (Gillette, 1976), in 1975 some 12 million workers were employed in countries other than their countries of origin.

Another voluntary and planned type of mobility is the so-called brain drain, for example, in the field of health. According to a study of the WHO (Mejiia & Pizurki, 1976), 6% of the world's supply of physicians and 4% of the nurses do not work in their own country or where they were trained. This is partly due to the fact that in certain countries too many medical and nursing students have been trained for all of them to find a possibility of using their training in their own countries. They are forced to look elsewhere for employment. In addition to these findings, the WHO study revealed that the type of training offered in certain developing countries to medical and nursing students does not meet the real needs of the country, being too specialized and sophisticated for health delivery on a country-wide scale.

Another reason for the brain drain, putting aside the above mentioned "push factor" in the country of training, is the "pull factor." The higher standard of living and the educational level of Western countries are the cause for the graduated to look for employment there. Medical curiosity and scientific interest are, of course, also pulling factors. The interest of a young physician or nurse may be directed toward a more specialized field of medicine than the ones available in their own country.

Planned but forced relocation is another kind of migration. Large population groups have been moved because of irrigation and dam construction. On a rather small scale this happened in Switzerland, and on a large scale in Africa, China, and the Soviet Union.

Finally, there is the *forced mobility of the persecuted* who flee from the aggressor. About twenty million people of the northern hemisphere have been forcibly separated from their home countries during the past 50 years. There have to be added those who fled for physical reasons like earthquakes, famine, floods, and drought. In Africa, millions have sought refuge in neighboring countries during the last decades, and in this year, 1977, about 65 citizens of mainland China are looking for an asylum in Hong Kong every day. About 100,000 Tibetans fled from the Chinese invader to Nepal and India, and since 1975 over 200,000 Vietnamese, Laotians, and Cambodians have migrated to the United States and Canada. There does, indeed, exist a vast sea of sorrow, grief, and despair all over the globe! The case of Anna Maurer which I have described is one example.

Psychological Reactions to Migration and Flight

It is almost unavoidable that a person or group meeting an alien civilization will experience a *culture clash,* if not a *culture shock.* As a long-lasting result of this collision, the immigrant may develop a nagging distrust against, for example, the type of health delivery in the new country. Zeldowicz (1978) describes the psychological reactions of southern Italians who had emigrated to western Canada. Not only did they show considerably more *psychosomatic* symptoms than the

Canadian population, but they were often even unable to accept psychotherapeutic treatment. Instead, they insisted on receiving visible help, "medicine," for their ailments. This attitude is similar to the one sometimes observed in lower social strata within a country. Patients who are not used to verbalizing their complaints in a more abstract way want concrete, visible help—"a bottle," not just "words."

A culture clash can occur within the migrant family. Such conflicts have been observed, for instance, in Israel between parents and the next generation. An interesting case of such a clash between husband and wife came to my attention in 1976. The partners were Turks, but got acquainted in Switzerland where they were staying as foreign workers. They got married fairly soon after they had met. The woman, however, had arrived in Switzerland some years earlier than her husband. She had not only adjusted herself well to the central-European way of life; she had even come to like it better than the pattern prevailing in her own country, where women play a more secluded role in social life. The man was still fully identified with the patriarchal family structure prevailing in his home village, and he expected his wife to be his obedient companion. She had already made some friends in the previous time of her stay in Switzerland when she was single, and she refused to fulfill his demands. The differences between them deepened, each one insisting on his or her style of life, respectively. A gap sprang up between them, and finally the husband, enraged and torn into pieces by jealousy, made a murderous assault on his wife. During the trial, a psychiatric opinion demanded by the court[1] revealed the culture clash between the woman, who had discovered her emancipatory potentials, and her partner, who was still bound to a strictly patriarchal civilization.

Another difficult consequence of migration and flight is the *identity crisis* frequently observed among children of immigrants and refugees. This has been experienced with the Tibetans in Switzerland, though they came on the invitation of the Swiss government, and were cordially welcomed and received by the people, who felt deeply for these refugees coming from faraway mountains. The adults adjusted themselves well to the Swiss way of life, but they preserved their own culture and religion. As for the children, they also got on well in the beginning. However, those who by now are reaching adolescence have to face the difficult question of which culture they belong to. "What am I, a Swiss or a Tibetan?", some of them ask, and, not finding an answer and not being able to decide where they belong, they may get

[1] This case was brought to my attention by my husband, H. O. Pfister, who as a psychiatrist had been asked by the court to examine the couple and write an opinion.

depressed. This does not mean that they are uprooted, but they do vacillate between two cultures. Some of them prefer to return to Asia, to join their people there, who live in India in a large Tibetan community as a closed minority group.

With the Tibetans we have already reached the problem of the *refugee*. These people often have to undergo severe hardships, such as persecution, isolation, and, in the country of asylum, internment in hastily built camps. Many remain uncertain about their future and their definite resettlement for years. In this uncertainty they frequently show psychological reactions which belong to the category of *traumatic neurosis*, that is to say, a mental disturbance which develops shortly after unexpected and shocking experiences. I have grouped these reactions of refugees under the headings flight reactions, internment neurosis, and uprooting neurosis, and have described them in detail (Pfister-Ammende, 1945, 1946, 1952, 1973).

The *flight reactions* appear in four different ways. One is the fear of the persecutor. When the danger to life is over, this fear changes into a continuous anxiety. If in the new surroundings a situation recalls the previous experience of persecution, the fear is projected upon it. I have described such a mechanism in the case of Anna Maurer. Another reaction is the hypertrophy of the instinct for self-preservation. At the same time, the moral values of the individual can deteriorate. This seems to be the case particularly with regard to moral behavior which is based on superego control of the drives. Furthermore, we can observe that refugees often cling to the values from the time before the flight, particularly to the lost home country. This attitude may even develop into a real fixation on previous values. Finally, there is the overestimation of the new country and its authorities, as by fantasies about a savior, which were not fulfilled, and could not be cathected onto a new object, since no savior appeared, and the individual had to escape by his own efforts.

Internment and uprooting neuroses show an initial period of depression, and then a phase of aggression, which is expressed by restlessness and hate against the previously often overestimated authorities. If there is no discharge for these aggressive impulses, they may turn into regressive avoidance reactions, such as mass psychosis of flight or mass anxiety. Of course, the choice of the symptoms (*Symptomwahl*) depends on the individual character and its modes of reaction. The final phase is apathy, as a kind of self-destruction. It is very difficult to help a person who has reached this phase of uprooting neurosis. These patients need most careful, long-lasting psychotherapeutic treatment.

These are my observations about reactions of refugees. Once more

I want to emphasize that migration and flight *may* lead to uprooting, but need not do so in every case. There are individuals who cannot become uprooted; for example, those refugees who have already established firm ties in a new country and anticipate a new life in this society. A deep commitment to a political, religious, or humanitarian idea also prevents uprooting. Those people find their task and fulfillment in life anywhere, in transitional periods as well as in a firmly settled way of life. Of course, belonging to a national or ideological group means feeling at home instead of feeling lonely and isolated. On the other hand, merely personal albeit strong ties to other persons do not always give the refugee the support he needs in order to feel at home. Sometimes these personal bonds suffer too much from the hardships of camp life and an uncertain future.

Only in extremely rare cases do we find individuals who have never been rooted, but have been drifting along all their lives. They never build up a real relationship, and live in a world of their own. These people get along quite well as migrants.

There also exist certain *high-risk groups* among migrants and refugees. In a survey of refugees and Swiss repatriates which I undertook after World War II (Pfister-Ammende, 1955), a high incidence of mental disturbance was discovered among boys under the age of 14 years. Similar discoveries were made in investigations among other refugee populations (Lin, Masuda, & Tazuma, 1979) and immigrant groups. For instance, in 1977 a study in the Canton Zurich showed that boys from immigrant families meet considerably greater difficulties in school than girls (Haefeli, 1977). It may be that girls of European civilizations can adjust themselves more easily to new conditions, and are more flexible, than boys. Widows, old people, and males who in their original country held comparatively high positions are also especially vulnerable. Lin, Tazuma, and Masuda made investigations among 239 Vietnamese who were brought to the United States in 1975. They found out that the divorced or widowed female heads of households were the least resourceful and most distressed of all.

Preventive Social and Mental Health Measures

Preventing every kind of migration is not only utopian, but it is not even always desirable. Civilizations at all times have mixed, and this has often had great advantages. However, the social and economic development of a country can be seriously hampered by emigration, as in the case of the brain drain from developing countries to the industrialized world. The loss of future cadres and then of developmental potentials is a negative aspect of emigration for a country. This chain of push and pull factors should be interrupted. Young people should be trained in those skills which are actually needed, and such training should be offered in the right proportion, so as not to create a surplus in the respective field. These conclusions demand a *political* decision which does not belong to the field of mental health. It represents a measure of truly primary prevention of undesirable emigration from a country.

What kinds of measures should be undertaken in order to prevent the uprooting of migrants and refugees? Measures with this aim can be summarized under the heading "planning." Wherever possible, relocation and resettlement should be carried out with the active participation of the migrants and the inhabitants of the receiving area. It also needs intermediate liaison teams. These act as catalysts, and should be composed of migrants and people from the new area, with social and mental-health workers. It sometimes happens that so-called migration offices act only as the long arm of a central office which regards immigration as a contractual arrangement between a much-needed labor force and the government. In that case, the authorities certainly do not take into account that in the migration process and its handling the prevention of uprooting is an important mental-health issue.

The staff of the liaison teams should be trained in the psychological problems of migrants and refugees. It might be possible and advisable in some instances to set up consulting services for migrants in their original countries. In any case, the immigrants should know that those services do exist, and where they can be reached.

Epidemiological surveys are recommended in order to find out which categories of the migrating population are particularly endangered regarding mental health during the migrating period, and whether there are mentally sick people who need treatment. For example, a large Nubian population group in the Sudan was relocated because their area was to be inundated in connection with the erection of the Aswan Dam. Before the move, in 1957, the Sudanese psychiatrists carried through an epidemiological survey (Baasher, 1961). It revealed the prevalence rate of mental disease in the migrant population, and the rather unexpected fact that the subgroup with the highest risk for their mental health consisted of mothers whose husbands worked most of the time in other parts of the country. This finding of the survey is very similar to that of Lin, Tazuma, and Masuda in 1979. The preventive measures taken by the Sudanese authorities consisted of uniting these families shortly before and during the relocation process, whenever possible.

Not only the migrants, but also the population of the area or country receiving migrants and/or refugees must be prepared for the move. Their attitude toward the newcomer is of immense importance in the resettlement process. Information about the newcomers and discussion of their needs in small groups would help.

If migrants or refugees first have to be accommodated in camps, certain protective mental-health measures should be taken. The guiding principle is that the inhabitants of the camp should not live in a social and spiritual vacuum. In order to obviate this, the following factors must be carefully examined: the type of dwelling unit, the number of inhabitants, the degree of organization of the camp life, the contact with the outside world, and, last but not least, the atmosphere of the camp. More details are given elsewhere. (Pfister-Ammende, 1973b).

These suggestions represent some major precautions to be taken during migration, after flight, and during the resettlement period. If, in spite of them, deep mental suffering, depression, and apathy, or major psychosomatic symptoms, continue among uprooted individuals, they should be given psychotherapeutic treatment. This would represent a kind of secondary prevention: attempting to avoid the danger that the patient will sink into the final state of uprooting neurosis, which is apathy. Unfortunately, attempts at psychotherapeutic treatment in the final chronic stage of uprootedness are not very promising.

Concluding Remarks

To summarize: we cannot and should not prevent all migration; and we cannot in all instances prevent the sufferings from uprootedness. The misery of individuals like Anna Maurer will occur again. It may even be that it will stay with us forever. Faced with this suffering, we should do everything possible and feasible to help these lonely people in their endeavor to find new roots, new courage, and trust in their fellowman.

References

Baasher, T. A. Survey on mental illness in Wadi Halfa. Techn, Sess. IV B,*1*. *International Congress on Mental Health*, WFMH, 1961.

Berger, D. M. The survivor syndrome. *American Journal of Psychotherapy*, 1977, *31*, 238–251.

Coelho, G. V., & Stein, J. J. Coping with stresses of an urban planet: Impacts of uprooting and overcrowding. *Habitat*, 1977, *2*, 379–390.
Eitinger, L. Concentration camp survivors in Norway and Israel. The Hague: *Martinus Nijhoff*, 1972.
Gillette, A. Die Quadratur des Kreises: Das Verlangen junger ausländischer Arbeitnehmer in Europa. *UNESCO-presse*, 1976, VII, *8/9*.
Haefeli, K. Schulische Probleme von Fremdarbeiterkindern. *Schweizerische Lehrerzeitung*, 1977, *33*, 1035–1037.
Lin, K.-M., Masuda, M., & Tazuma, L. Adaptional problems of Vietnamese refugees. Part I: Health and mental health status. *Archives of General Psychiatry*, 1979, *36*, 955–961.
Lin, K.-M., Tazuma, L., & Masuda, M. Personal communication, 1975.
Mejia, A., & Pizurki, H. World migration of health manpower. *WHO Chronicle*, 1976, *30*, 455–460.
Pfister-Ammende, M. Entwurzelung und Wiederverwurzelung. *Ueber die Grenzen*, 1945, *12*, 4–5.
Pfister-Ammende, M. Vorläufige Mitteilung über psychologische Untersuchungen an Flüchtlingen. *Bulletin der Schweizerischen Akademie der Medizinischen Wissenschaften*, 1946, *2*, 102–120.
Pfister-Ammende, M. Zur Psychopathologie der Entwurzelung. *Bulletin der Schweizerischen Akademie der Medizinischen Wissenschaften*, 1952, *8*, 338–345.
Pfister-Ammende, M. The Symptomatology, treatment, and prognosis in mentally ill refugees and repatriates in Switzerland. In H. B. M. Murphy (Ed.), *Flight and Resettlement*, *UNESCO*, 1955, 147–172.
Pfister-Ammende, M. Displaced Soviet citizens in Switzerland. In C. A. Zwingmann & M. Pfister-Ammende (Eds.), *Uprooting and after*. New York: Springer, 1973. (a)
Pfister-Ammende, M. Mental hygiene in refugee camps. In C. A. Zwingmann & M. Pfister-Ammende (Eds.), *Uprooting and after*. New York: Springer, 1973. (b)
Zeldowicz, H. Some psychiatric aspects of Italian immigrants in Canada. *Proceedings of the 1977 World Congress on Mental Health*, WFMH, August 1977.

III

Stressful Situations of Children and Adolescents in Transition: The Role of Attachments and Social Supports

Introduction

Usually, a child grows up in a nurturant environment of friendly and familiar caretakers. From the secure home base of a caring, intimate world the child learns to form new bonds and to cope with changes in its physical and social world.

In this section, uprooting issues that involve attachment and separation are considered from a developmental point of view, with special reference to children and adolescents in transitional situations. In Chapter 9, Bretherton reviews attachment theories and formulates a model of the child's early internal representation of reciprocal exchanges between himself and the caregiver. Social bonds are organized early in life, according to the author, in reciprocal shared programs of interaction between partners. The child has an internal representation of the reciprocal play which he expects and, in a sense, effectuates. Parents and other primary caregivers are therefore attachment figures for a child because they interact in providing expectable support and security.

In many postindustrial societies, a child needs to be able to form attachment bonds with substitute caregivers or subsidiary caregivers, such as adoptive parents, foster parents, or day-care personnel. Since children often resist attempts to make them transfer allegiance from one figure to another when changes are made in their intimate social living arrangements, the ways of enhancing the development of attachments need to be explored. An unfamiliar caregiver needs to establish joint activities with a child before he can foster trust and acceptance. Previous sharing of information with primary caregivers, and establishing contact in the presence of the primary attachment figure, appear to be two means of easing the transition. An important behavioral hypothesis is that the basis for security in young children is localized in other persons, and this allegiance is not easily transferred. By studying the reactions of the young to unfamiliar persons, we may be able to recognize conflicts which appear in children who are facing new situations, and also the development of new attachments. This development may take some time, and no specific behaviors can be prescribed. Bretherton's caution is also applicable to those who deal with adult separations and attachments. Transitional situations may produce conflicts within uprooted individuals, and between such individuals and those closest to them.

In Chapter 10, Goldberg examines the conflicts experienced by adolescents and their families when they relocated from one country to another. Her report of interviews with over 200 Americans who have lived overseas and with mobile Americans in the United States supports Coelho and Stein's hypothesis (Chapter 2) that individuals experience considerable stress when relocation takes place during critical developmental periods in the life course. Families of multinational corporation executives, of foreign-service officials, and of personnel of international organizations are exposed to the stresses of living in unfamiliar cultures. They are challenged to make new efforts to learn how to adapt to a new situation, as a family group, since their young dependents are uprooted along with them. Younger people are more vulnerable because they may be unable to perceive the importance or necessity for the move. Stressful conflicts can result from the ambivalence which adolescents and their parents feel over the expression of independence in a new and unfamiliar situation. This ambivalence is intensified when families are resettled in a culture which is perceived as being strange and potentially dangerous. Adolescents wish to be more independent of their parents, yet they are faced with an unknown environment that makes them more dependent on those with similar cultural backgrounds, such persons often including their own families. As natural guardians, parents often feel guilty for having uprooted their children. Many parents go to great lengths to provide the amenities of home, which are often quite inappropriate and cause an additional source of tension in the new setting. Some parents restrict their children because of their own fears and insecurity in a strange environment.

The adolescent may also have difficulties in accepting the norms and mores of his new peers, and thus may become a marginal observer in school and local activities. These experiences, though not permanently disabling or damaging, appear to be vividly remembered by the respondents in the study. This finding gives additional evidence of the long-term effects of such crises of relocation and adaptation. On reentry into the home culture, the adolescent may experience conflicts and feelings of disappointment and alienation similar to those experienced abroad.

The article by Werkman (Chapter 11) emphasizes that those who go abroad and later return to their home country need a period of "decompression" in which the individual reconciles his expectations with the current reality of the situation. The coping task for these adolescents is, as Marris (Chapter 5) and Back (Chapter 6) have suggested, to evaluate past experience and integrate it with the present in order to manage the new environmental stresses. Werkman also finds that, though most American adolescents and young adults are able to adjust to new peer groups and societal values, many no longer feel comfortable "at home." They are nostalgic for their former way of life abroad. They are resentful of the indifference expressed by their compatriots and their peers. Many adolescents feel unable to share the special experience which they gained abroad, because it often means very little to those who have stayed behind. Many of these disappointed and alienated adolescents look forward to relocating again. Thus it appears that many individuals value their experiences of uprooting, and are unable, or unwilling, to resettle in their home countries even when they have the opportunity. Such individuals then may "feel" much more at home with peers who have similar multiculture experiences.

In Chapter 12, Gordon, on the basis of his study of runaways, suggests that self-uprooting may result in further maturity for the adolescent and his family. Gordon provides new insights into historical responses of social institutions to youthful runaways. During the 18th and 19th centuries, according to the author, runaways were considered to be social deviants who needed to be incarcerated in reform schools or in punitive institutions. From the beginning of the 20th century, however, the pressures of poverty and familial disputes were recognized as legitimate causes for running away, and runaway youths were recommended for treatment rather than punishment. In the 1960s running away was often recognized as a protest against a restrictive society and an oppressive political system, and an array of "counterculture" support groups emerged to offer refuge and assistance to runaways. Although political protest is not currently a major factor in running away, many waystations and runaway homes have been formed, and provide a mental-health buffer for youth who seek to redefine themselves, and to reexamine family situations that have provoked abuse, mistrust, and depression.

Running away can be seen as an attempt on the part of the adolescent to uproot himself from an unsatisfactory primary group. The intervention of support groups can provide an opportunity for uprooted adolescents to increase their self-understanding as well as their understanding of other relationships. It can further result in the reconciliation of families when parents gain insight into their own growth possibilities and recognize that they must also change in response to the adolescent's needs that have been dramatized by running away.

9

Young Children in Stressful Situations: The Supporting Role of Attachment Figures and Unfamiliar Caregivers

Inge Bretherton

Introduction

Attachment, affectional bond, emotional tie: when describing the relationship of a young child to his or her major caregiver, we tend to fall back on the same metaphor. Although it has been notoriously difficult to define this bond in behavioral terms, there are obviously aspects of caregiver–infant interactions which lead to the impression that the two partners are, in some way, linked by psychological processes.

A metaphor indicates that we have an intuitive notion about the existence of such processes, but a metaphor is in and of itself neither an explanation of attachment nor an explanation of how attachment to a caregiver may provide emotional support to a young child in times of stress. Only by going beyond the metaphor and attempting to

The first version of this paper was presented at the Annual Meeting of the Animal Behavior Society, Boulder, Colorado, June 1975.

Inge Bretherton • Department of Psychology, University of Colorado, Boulder, Colorado 80309.

uncover the underlying processes which regulate the interpersonal behavior that appears as "surface structures" (Chomsky, 1965) will we ultimately become able to understand the nature and functioning of social bonds.

In the recent past there have been several attempts at conceptualizing the processes to which we give the name "bond" or "attachment." In fact, there are currently two models of bonding or attachment which lead a side-by-side existence. Model 1 may be termed "attachment as interaction," Model 2 "attachment as mediated by interaction." Hinde (1974) postulated that attachment *is* the nature, quality, and patterning of interaction. The important point here is that the bond is seen to lie in the way in which two partners interact, and not in some structure which exists outside of the interaction. Brazelton, Sander, and Stern belong to the same school of thought. Bowlby (1969, 1973) and Ainsworth (1973), on the other hand, have consistently maintained that attachment is *mediated* by interaction. According to these two authors, the child *becomes* attached to the mothering figure during the second six months of life as the result of prior social interaction with her. In Model 1, the behavior is viewed as inseparable from the bond, which cannot even be discussed without also discussing the dyadic interaction. In Model 2, attachment behavior is clearly distinguished from attachment *qua* bond (a term used by Lamb, 1974). Although there is a contradiction between these two positions, it is a contradiction which, on closer inspection, turns out to be more apparent than real.

In this paper, I shall first outline the two theoretical models of attachment in some detail, and then show how a more general theory of social bonding can be constructed by integrating Models 1 and 2. Theories influence the way in which we behave and the way in which we view the behavior of others. Both of the two theoretical models of attachment have sufficiently influenced everyday thinking about social bonding and infant–mother relations that I believe a theoretical clarification to be useful before considering key practical implications. I am obviously also hoping that your thinking will be influenced by my integration of Models 1 and 2.

In Part II of this paper I shall make use of the ideas presented in Part I in order to discuss how adults, be they familiar caregivers or unfamiliar substitutes, can offer psychological support to infants and toddlers in situations which tend to be fear- or stress-arousing. These are situations which are associated with change and uprooting: entering a day-care situation, being hospitalized, or being temporarily or permanently separated from the primary caregiver.

Part I: Theories of Attachment

Model 1: Attachment or Bonding as Interaction

There is mutual regulation and adaptation in the transactions between caregiver and infant from the very earliest days of life, as Sander (1977) and his colleagues have so elegantly shown. Ten days after birth, a change of the primary caregiver (in which one experienced adult is substituted for another) had the effect of disrupting a baby's eating, sleeping, and crying rhythms. A considerable period of readaptation was necessary before the transactions between infant and foster mother became restabilized. Even when only *one* critical aspect of caregiver behavior was altered, the normally smooth interactions between a caregiver and a newborn infant became disequilibrated. If, for example, the mother's face was masked during an infant's waking period on the seventh day of life, the infant reacted with startled surprise to the mother when looking up at her at the initiation of feeding. Brazelton's (1978) finding that very young infants become perturbed when, in the midst of a play interaction, the mother presents a sober, immobile face for a few minutes, further substantiates the notion of mutual regulation. Additional support comes from an experiment conducted by Papoušek and Papoušek (1975). These investigators discovered that four-month-old infants were upset by unusual leave-taking behavior on the part of the mother. If the mother left the laboratory playroom in her normal, everyday fashion (with gradual disengagement), the infants greeted her return with delight. But if the experimenter switched off the lights just before the mother's departure, the infants abruptly discovered the mother's absence when the light came on again. Under these circumstances, they began to turn away from the mother when she reentered the room. Merely switching the lights on and off in the presence of the mother did not affect the infants.

Specific caregiver–infant adaptations, then, occur very early, a fact which can be most easily demonstrated by their disruption. Such mutual adaptation illustrates initial bonding (Sander, 1977). Sander looks at bonding as the interfacing of two adaptable systems: each partner has attuned (and continues to attune) his or her behavioral programs to the behavioral programs of the other. Each—to use the language of learning theory—has been shaped by and has shaped the other. In so doing, each partner has developed expectancies of the other's behavior. As Hinde (1976) puts it, the behavior of the partners has become meshed; Stern (1977), thinking along the same lines, likens the interaction between the partners to dancing with each other; Brazelton repeatedly used such words as affective synchrony and reciprocity. For all of these investigators the bond lies in the mutual adaptation and regulation of dyadic behavior.

Model 2: Attachment as Mediated by Interaction

In his book *Attachment,* Bowlby (1969) points out that human infants seem to be genetically biased to respond to stimuli which, in the ordinary expectable environment, are most likely to come from human companions. Moreover, human infants also emit signals which tend to be effective elicitors of adult caregiving activities. For example, newborns are especially sensitive to the sound of the human voice, and to being held, rocked, and patted. Human caregivers, in turn, are sensitive to infant signals such as crying, orienting, and molding into the caregiver's body.

In the course of the first few months of life, infants begin to discriminate between familiar and unfamiliar adults, engaging in the most lively social interaction with the familiar caregivers. A five-month-old may greet the return of a somewhat familiar person with a smile. When the mother returns after a brief absence, on the other hand, she tends to receive a smile accompanied by bouncing and vocalizing (Stayton, Ainsworth, & Main, 1973). Infants also come to respond more quickly to being soothed by the principal caregiver, and to protest his or her departure more vocally than that of less intimate companions (*ibid.*).

A further focusing on the primary caregiver occurs during the second half of the first year of life. Research findings show that at this period separation from the mothering figure becomes a much more traumatic event for a baby (Ainsworth, 1973; Bowlby, 1969; Yarrow, 1967). Bowlby ascribes this change to the establishment of a control system *within the child,* a system which regulates proximity maintenance to the mothering person. The organization of Bowlby's control system is accompanied by two developmental milestones. The first of these is a motor milestone, namely the capacity to move about independently by crawling and later by walking. The second is a cognitive milestone, namely the capacity to hold objects in mind even when they are not present to perception, and hence the capacity to make simple plans calculated to regain the absent object. Of course, a baby's proximity-promoting behavior has for the past several months been preferentially directed toward the primary caregiver(s). Nevertheless, before the attainment of locomotion these behaviors consisted mainly of signaling, such as crying, vocalizing, and postural cues.

Bowlby proposes that during the second six months of life all the proximity-promoting behaviors become organized into one goal-corrected system.

This control system prevents the infant or toddler from straying too far from his or her protective figure, and leads him/her to draw closer to the protective figure in stressful situations. The system generates plans by utilizing the baby's at first primitive and later more sophisticated internal working models (symbolic representations) of the environment, of the mothering person, and of the self, models which can be used to predict the behavior of the caregiver. The function of this control system, in evolutionary terms, is protection of the infant from potentially dangerous situations. *When Bowlby and Ainsworth refer to attachment as being mediated by interaction, they mean that the child's proximity-regulating control system becomes organized around that person with whom the child has previously had the most lively and intimate social interaction.*

Because Bowlby's ideas have so frequently been misunderstood, I have decided to present them in diagram form. I hope that Figure 1 will make it easier to comprehend the functioning of the system as envisioned by Bowlby. A term which may need some clarification at this point is the word "goal-corrected", a label which Bowlby uses to describe the attachment system. Goal-directed systems do not make use of continuous feedback, goal-corrected systems do. An arrow aimed at a target is goal-directed once it is released: its direction can no longer be changed if the target continues to move. A hawk in pursuit of a mouse is goal-corrected: it continually corrects its own movements to keep up with the escape movements of the prey.

A goal-corrected system requires a number of subsystems: sensors which respond to input, but which can also seek input; monitoring devices which detect relevant portions of the total sensory input and appraise it; control or decision-making devices which perform a running integration and evaluation of all available information from the monitoring devices and also from the representational model of the environment. The control device resets the goal, in this case a greater or lesser degree of proximity to the caregiver, and then selects an optimal course of action from a number of possible alternatives while keeping track of any changes in the environment which might necessitate a change of plan. Finally, the goal-corrected system must have access to an effector system, the final common path for which other behavioral systems may also be competing.

The system shown in Figure 1 is explicitly restricted to the regulation of proximity-seeking behaviors which are motivated by a desire for security (as opposed to playful interaction, for example).

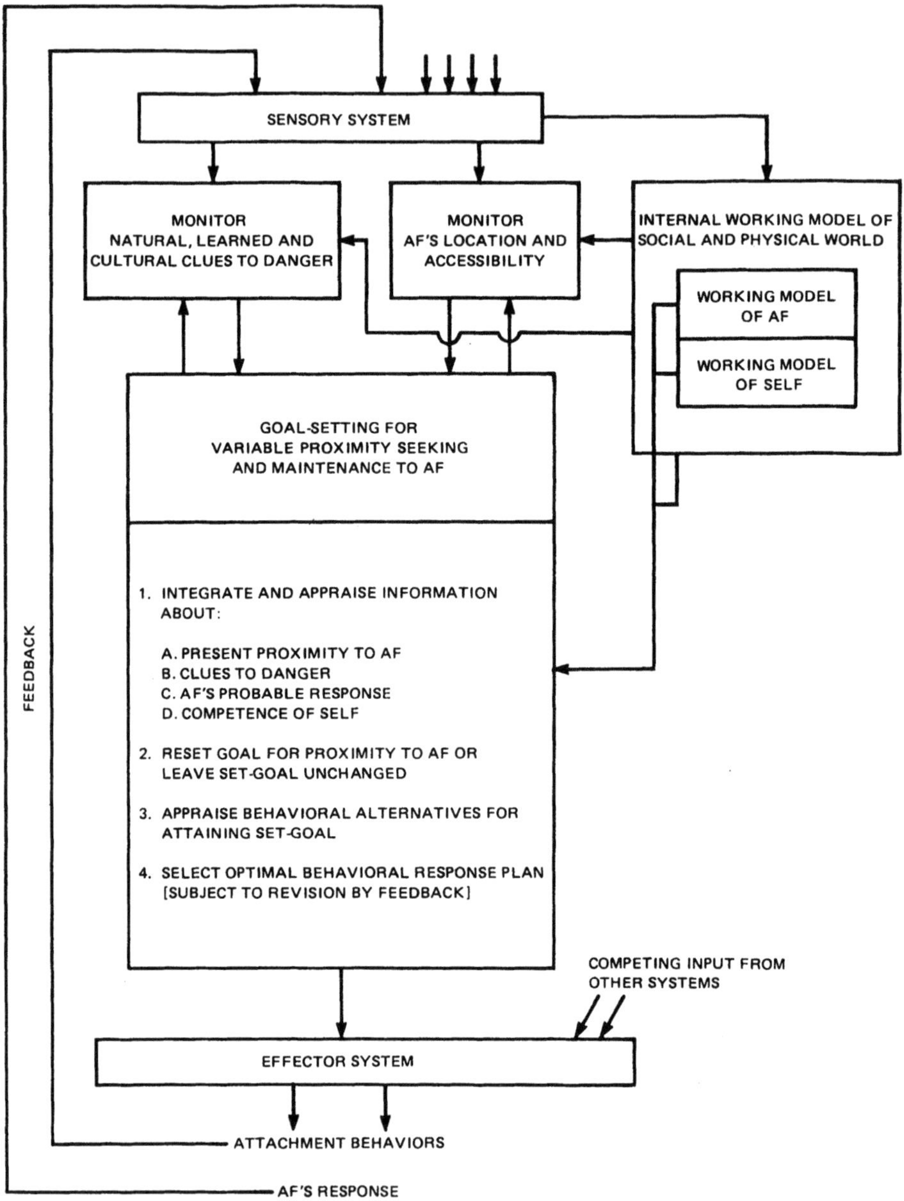

Figure 1. The security-regulating system: A simplified model of a control-system for the regulation of proximity to and contact with an attachment figure (AF).

This distinction is implicit in much of Bowlby's writing, a point to which I shall return later. Bowlby does draw a clear distinction between *feeling secure* and actually *being safe*. Although feeling secure is no guarantee of actual safety, feelings of security tend to be correlated with situations that are also objectively safer. Being close to a protective attachment or security figure is likely to ensure that a young child will actually be safer than were he or she alone (Bowlby, 1973).

Let me now use Figure 1 in order to describe in more detail the functioning of the proximity- and contact-regulating system along the lines suggested by Bowlby. The individual boxes of the diagram stand for specific sets of programs, analogous to very sophisticated and flexible computer programs. The arrows represent the direction of informational flow. Two classes of stimuli are monitored by the system: stimuli which indicate the presence of potential danger or stress (internal and external), and information concerning the whereabouts and accessibility of the attachment figure(s). For the sake of simplicity, Figure 1 shows only programs monitoring the presence of clues to danger, and ignores other types of stress-related input (see Bowlby, 1973, for a detailed account of his views on clues to danger). Bowlby assumes that some clues to danger are recognized by the organism without the need for environmental experience, such as fear of abrupt stimulation of any kind. Very rapidly, however, a young child begins to respond no longer just to these "natural clues to danger," but also to others which are learnt through direct experience, through observation of companions, or through instructions from caregivers (learned and cultural clues to danger). The baby interprets the meaning of such clues in the light of internal working models or symbolic representations of the social/physical environment and of the self.

A number of studies have shown that the mere presence of a familiar caregiver acts as a buffer to stress. In other words, in the company of an attachment figure a baby responds with lesser distress or no distress to stimuli which would give rise to more intense fear responses were the baby alone or in the company of a stranger (Campos, Gaensbauer, Sorce, & Henderson, 1973). Stimuli which tend to elicit only very mild fear responses in the absence of the attachment figure may be investigated with alacrity in that person's presence. The proximity- and contact-regulating system, in other words, integrates two different kinds of information, and appraises them jointly: information about potential danger, and information about the availability of the caregiver.

In making an appraisal, the system takes into account not only the actual proximity of the attachment figure, but also whether that person can ordinarily be relied upon to respond promptly to the child's bid for proximity and contact. If the caregiver tends usually to respond when needed, the child can afford to move further away than if the attachment figure is habitually inattentive. Moreover, the system must take into account the child's own ability. A child who is competent at locomotion (and therefore able to return to his or her secure base quickly if danger threatens) can afford to move further away as long as no clues to danger are present. It may therefore be seen that the behavioral output of such a system can change dramatically in the course of a child's development, even though the basic interrelationships between the subsystems remain unaltered. Let me illustrate this with a few concrete examples.

1. With increasing age, different types of stimuli are likely to arouse alarm. Whereas younger children tend to be frightened by sudden changes in stimulation (a bright flash of light, a slamming door, or rapid changes in *g* [gravitational pull]), older children are more often afraid of animals, unfamiliar people, imaginary creatures, and being alone in the dark. What children fear (and hence what activates attachment behavior) changes with cognitive development (Jersild, 1943).

2. With improvement of the child's locomotor ability, accessibility of the attachment figure takes on a different meaning. For an eight-month-old who can barely crawl, accessibility may mean a distance of several feet; for a three-year-old, it may mean several hundred yards.

3. As the child's spatial representation becomes more sophisticated, the child can afford to move out of the caregiver's sight, even in less familiar environments. If frightened, the child will know how to return to base.

4. Situations which at an early age require the aid of the attachment figure can later on be dealt with by the child on his or her own. For a one-year-old, a closed door is a much greater obstacle to regaining the attachment figure than it would be for a three-year-old.

5. As the child's symbolic model (working model) of the attachment figure becomes less egocentric (in the Piagetian sense), the child's strategies for ensuring proximity to the attachment figure become more subtle. The preschool child can increasingly take the caregiver's wishes and interests into account when formulating plans. For example, a little girl may ask her mother to take her along to a conference, instead of imploring the mother to remain at home.

Integration of Models 1 and 2

A number of differences between Model 1 and Model 2 must have become immediately obvious to the reader: Model 1 deals with the organization of *dyadic* behavior, Model 2 is much more concerned with the organization of attachment behavior *within the child.* Model 2 makes extensive use of the idea of symbolic representation (or working models), Model 1 does not. Model 1 is more general, Model 2 is restricted to a particular category of interaction. Model 2 defines attachment behavior as the output of a control system for proximity regulation, Model 1 is more concerned with the meshing and reciprocal timing of *any* dyadic behavior. My integration of Models 1 and 2 will proceed by applying the major ideas of each model to those of the other, and by adding a few new ideas of my own.

Revision of Model 1. I ended my account of Model 1 by stating that when the interactions between a dyad have become well meshed, each partner has come to have expectancies about the behavior of the other. The fact that such expectancies exist implies that each partner must have formed an internal representation of the other's behavioral programs. Without such representations there could be no meshing or interfacing of the dyad's interactions. To return to Stern's metaphor: in order to dance together in unison, both partners must know the steps, so that each can anticipate the other's next move.

It is not particularly daring to assume that the adult caregiver may be able to construct symbolic representations (internal working models) of his or her interactions with an infant. Such internal models can be used, as Bowlby has suggested, to interpret the partner's behavior, to forecast the partner's future behavior, and to construct plans. Yet none of our current theories grants the infant symbolic capacity until the end of the first year or so. Since it has been shown, however, that even very young infants respond to unpredictable changes in their caregiver's behavior with surprise or distress, we must presume that infants are able to construct internal models of their interactions with the caregiver, though these models are not symbolic. Such nonsymbolic or sensorimotor models cannot be used to make plans or predict the partner's future behavior in the partner's absence, but they are nevertheless available during actual interactions with the caregiver. That is, in the context of similar interactions and in the presence of the habitual partner the infant seems to be able to retrieve stored information about past interactions. I use the term nonsymbolic or sensorimotor model instead of Piaget's term "schema" because the latter has traditionally been employed to refer to the sensorimotor representation of very simple action patterns, such as the shaking of a rattle or the use of a stick to retrieve another object. The sensorimotor or nonsymbolic models of interactions with the caregiver are more complex than these action schemas, yet the concept is certainly very much akin to Piaget's (1952) concept of the schema.

One must assume that nonsymbolic or sensorimotor models of the baby's transactions with the caregiver become more sophisticated with development in the same manner as do the schemas which the child uses to act upon the world of physical objects. Once an infant has begun to understand that objects are permanent and can remove a cloth in order to find a hidden goal object, the first dim beginnings of a symbolic representation of the social and physical world become possible. From this point onward, an infant's representation of the mothering person (or indeed any partner) becomes less and less tied up in actual interactions. Gradually the infant becomes aware of the caregiver as a separate agent and as a person with his or her own interests and goals. Hence the child can, with increasing efficiency, influence the behavior of the partner and more accurately forecast the partner's likely response to an initiative for interaction. The more elaborate and accurate model of the caregiver and interactions with the caregiver is also useful in keeping the absent caregiver psychologically present, and even in interacting with him/her in fantasy.

The foregoing discussion implies that a social bond can be understood as the fact that each of the partners possesses internal representations (be they nonsymbolic or symbolic) of a range of interactions with the partner. It is then not strictly the patterning of the interactions themselves, but the underlying reciprocal organization of the dyadic behavior that constitutes the bond. Whereas in some sense it is appropriate to say that the patterning and quality of interaction is the bond, it would be more correct to say that the "shared plan" (Kaufman, 1976) or the shared internal programs which underlie the caregiver–infant interactions, are the bond. This organization (or internal working model) remains even when no interaction is taking place, just as a baby's grasping schema, to use a Piagetian analogy, remains when the baby is asleep. *The term "shared" refers to the fact that the infant has internal representations of interactions with the caregiver, and the caregiver, conversely, has internal representations of interactions with the infant. It does not, I repeat, not, mean that both representations have to have an equal degree of sophistication or accuracy.*

Of course, not every interaction is well meshed. If two members of a dyad cannot adjust their behavioral programs to one another, interaction between them will eventually cease. In relationships where some degree of mutual adjustment is achieved, the quality of the relationship or bond is determined by the quality of the meshing in their dyadic interactions. Hinde (1976) phrased this very succinctly:

> What the two partners do together may be less important than how they do it. They may, for instance, be talking, fighting, or kissing. In addition, we must specify how they are doing it. Are they talking in animated or

dispassionate fashion? What are they talking about? Are they fighting savagely? Kissing passionately, tenderly, or dutifully? (p. 3)

The quality of the dyad's interactions is based on the degree of mutuality in the "shared programs." Sometimes mutual adaptation or interfacing of behavioral plans is permanently hampered by either partner's inability or unwillingness to read the partner's signals. There are a number of reasons why this might happen: one partner's signals may be weak or deviant; alternatively, the partner may be deliberately or unwittingly "dyslexic."

Selma Fraiberg's (1974) studies of the cognitive and emotional development of congenitally blind babies showed that caregivers tend to feel disturbed by their infant's lack of eye contact during social interactions, and have to be taught to "read" their infant's gestures instead. Infants may show a similar response to the absence of eye contact and the nonresponsive, masklike face of a congenitally blind mother (Brazelton, Tronick, Adamson, Als, & Wise, 1975). These authors describe a one-month-old infant who habitually averted her face from her blind mother during social interactions. The same baby watched her blind father's face with interest: the father had not been blind from birth, and had a responsive face.

Temporarily disruptive effects can occur during interaction when either partner is unwilling to respond to the bids of the other because of, for example, emotional engagement in an antithetical activity, or because of depression. In such cases, the specific interaction may be inharmonious, though the capacity for engaging in mutually satisfying interactions is still present. One can draw an analogy here to linguistic competence and performance: the syntactic competence which underlies language production remains intact, even when it happens not to be reflected in a particular, ungrammatical performance. However, if a person's unwillingness to respond to the partner's signals becomes permanent, so that the quality of their interactions is affected consistently and not just intermittently, then the quality of the bond (or the meshing of their reciprocal interactive programs) may also be said to have deteriorated.

A large part of the work by Ainsworth and her colleagues (even though it is generally associated with Model 2) has centered on the investigation of the quality of interactions between infant and mother. Ainsworth places especial emphasis on how the interaction of the dyad is affected by the mother's ability and willingness to read her infant's signals and to respond to them appropriately and promptly. Ainsworth (1973) termed this capacity "maternal sensitivity." In reciprocal face-to-face interactions, some mothers allow the infant enough time to mobilize a response; others swamp their infants with stimulation without taking much heed of the infant's signals (Blehar, Lieberman, & Ainsworth, 1977). Similarly, in feeding interactions, some mothers consistently pace their behavior in response to the behavior of the infants; others do so much less (Ainsworth & Bell, 1969). In attempting to console a distressed infant, some mothers hold the infant until he or she is totally quieted or wishes to terminate physical contact; others attempt to put the baby down before he or she is ready for it (Bell & Ainsworth, 1972). Mothers who either cannot read their infants' signals appropriately, or who choose to ignore infant signals during the first three months of life, tend, at the end of the first year, to have less harmonious (or well meshed) relationships. The quality of the relationship or bond is thus dependent on both partners' ability and willingness to interface their behavior, to construct many well-adapted, shared interactive programs.

Interactions which do not mesh are not mutually satisfying or—to use Brazelton's expression (Brazelton *et al.*, 1975)—are not characterized by affective synchrony. The mutual satisfaction derived from harmonious interactions provides the motivation for further interactions, and thus for the development and elaboration of further shared plans. If, in initiating an interaction, one of the partners persistently fails to engage the other, there will be turning away and dejection (Brazelton *et al.*, 1975). During mutually satisfying interactions, on the other hand, both partners exhibit behaviors which are generally interpreted as expressions of pleasure and affection (eye contact, smiling, touching, positive vocalizations). Such behaviors serve as immediate positive feedback to the partner, indicating that affective synchrony is being established or maintained, that the interaction is going well. Indeed, in some of the early reciprocal games between mother and baby, achieving affective synchrony seems to be perhaps the major goal (Stern, 1974): such interactions consist almost entirely of behaviors which are generally labeled as bonding, bond-maintaining, or bond-servicing behaviors. (Kaufman [1970] used the term "bond-servicing" for the mutual grooming observed between friendly nonhuman primates).

The metaphor "bond-servicing" or "bond-maintaining" suggests that we have a notion about some structure (the bond) whose proper functioning is lubricated by the intermittent application of a specific category of behaviors. Another way of thinking about bond-maintaining behaviors is to regard them as motivators for further dyadic interaction because of their signal function. In one sense, behaviors such as touching and smiling are an integral part of the ongoing interaction, but in another they can be understood as reinforcers of the tendency to engage in further interaction—both in the short *and* in the long run. It is in this second sense—in the sense of providing positive feedback—that the metaphor "bond-servicing" or "bond-maintaining" applies.

Bonding or bond-maintaining behaviors can be observed during almost all harmonious dyadic interactions, although many other behaviors occur during these interactions as well. Not every relationship involves the same categories of interaction. A child may have developed a specific set of shared plans with one person (e.g., play) and another set of shared plans with another partner (e.g., providing and giving security). Different types of social bond could be characterized by the type of shared plans which exist in the relationship. This would make it easier to distinguish between the child's bond to mother, father, sibling, friend, and all the other people in a child's social network (a term used by Weinraub, Brooks, & Lewis, 1977). When Bowlby (1969) developed his theory of infant–caregiver attachment, he emphasized *one* particular category of interactive behavior: that which involves the child as the seeker and the mother as the giver of security. Much of Ainsworth's (1973) theorizing has also concentrated on this aspect of the relationship, though Ainsworth was, as we have noted, also concerned with investigating the degree to which harmoniousness in the total relationship affected the child's capacity to use the mother as a secure base for exploration and as a refuge in stressful situations. In the next section, I shall suggest how Bowlby's and Ainsworth's ideas can be incorporated into the revised version of Model 1.

Revision of Model 2. Can I substantiate my claim that Bowlby was primarily interested in describing that part of the mother–infant relationship which concerns the mother as the provider and the child as the seeker of security? In his book *Attachment* (1969), Bowlby defines attachment behavior as any behavior which has the child's proximity to the attachment figure (usually the primary caregiver) as its predictable outcome. He also emphasizes that attachment behavior is especially likely to become activated in situations of stress, such as perceived danger, fatigue, or illness. Bowlby did not, as I have done (Bretherton, 1978), explicitly distinguish between proximity seeking for the sake of playful interaction and proximity seeking for the sake of security (or protection). That he did have such a distinction in mind becomes clear in one short passage in *Attachment* (1969), where he contrasts the role of attachment (security) figure with that of playmate:

> It has already been remarked that we may need to distinguish more carefully than has hitherto been done between attachment figures and playmates. A child seeks his attachment figure when he is tired, hungry, or ill, and also when he is uncertain as to that figure's whereabouts; when

> the attachment figure is found, he wants to be held or cuddled. By contrast, a child seeks a playmate when he is in good spirits and confident of the whereabouts of his attachment figure; when the playmate is found, moreover, the child wants to engage in playful interaction with him/her . . . the roles are not incompatible . . . thus a child's mother may at times act both as a playmate and as a principal attachment figure. (p. 307)

The above passage strongly suggests that what Bowlby had in mind when defining attachment behaviors was the output of the security-regulating system, and not *all* proximity-promoting behaviors. Table I lists a variety of proximity-maintaining and proximity-seeking behaviors which may be activated by the security-regulating system in

Table I. Examples of behaviors directed to an attachment figure (AF) and activated by the security-regulating system

A.	AF stationary—no clues to danger present		
	1.	Looking (keeping visual tabs)	
	2.	Listening (keeping auditory tabs)	
	3.	Locomotion (checking on AF's whereabouts)	
B.	AF moving away—no other clues to danger present		
	1.	Looking (tracking AF's movements)	
	2.	Listening (tracking AF's movements)	
	3.	Crying	
	4.	Speaking ("Where are you going?", "Don't go away!")	
	5.	Locomotion (following AF)	
	6.	Physical contact (trying to restrain AF, hanging on to AF so as not to be left behind, holding AF's hand while moving with AF)	
C.	AF stationary and present—clues to danger present		
	1.	Looking (checking AF's reaction to event)	
	2.	Crying	
	3.	Speaking ("Come here!")	
	4.	Posture (arms up)	
	5.	Locomotion (run, walk, crawl to AF)	
	6.	Physical contact (cling to AF, bury head in AF's lap, hide behind AF)	
D.	AF absent		
	1.	Looking	Search
	2.	Listening	Search
	3.	Locomotion	Search
	4.	Crying	
	5.	Speaking ("Where are you?", "Come back!")	
E.	AF returns		
	1.	Looking (to check that it's AF)	
	2.	Facial expression (smile, frown)	
	3.	Crying	
	4.	Speaking ("Why did you go away?", "I'm glad you're back!")	
	5.	Locomotion (approaching AF)	
	6.	Physical contact (cling, clutch)	

a number of different contexts. Table II contrasts superficially similar behaviors such as approaching, looking, and vocalizing in order to demonstrate that these behaviors can take on a completely different meaning according to the context in which they are embedded. Column 1 of Table II lists security-motivated behaviors (which I believe to correspond to attachment behaviors in Bowlby's sense), column 2 lists proximity-seeking and maintaining behaviors which might occur in the course of playful and affectionate interactions. I do not mean to imply that any given proximity-seeking behavior can always be unambiguously classified as security- or interaction-motivated. Often both motivations may be at work; in other instances an observer may simply not be able to make an unambiguous judgment. I do think, however, that it is a mistake to equate running to the mother in fear with running to her to engage her in a game of peekaboo, and lump both together under the rubric "attachment behavior." Indeed, I would like to draw your notice to the fact that the behaviors which I have labeled "interaction-motivated" overlap (but are not quite congruent with) the bond-maintaining behaviors which I described in the previous section.

Without proposing any change in the organization of the security-regulating system as it is presented in Figure 1, I would now like to propose a change in how to interpret its functioning. Bowlby did not view this system as a continuously active one, postulating instead that the system is switched on (and then generates attachment behaviors) when the child is experiencing stress or is threatened in some way. Once the set-goal of the system has been attained (e.g. the appropriate degree of proximity to the attachment figure), the attachment system

Table II. Comparison of security-motivated and interaction-motivated attachment behaviors directed to the mother (M)

Behavior	Security-motivated	Interaction-motivated
1. Looking	Keeping visual tabs on M, Checking on M's reaction to an alarming event.	Gazing fondly at M, Looking up to M for approval.
2. Vocalizing	Protesting M's departure.	Delighted squeals during peekaboo game with M.
3. Speaking	(To third person) "Please open that door. I want to see where my M has gone."	"Mommy, I love you."
4. Approaching	Rushing to M when a stranger enters.	Approaching M asking to play horsey.
5. Physical contact	Clinging to M when frightened by a dog.	Climbing on M's lap to hug and kiss her.

can be switched off again. I suggest that it would be more useful to regard the system as one which remains continuously active. Whereas the behaviors which the system regulates are switched on and off according to changing circumstances, the system itself continues to perform its monitoring function, even when no overt proximity-seeking behaviors are activated.

Viewed in this way, the relationship between the role of the attachment figure as a secure base and as a haven of safety (or, as Bowlby would prefer, a secure haven) becomes much clearer. In a familiar environment in which the attachment figure is accessible and in which no alarming stimuli are present, a young child may engage in a variety of activities: exploration of toys or other objects, play with a familiar playmate, or play with the attachment figure. In a less familiar environment, a baby may also wish to engage in exploration or interaction with unfamiliar people. However, the distance which an infant is willing to tolerate between herself and her mother is likely to decrease under these circumstances. As one mother phrased it: "There seems to be a magic zone." In the case of her one-year-old, the magic zone extended to a distance of three feet from the mother. An unfamiliar adult was unable to entice this baby to move beyond the magic zone, but within it the baby was quite willing, even eager, to engage in play with the unfamiliar person.

In any one situation, a gradual shift from using the mother or other attachment figure as a secure base to using her as a secure haven, and vice versa, may be observed. For example, a baby may first inspect an unfamiliar adult while holding on to her mother's knee. After some coaxing, the baby may let go of her mother with one hand, using the free hand to play ball with the stranger. Finally, the baby may let go of the mother altogether, and interact with the unfamiliar playmate while sitting at the mother's feet. This sequence illustrates a gradual expansion of "the magic zone" as the baby's appraisal of the situation changes. Under other circumstances, one may observe a contraction of the magic zone. However, even when the child seems to be exploring with apparent unconcern for her mother's whereabouts, the magic zone still restrains the baby's movements—it has merely become very large (as illustrated by Anderson's [1972] findings showing that children in a park did not travel more than 200 feet away from their mother).

The functioning of the security-regulating system appears to the external observer as the regulation of proximity and physical contact to an attachment figure. It is presumably experienced by the child as a feeling of security when the system's set-goal (being close enough to the figure) is attained, as a feeling of anxiety and sometimes anger when the set-goal cannot be reached, and as a feeling of pleasure when the set-goal is attained after overcoming some obstacle (Bowlby, 1977). The bond itself is the organization of the child's security-regulating system around a specific person or persons and, the organization of the attachment figure's protective system around the specific child (which I shall discuss in more detail shortly). Although affect is strongly involved in the operation of the bond, affect itself is not the bond. Attachment is a special shared interactive program a la Model 1. It is so special because it tends to be much more exclusively organized around a small hierarchy of major caregivers than are other shared dyadic programs. Infants who are distressed tend to seek the mother, if she is available. If a crying baby is picked up by the father, for example, he may reject being comforted by him if he can see his mother (the reverse can also occur, if the father is the principal attachment figure). If a baby wants social interaction, on the other hand, he seems to be more willing to approach a much larger number of persons. Interaction-motivated behaviors tend also to be preferentially directed to the major caregivers, but the preference is much less pronounced than it is for the security-motivated behaviors (Tracy, Lamb, & Ainsworth, 1976). When Bowlby (1969) proposed the term "monotropy" to refer to a young child's tendency to prefer one caregiver to all others, he was again—I think—concerned with security-motivated attachment behavior.

One interesting aspect of the attachment system is that we know so much more about how the functioning of the system is organized within the child, about the circumstances which result in the activation of security-motivated attachment behavior. The specific functioning of the system is of course partially determined by the reciprocal behavior of the attachment figure, by temperamental differences, and individual experience with alarming events. However, we can predict much more accurately when a child will exhibit security-motivated behaviors than when he will tend to engage in playful interaction (except that this will occur only in situations where he feels at ease). Some infants and toddlers whom I have observed (Bretherton, Stolberg, & Kreye, 1980) interspersed short bouts of interaction with the mother with short bouts of exploration; for other children, long bouts of exploration alternated with long bouts of interaction. Thus, although I could at least conceive of writing a computer program which would simulate the functioning of the security-regulating system, it is not at all obvious to me how I could even begin to write a program which would simulate other categories of interaction with a fair degree of realism. The system for security regulation is more immediately implicated in survival than are some of the other shared dyadic systems, and it may be for this reason that security-regulated attachment behavior is organized in a somewhat more standard fashion across individuals than, say, the regulation of play interactions. For the latter there appears to be much more room for variation from dyad to dyad.

Redefining Social Bonds and Attachment. The major points of the preceding discussion can be summarized as follows:

Social bonds can be said to exist between two human beings when each partner has constructed shared programs of interaction with the other. Although all relationships involve a number of such shared programs, the relationship comprises the *joint* functioning of all of these programs. A bond is more than the sum of its shared dyadic programs, since the functioning of one program may influence the functioning of all the others. Nevertheless, it may be helpful to characterize different types of bond by the major category of shared programs which are involved in the relationship. A child may, for example have a more intense playmate relationship with his father, and yet strongly prefer the mother as a security figure. Indeed, research shows that this is quite often the case (Lamb, 1977). I therefore propose that we describe *the type of bond* which exists between two partners by the most important dyadic programs they have constructed; that we describe the *uniqueness* of each bond by the particular, idiosyncratic adaptations each member of the dyad has made to the other (one play bond is not the same as any other); that we describe the *quality of a bond* by how well meshed the dyadic programs are, and, finally, that we describe the *exclusiveness of a bond* by how reluctant each member of the dyad is to form a new, similar bond with another partner.

One might consider restricting the term "attachment" to that component of a relationship which has to do with the regulation of security and protection. However, the effectiveness of a person as a security giver and seeker is no doubt influenced by the smooth—or not so smooth—functioning of other shared dyadic programs, as the work of Ainsworth and her colleagues has shown (Ainsworth, Blehar, Waters, & Wall, 1978). Therefore, I advocate that we retain the word "attachment" as a technical term for those relationships in which one partner serves as a major security figure for the other, always bearing in mind that more is involved in such a relationship than the regulation of security and protection.

Using this definition of attachment we can now also define the term "attachment figure." *Attachment figure* is a label for any figure around whom a child's security-regulating system is organized, the principal attachment figure being that person whom a baby consistently seeks out in times of stress. The subsidiary attachment figures are those who can be effective security providers in the absence of the principal figure. Although a young child may have formed different types of social bonds with a range of people in his or her social network (Weinraub, Brooks, & Lewis, 1977), he or she will generally only be attached to a very few figures. Research suggests that to establish a new play-relationship may be fairly easy, but for a young child to enter into a new, major attachment (once the security-regulating system has become organized) is a much more arduous enterprise.

Part II: The Supportive Role of Adults

The Supportive Role of Attachment Figures

In the preceding sections, I have discussed the processes whereby two persons may be said to be bonded or, more rarely, attached, but I have not yet explicitly described the role of the caregiver in the functioning of the security-regulating system.

The functioning of this system within the child is dependent on some cooperation from the caregiver, but not to the same extent as are some of the other dyadic transactions, such as cooperative play. An infant rhesus monkey (Harlow, 1961) can become attached to a surrogate mother made of a wire frame covered in soft terry cloth. This surrogate has two essential characteristics: she is always accessible, and she permits the infant to cling whenever it wishes to do so. Although the surrogate mother neither grooms, nor plays with, nor disciplines her infant, the infant nevertheless organizes its security-regulating system around her. When alarmed, the infant runs to the ever-available but passive surrogate for comfort. In an unfamiliar environment it may at first cling to her, but it soon quietens and begins to use her as a secure base for exploration. The security blankets and cuddly toys of young human children perform much the same function.

A human mother, unlike Harlow's surrogate, plays diverse roles in her young child's life, be it as social companion, playmate, teacher, or security figure. Even in her role as security figure a human mother differs from a cloth surrogate who merely permits a frightened infant to approach and cling.

In fact, the human caregiver must be presumed to have an internal control system that is complementary to the child's security-regulating system—a system which is organized around the protection and retrieval of a specific child. Maternal protective soothing and retrieval behavior has been studied in other mammals, but not to any great extent in human beings. It too can be contrasted with maternal sociable interaction (analogous to the distinction we have already made for the child):

> Whereas crying leads a mother to protect, feed or comfort her baby, smiling and babbling elicit behavior of a very different kind. When her baby smiles a mother smiles back, "talks" to him, strokes and pats him, and perhaps picks him up. In all this each partner seems to be expressing joy in the other's presence. (Bowlby, 1969, p. 246)

Like some of the other programs a caregiver and child share, the dyad's security behavior is also reciprocally organized. Moreover, the reciprocal system has more survival value than the separate systems operating within caregiver and child taken alone. Should a young child, one some occasion, be so entranced by a novel environment that his or her security-regulating system is completely overridden by exploratory behavior, then the caregiver is available to retrieve the child because the set-goal in the caregiver's protective system has been exceeded—not because the child signaled a desire to be closer to the caregiver. Even in situations where a frightened child approaches the caregiver, the caregiver usually does more than merely allow the child to remain in proximity and seek physical contact. If the caregiver does not consider the situation particularly dangerous, he or she will usually take active measures to soothe the child by patting, rocking, or talking. In a recent study (Bretherton *et al.*, 1980) of 12-month-olds, I observed that every mother engaged in such behavior when her baby sought physical contact. Thus the child, in a situation in which he or she feels insecure, is not merely reassured by proximity to and physical contact with the caregiver, but by the caregiver's active soothing behaviors and attempts at familiarizing the child with the alarming stimulus, as by stroking a dog of which the child was frightened, or by encouragement: "Go and play with the nice lady!" It is in this way that an attachment figure may use the security-regulating system to teach a child what should be feared and what is harmless.

If the caregiver him- or herself is alarmed by the event which triggered the child's proximity-seeking, both partners will normally attempt to leave the situation together. Should this be neither possible nor appropriate, the caregiver's ability to play the role of security figure will be severely hampered. Hence the reason why frightened mothers are frequently regarded as a nuisance by medical personnel.

We need some systematic studies on how to make it easier for the mother or primary caregiver to perform her or his natural role as an emotional support system. This could possibly be achieved by developing better techniques for giving information about medical procedures, and by building a relationship of mutual trust between the child's security figures and those who will be the temporary caregivers, rather than by resorting to bland exhortations "not to worry." It is interesting that learning-to-swim programs for toddlers already make use of this concept: The nonswimming mother is encouraged to learn to swim herself, lest the young child take the cue from the mother and come to fear the situation.

Not all caregivers are equally effective security givers. In fact, the caregiver's responsiveness to a young child's bids for social interaction is not necessarily related to how that same caregiver reacts to the child's security-seeking behavior. Mothers who are very responsive to their infants' positive vocalizations are not always equally responsive to distress vocalizations (Yarrow, Rubenstein, & Pederson, 1975). Unresponsiveness to distress vocalizations, interestingly, was found to have pervasive effects on the infants' gross and fine motor control (associated with the capacity to explore the environment), and on several other motivational and cognitive variables. The mother's responsiveness to positive vocalizations, on the other hand, was merely related to the infant's subsequent positive vocalization. Mothers who do not respond fairly promptly and consistently to distress signals from their young infants (Bell & Ainsworth, 1972) and who do not enjoy physical contact with them (Main, 1976) seem to be less effective as security figures. In an unfamiliar environment, these infants tend to be either very passive, unable to use the mother as a secure base from which to explore the new room with its array of toys, or they seem unalarmed by the new situation, but pointedly snub the mother or avoid her when she returns after a brief absence from the room. These patterns have been extensively documented by Ainsworth *et al.* (1978).

Avoidance of the mother can be seen in a much more extreme form in abused or severely neglected infants. In a pilot study (Gaensbauer, 1977) such infants aged between 12 and 18 months showed little interaction-seeking when they were observed in the company of their mothers, nor did they protest the mother's departure from an unfamiliar room. On the mother's return after a brief absence, the abused infants avoided interaction, and certainly did not approach or cling to the mother. On occasion, some of the infants cried when the stranger left the room, and even displayed searching behavior for him. It seems that these infants had never organized their security-regulating system around their major caregiver. After some weeks in foster care, some of the infants began to show both interaction- and security-motivated attachment behavior to the foster mother, though their initiative was still relatively weak compared to that of nonabused infants. In less severe cases, where the mother is only intermittently rejecting or ignoring, a child may still become attached to her. Yet, because the child must in such cases take the major responsibility for maintaining proximity to a somewhat reluctant and uncooperative security figure, he or she may appear extremely clingy and overdependent. Bowlby (1973) coined the term "anxious attachment" for this kind of relationship.

When an effective (principal) attachment figure accompanies a young child in a potentially stressful situation (e.g. hospitalization), that person is supportive merely by dint of his or her presence, and by being able to reassure the child, as Bowlby repeatedly points out. But, over and beyond this, the familiar caregiver can, at such times, act as a mediator between the new social and physical environment and the child. The caregiver can provide information about the child's customary routines, food preferences, and sleeping habits, and in many other ways ease the child's adjustment to the new environment.

When a young child is separated from the mother or primary caregiver, on the other hand, he or she is not only separated from the security figure in his or her role as mediator of the new, but from all the other roles which the primary caregiver normally plays in the child's life: as provider of physical care, as playmate, helper, and teacher. Thus, a substitute caregiver may be faced with having to assume *all* these roles at once, without first having had an opportunity to develop any "shared interactive programs" with the child. How can an unfamiliar person facilitate the establishment of a bond between him- or herself and a young child? The next section will address itself to that point.

Unfamiliar Persons as Supportive Figures

Unfamiliar Persons as Friendly Companions. During the second half of the first year, some infants begin strongly to protest the encounter with an unfamiliar adult, even when the child's principal attachment figure is present (see Ainsworth, 1973, and Bowlby, 1969, for reviews). Although not all infants show intense fear of strangers, a majority tend to become somewhat cautious with persons they have not seen before. How then does it make sense to discuss the role of an unfamiliar person, not as an alarming stimulus, but as a support system in stressful situations?

Although many older infants turn away, withdraw to the attachment figure, and occasionally even cry when first faced with a stranger, these behaviors do not as a rule persist for more than a few minutes if the child's mother is present. In some of the earlier studies investigating stranger-anxiety, a sizable number of infants aged eight months and over responded to an unfamiliar adult's approach with frowning, crying, and leaning away. However, these studies were generally conducted under conditions where the infant was confined in a seat at some distance from the mother, the stranger's behavior was not contingent on the infant's response, and physical contact was instituted almost immediately (Morgan & Ricciuti, 1969; Rand & Jennings, 1974). But if the infant either is not confined or is seated close to the mother, and if, in addition, the stranger engages the child in play with toys and gears her or his behavior to that of the child, then the initial response to unfamiliar persons tends to be much more positive. The rapidity with which overt wary behaviors disappear (in most though not all circumstances) when an adult, female stranger invites one-year-olds to play with her is remarkable. I found (Bretherton, 1978) that after four minutes of interaction, infant behaviors such as gaze aversion, turning coyly away, and especially crying, disappeared completely. Although the absolute frequency of positive behaviors such as smiling, or giving the stranger a toy, did not increase over the eight-minute observation period, the positive behaviors seemed intensified once they were no longer accompanied by wariness. Almost all of the 48 one-year-old infants I observed entered into some cooperative play with the unfamiliar person within eight minutes of being invited to play.

When, instead, the stranger remained relatively passive—only responding to the infants' social bids without herself making overtures—very few one-year-olds made approaches to the stranger during a four-minute observation period, a finding which has been corroborated by others (Eckerman & Rheingold, 1974). In an unfamiliar environment, one-year-olds tend to approach novel toys and to investigate them with alacrity, but they generally delay proximal interaction with unfamiliar adults. If given sufficient time, however, many infants will seek out strangers in order to interact with them at close range, by giving or offering toys, or by vocalizing to and touching the stranger, though in most cases these attempts at interaction are at first not maintained for very long (Bretherton *et al.*, 1980). In all three age groups (12-, 18-, and 24-month-olds) we observed fewer *systematic* attempts at "making friends" with the strangers. One or two infants seemed to fall in love with the stranger almost at first sight, claiming her as a playmate for almost the whole 40-minute observation period. Yet, in another case, an infant climbed into the unfamiliar person's lap after a 10-minute exposure to her and reclined there for two minutes, but paid little heed to her for the remainder of the session. Relatively few children, after having approached the stranger for the first time, systematically increased their interaction bouts with her. The findings suggest that, if an unfamiliar person needs to establish rapport with a young child within a limited period, he or she might be better advised not to leave the pace of the interaction, or the responsibility for maintaining an interaction once started, entirely up to the child. We do not know, however, to what extent this claim holds for children over two years of age.

Although the *presence* of the mother tends to have a positive effect on how infants respond to the encounter with a new person (Campos *et al.*, 1973), the behavior of the unfamiliar person toward the mother seems to have little influence on a young child's willingness to engage in interaction with the stranger. Clarke-Stewart (1978) staged a quarrel between mother and stranger over possession of a magazine. This event, observed by the toddler, had surprisingly little effect on how the child subsequently responded to the stranger's overtures. In our own work (Bretherton *et al.*, 1980) I found that when one- to two-year-old infants were offered the opportunity to interact with two strangers in the presence of their mother, they chose to interact more with the stranger who sat quietly beside the mother, and less with the stranger who conversed with her. Rather than making more overtures to the stranger who was friendly with the mother ("she must be o.k., since she's talking to my mommy"), these children directed more social bids to the stranger who was more accessible (that is, who was not socially engaged with another person). But if a stranger engages in deliberately frustrating behavior toward the *child* (Clarke-Stewart, 1978), it becomes very difficult to establish rapport once the frustrating behavior is discontinued. Infants and toddlers, then, seem to take their cue *directly* from the person with whom they are interacting, rather than from observation of a transaction between that person and their attachment figure. These findings may not apply to children over two years of age; research on this point is not yet available.

In real-life situations, mothers rarely play the rather noninterventive role assigned to them in many of our experiments. I have observed that many mothers tend to encourage their young children to interact with visiting adult friends, but so far no systematic studies have been conducted on how mothers try to facilitate social contact between their infants and unfamiliar persons, and on how successful these techniques generally are.

By returning to the scheme I presented in Part 1 of this paper we can now ask what happens when an infant and an unfamiliar person begin to engage in a mutual transaction. As the interactions between a stranger–infant dyad proceed, the partners are establishing shared programs of behavior, and as these programs become increasingly reciprocal or meshed, the firmness and quality of the bond between them also grows. Despite the fact that not every interaction *is* a bond, continued interaction will lead, for both partners, to a joint reorganization of behavior which—in my scheme—constitutes a bond.

Older infants who have already formed attachment relationships and who, in addition, may be participants in a sizable social network (Weinraub, Brooks, & Lewis, 1977) can make use of the skills acquired during their interaction with others in order to establish new relationships. Thus, a stranger may build on an infant's capacity to play peekaboo with the mother by attempting to play that very game with the child also. Rafman (1974) used this technique with depressed, hospitalized infants. After observing the ritualized games through which a particular mother–child couple usually interacted, Rafman sought herself to engage the child in interaction by faithfully imitating the mother's part in the game. This method proved successful for making social contact with young children who had hitherto resisted it. In similar experiments with healthy infants, Rafman found that the infants tended at first to respond with puzzlement to this procedure, then by beginning to enact (without much enthusiasm) their own part of the game. Half the infants subsequently continued the game eagerly, the other half—after the initial response—became distressed. Having learned to waltz with one person makes you better able, but not necessarily more willing, to learn to waltz with another. Much of what is generally labeled "fear of strangers" seems to me to be not fear of what the stranger might do to the child (that is, fear of being harmed by the stranger), but fear of and resistance to engaging in the process of mutual adjustment and regulation (bonding) with a new person. Nevertheless, once the initial resistance or hesitation is overcome, skills acquired during interaction with attachment figures and friends will be used to establish bonds with new people.

Unfamiliar Caregivers as Security Figures. Can a person who has a playmate relationship with a young child become a security figure to that child also? It is an interesting fact that, in the absence of the principal attachment figure, distressed infants tend to display elements of security behavior toward adults whom they have known for but a few minutes. Rheingold (1969) reports that she was unable to "peel off" infants whose reactions to an empty, unfamiliar room she was trying to test. Ainsworth and her group also showed that infants tend to cling to strangers when they are left briefly by their mothers (Ainsworth *et al.*, 1978). It appears that, in the absence of major attachment figures, the security-regulating system can also be activated by others, even by unfamiliar people. The infants' behavior tends to be somewhat ambivalent under these circumstances, however. Young children may cling desperately to a stranger after the mother's departure from the room, but they may simultaneously continue to cry in the stranger's arms, and intersperse efforts at clinging to her with efforts at getting away. When the mother returns, on the other hand, she can generally soothe the child swiftly. Thus, despite the fact that behaviors associated with the security-regulating system are often shown to strangers, the effectiveness of strangers as security figures is at first very limited.

Once a stranger has established a friendly interaction with a young child, a change seems also to occur in that person's ability to play the role of substitute security figure. Fleener (1973) compared the behavior of 10-month-olds to two persons. One of the persons had interacted (played and performed routine care) with the child in the mother's absence for several hours, the other had interacted with the infant but briefly. When the substitute caregiver left the infant alone with the almost total stranger, many infants cried, whereas no distress occurred when the roles were reversed. In a similar study (Bretherton, 1978) I compared the behavior of infants who had interacted with a stranger in play for one minute with that of infants who had experienced a similar interaction for eight minutes. After the more prolonged interaction, infants cried substantially less when their mother left the room for a brief period. In addition, those infants who began to cry tended to be more effectively soothed by the stranger than the group who had only interacted with the unfamiliar adult for one minute.

Although the young child has a very strong propensity to seek to be near the principal attachment figure when some stressful event occurs, others can at least partially fulfill the role of security figure when the preferred persons are not present.

Since the principal attachment figure is usually so exclusively preferred when he or she is available, one must ask whether, or to what extent, a young child may still be experiencing anxiety, even when the substitute security figure seems to be effective (that is, whether the security-regulating system is not still activating plans for regaining the principal attachment figure). We do know (Ricciuti, 1974) that a subsidiary figure, such as a substitute caregiver in a day-care situation, is not so well accepted as a security figure by one-year-olds as is the principal attachment figure when the observations take place in an unfamiliar environment. It is possible that the effectiveness of the substitute is more linked to the particular context in which he or she gives substitute care than is the effectiveness of the principal attachment figure.

Other factors may also affect the degree to which a substitute security figure is accepted; in one of my studies (Bretherton, 1978) the stranger continued to interact with the baby in play throughout the mother's brief absence. It may have been that fact, in addition to the prolonged opportunity to establish rapport through play before the mother's departure, which led to a lesser incidence of crying during the mother's absence.

Although infants, even those infants who greet a stranger's entrance with friendly behavior, may occasionally snub the unfamiliar person's overtures, the tendency to reject or avoid a stranger is likely to become much more pronounced when the baby's attachment figure is absent. An infant may accept *partial* comfort from a person he or she barely knows, but at the same time push away, kick, or hit a stranger who attempts to play the role of security figure. To an unfamiliar person attempting to soothe a distressed child, such rejecting behavior is painful (as I have discovered from the various persons, including myself, who played the part of stranger in laboratory studies). We need more studies of the form and intensity such behaviors take in real-life situations, especially in situations where the child cannot make friends with the new caregiver in the presence of the attachment figure (for example, when an infant or toddler is transferred from a foster mother to an adoptive home).

The most systematic study of how young children can be given emotional support during a separation from their families was undertaken by James and Joyce Robertson (1971). After observing and describing the sometimes devastating effects of hospitalization on young children who were allowed only very limited visiting from their parents, Robertson (1953) showed that this emotional upset could be almost entirely prevented if the mother accompanied the child and remained with him or her during the period of hospitalization. Sometimes, however, separation of young children from major attachment figures is unavoidable.

With a small sample of four toddlers, the Robertsons (1971) therefore decided to investigate in depth how the deleterious effects of separation could be lessened. To this end they developed a number of strategies, some of which have already been mentioned, which fell into two related categories: First, to preserve as much as possible the ties with the child's natural parents and the familiar home environment, and second, to establish rapport with the child prior to separation, and to familiarize the child with the new environment. Some weeks before the impending separation (due to the expected birth of a sibling) the Robertsons collected information from the child's parents about the child's eating, sleeping, and toilet habits. Joyce Robertson (the foster mother) also began to make repeated visits to the foster child's home, and invited the child to visit her home before the separation took place. In addition, the Robertsons encouraged the child to bring familiar objects (toys, a bed, and a picture of the mother) to the foster home. The picture was used to talk about the child's family during the separation.

Yet, with all these precautions, the Robertsons (like the strangers I previously mentioned) had to withstand occasional rejections from their foster children, rejections which might have angered persons less prepared for and understanding of them. One two-year-old boy, after his father's daily visit, would say, "No, that is my Daddy's chair," when one of the Robertsons attempted to occupy the chair the father had used during his visit. Another child said to Joyce Robertson, "Don't cuddle me. Only my Mummy cuddles me." Paradoxically, the same child at other times told her foster mother, "You are my Mummy."

The Robertsons' study poignantly illustrates young children's loyalty to their principal attachment figures and their resistance to relinquishing them even though the beginnings of a new attachment to the substitute caregivers are already being formed. We desperately need further research on how to handle transitional situations of this nature. Robertson & Robertson (1971) describe the toddlers' reactions to the father's daily visit during which increased anger tended to be directed toward him, so that toward the end of the 10-day separation period some fathers curtailed their visits because of the painful feelings they aroused. We do not know to what extent and how these responses could be alleviated. Such situations are, however, sufficiently common (for example, when foster children are visited by their natural parents) to warrant further study.

Less conflict seems to arise when a young child spends only part of the day with a substitute parent (as in day care), but it is much more difficult to maintain the attachment to the natural parents while also making a transition to new attachment figures. With continual transitions from one set of major attachment figures to another, it may ultimately become extremely difficult for the child to enter into any new attachment relationships, as the writings of Burlingham & Freud (1944) indicate. Thus, to require a young child to reorganize his or her security-regulating system continually around new *principal* figures may finally make the child unwilling to engage in this task yet again—each effort brings only short-term results, and each new loss engenders renewed pain, leading finally to what Bowlby termed a state of "detachment," or the inability to form a loving relationship with anyone.

It seems plausible to suppose that not all children find it equally difficult to accept a new security figure. When speaking of strength of attachment, we may therefore have in mind two separate issues which are not necessarily related: first, the extent to which the child prefers one attachment figure to all others, especially in stressful situations, and second, the extent to which a child resists transferring his or her allegiance from one primary attachment figure to another. In-depth interviews with those who come to serve as subsidiary and primary substitute caregivers (foster parents, adoptive parents, day-care personnel) on the difficulties of fulfilling this role, and on how to cope more or less successfully with the challenges that arise, might lead us to a more systematic body of information, on which observational studies could then more fruitfully be based.

Conclusion

At a time when large numbers of families have to move or choose to move in order to improve work and living conditions, and when, moreover, there is a growing tendency for families to split up and regroup into new units because of divorce and remarriage, young children's attachments and friendships with others are also more frequently disrupted. It therefore becomes more important than ever to learn how to keep the stress resulting from these disruptions within levels a small child can manage. Our recent research efforts have concentrated on understanding the development of attachment. These efforts were a direct response to many studies undertaken during the 1940s, 1950s, and early 1960s which found that separation of a young child from his or her major caregivers without adequate substitute care could have devastating results. I suggest that we now begin a new series of systematic field studies of separation and the formation of new attachments, based on the insights gained from our recent studies of attachment.

Attachment figures are in an ideal position to support a young child's ability to adjust to new situations and people. Because young children feel more secure in the presence of their attachment figures, those figures are also privileged when it comes to encouraging a child to try out new coping behaviors, or to teaching such behaviors to the child.

It is not possible in all situations for attachment figures and new caregivers to collaborate, to share information which will make the transition to a new setting, but this should be the aim wherever it does prove to be at all feasible. For the unfamiliar caregiver, the first order of business is to establish rapport with a young child, a task which will be easier when the child's attachment figure is present, though I can think of many situations in which this cannot be done (as when a child is forcibly removed from neglecting parents). Under such circumstances, the new caregiver's task will be more difficult, and he or she may have to be prepared to endure more rejecting behavior.

I think that it would be a mistake to suggest that the substitute caregivers engage in *specific behaviors*, that is, smile frequently, speak to the child, or offer much physical contact. The primary goal should be to establish a social bond through the *meshing of the two people's interactions*. The quality of this meshing is more important than the mode of interaction which is chosen. As Hinde has said, what the two interactants do together is less important than how they do it. It does appear, that for the new caregiver to wait until the child is ready to seek him or her out for social interaction may not be the most promising approach. Many young children accept overtures from adults, and respond by entering into reciprocal interactions, even though they were not willing to make the first move on their own. Moreover, to use the intermediary of a toy when trying to make friends is especially likely to lead to successful interactions. Once a relationship has been established, and especially once the child has begun to form an attachment to a new caregiver, that person can build on the child's trust to help him or her to find ways of coping with the new environment or situation. To discourage clingy behavior in a child who feels patently insecure and anxious does not usually result in greater independence, but rather in heightened anxiety. It is vital for familiar and unfamiliar caregivers to realize that a child's attachment to another person, though it implies reliance on that person, can function and be encouraged to function in such a way as to lead ultimately to more autonomous coping, to more self-reliance.

Acknowledgment

The author would like to thank Mary Ainsworth, Vicki Carlson, Mary Main, and Marcia Rosser for their helpful comments on a draft of this paper.

References

Ainsworth, M. D. S. The development of infant–mother attachment. In B. M. Caldwell & H. W. Ricciuti (Eds.), *Review of child development research 3*. New York: Russell Sage Foundation, 1973.

Ainsworth, M. D. S., & Bell, S. M. Some contemporary patterns of mother–infant interaction in the feeding situation. In J. A. Ambrose (Ed.), *Stimulation in early infancy*. London: Academic Press, 1969.

Ainsworth, M. D. S., Blehar, M. C., Waters, E., & Wall, S. M. *Patterns of attachment*. Hillsdale, N.J.: Lawrence Erlbaum, 1978.

Anderson, J. W. Attachment behaviour out of doors. In N. B. Jones (Ed.), *Ethological studies of child behaviour.* Cambridge: Cambridge University Press, 1972.

Bell, S. M., & Ainsworth, M. D. S. Infant crying and maternal responsiveness. *Child Development,* 1972, *43,* 1171–1190.

Blehar, M. C., Lieberman, A. F., & Ainsworth, M. D. S. Early face-to-face interaction and its relation to later infant–mother attachment. *Child Development,* 1977, *48,* 182–194.

Bowlby, J. *Attachment and loss.* (vol. I) *Attachment.* New York: Basic Books, 1969.

Bowlby, J. *Attachment and loss.* (vol. II) *Separation.* New York: Basic Books, 1973.

Bowlby, J. Attachment theory, separation anxiety, and mourning. In D. A. Hamburg & H. K. Brodie (Eds.), *American handbook of psychiatry IV.* New York: Basic Books, 1977.

Brazelton, T. B. Personal communication, October 1978.

Brazelton, T. B., Tronick, E., Adamson, L., Als, H., & Wise, S. Early mother–infant reciprocity. In CIBA Foundation Symposium 33, *Parent–infant interaction.* New York: Associated Scientific Publishers, 1975.

Bretherton, I, Stolberg, U., & Kreye, M. *Engaging strangers in proximal interaction: Infants' social initiative.* Manuscript submitted for publication, 1980.

Bretherton, I. Making friends with one-year-olds: An experimental study of infant–stranger interaction. *Merrill-Palmer Quarterly,* 1978, *24,* 29–51.

Burlingham, D., & Freud, A. *Infants without families.* London: Allen & Unwin, 1944.

Campos, J. C., Gaensbauer, T., Sorce, J., & Henderson, C. *Cardiac and behavioral reactions of infants to strangers: Effects of mother's presence and absence and of experimental sequence.* Paper presented at the Biennial Meeting of the Society for Research in Child Development, Philadelphia, April 1973.

Chomsky, N. *Aspects of the theory of syntax.* Cambridge: M.I.T. Press, 1965.

Clarke-Stewart, K. A. Recasting the lone stranger. In J. Glick & K. A. Clarke-Stewart (Eds.), *The development of social understanding.* New York: Gardner Press, 1978.

Eckerman, C. O., & Rheingold, H. L. Infants' exploratory responses to toys and to people. *Developmental Psychology,* 1974, *10,* 255–259.

Fleener, D. E. *Experimental production of infant–maternal attachment behaviors.* Paper presented at the Annual Meeting of the American Psychological Association, Montreal, 1973.

Fraiberg, S. Blind infants and their mothers: An examination of the sign system. In M. Lewis & L. A. Rosenblum (Eds.), *The effect of the infant on its caregiver.* New York: Wiley, 1974.

Gaensbauer, T. Personal communication, March 1977.

Harlow, H. F. The development of affectional patterns in infant monkeys. In B. M. Foss (Ed.), *Determinants of infant behaviour I.* London: Methuen, 1961.

Hinde, R. A. *Biological bases of human behavior.* New York: McGraw-Hill, 1974.

Hinde, R. A. On describing relationships. *Journal of Child Psychology and Psychiatry,* 1976, *17,* 1–19.

Jersild, A. T. Studies of children's fears. In R. G. Barker, J. S. Kounin, & H. F. Wright (Eds.), *Child behavior and development.* New York: McGraw-Hill, 1943.

Kaufman, I. C. Biologic considerations of parenthood. In E. J. Anthony & T. Benedek (Eds.), *Parenthood: Its psychology and psychopathology.* Boston: Little and Brown, 1970.

Kaufman, I. C. Personal communication, April 1976.

Lamb, M. E. In defense of the concept of attachment. *Human Development,* 1974, *17,* 376–385.

Lamb, M. E. Father–infant and mother–infant interaction in the first year of life. *Child Development*, 1977, *48*, 167–181.

Main, M. Security and knowledge. In K. E. Grossman (Ed.), *Soziale Grundlagen des Lernens*. München: Kinder-Verlag, 1976.

Morgan, G. A., & Ricciuti, H. N. Infants' responses to strangers during the first year. In B. M. Foss (Ed.), *Determinants of infant behaviour IV*. London: Methuen, 1969.

Papoušek, H., & Papoušek, M. Cognitive aspects of preverbal infant–adult interaction. In CIBA Foundation Symposium 33, *Parent–infant interaction*. New York: Associated Scientific Publishers, 1975.

Piaget, J. *The origins of intelligence in children*. (M. Cook, trans.). New York: International Universities Press, 1952. (Originally published, 1936).

Rafman, S. Reactions to strangers' imitation of mother. In T. G. Décarie (Ed.), *The infant's reaction to strangers*. New York: International Universities Press, 1974.

Rand, C. S. W., & Jennings, K. D. *Reactions of infants and young children to a stranger in an unfamiliar situation*. Report prepared for the William T. Grant Foundation, 1974.

Rheingold, H. L. The effects of a strange environment on the behavior of infants. In B. M. Foss (Ed.), *Determinants of infant behavior IV*. London: Methuen, 1969.

Ricciuti, H. N. Fear and the development of social attachments. In M. Lewis & L. A. Rosenblum (Eds.), *The origins of fear*. New York: Wiley, 1974.

Robertson, J. Some responses of young children to loss of maternal care. *Nursing Times*, 1953, *49*.

Robertson, J., & Robertson, J. Young children in brief separations: A fresh look. *The psychoanalytic study of the child*, 1971, *26*, 264–315.

Sander, L. W. The regulation of exchange in the infant–caregiver system and some aspects of the context–content relationship. In M. Lewis & L. A. Rosenblum (Eds.), *Interaction, conversation, and the development of language*. New York: Wiley, 1977.

Stayton, D. J., Ainsworth, M. D. S., & Main, M. Development of separation behavior in the first year of life. *Developmental Psychology*, 1973, *9*, 213–225.

Stern, D. N. Mother and infant at play: The dyadic interaction involving facial, vocal, and gaze behaviors. In M. Lewis & L. A. Rosenblum (Eds.), *The effect of the infant on its caregiver*. New York: Wiley, 1974.

Stern, D. N. *The first relationship: Infant and mother*. Cambridge, Mass.: Harvard University Press, 1977.

Tracy, L. R., Lamb, M. E., & Ainsworth, M. D. S. Infant approach as related to attachment. *Child Development*, 1976, *47*, 571–578.

Weinraub, M., Brooks, J., & Lewis, M. The social network: A reconsideration of the concept of attachment. *Human Development*, 1977, *20*, 31–47.

Yarrow, L. The development of focused relationships during infancy. In J. Hellmuth (Ed.), *Exceptional infant* (vol. 1). Seattle: Special Child Publications, 1967.

Yarrow, L., Rubenstein, J. L., & Pedersen, F. J. *Infant and environment: Early cognitive and motivational development*. New York: Wiley, 1975.

10

Relocation and the Family: A Crisis in Adolescent Development

Elaine Ruth Goldberg

Adolescent Transitions: A Challenge to Parents and Teachers

When adolescents move, they face the combined stresses of adaptation to a new environment and the pressures of adolescent development. Although either transition can produce dramatic upheavals in individuals and family systems, the double stress of moving and growing up is an extreme challenge to an adolescent's security and well-being The long-term effects of academic and social status on his or her self-concept and the urgency of college and career choices raise the stakes of uprooting.

Parents and others responsible for promoting and enhancing the growth and development of adolescents are concerned about the effects of mobility. Schools, preoccupied with predicting and measuring success, magnify these concerns.

The research this paper is based on was stimulated by the following questions:

How does relocation affect adolescent development?

What psychological resources (i.e., habits or learned patterns of thinking and acting) enable some adolescents to view relocation as growth-enhancing, where others suffer intense loss, failure, and depression?

What role can school staff, mental-health and medical professionals, religious groups, community agencies, organizations, and employ-

Elaine Ruth Goldberg • Educational Consultant, Brighton, Massachusetts 02135.

ers play in lowering the vulnerabilities of adolescents and others affected by the stresses of mobility and a mobile society?

How can professionals adapt their skills, attitudes, and policies to meet the needs of relocated adolescents and the needs of an increasingly mobile society?

The Study Group

To address these questions, this investigator drew upon the recorded conversations and written accounts of more than 200 mobile adolescents, their parents, and men and women (ages 24–60) who had been mobile adolescents.

This study group represents adolescents who had moved many times and those who had moved only once, those who had moved only within the United States and those who lived overseas in one or more locations. In many families, one or both parents had moved as adolescents, and in about half the families the adolescents had moved one or more times before age 12.

In most families, "father's job" was the principal reason for moving. Fathers were typically corporate, academic, government service, or military personnel. All families were middle-class or upper middle-class by income, father's occupation, and/or family background. This investigator met the people interviewed for this study in the United states and in Europe through schools, clubs, formal groups, and professional affiliations, and through friends and acquaintances.

Both current and retrospective views reveal the impact of mobility on the adolescent's self-concept. The major topics of the interviews were family relationships, social status and peer relations, school performance, and life choices. Individuals felt different degrees of success or satisfaction in these topic areas based on widely ranging standards that reflect varying needs for approval, recognition, and acceptance.

In general, parents described family relations in terms of adolescent participation and responsibility. Parents had more positive feelings about moving if their adolescent son(s) or daughter(s) maintained or improved cooperative attitudes, and if their academic and social achievements conformed to parental expectations.

Early adolescents tended to appreciate improved family closeness and communication, especially if the family spent more time together than they had before moving. In the following quote, a 13-year-old girl described how moving improved her family climate.

> We were all experiencing something in common: facing the problem of how to survive in this strange new place. We would go around the dinner table

> and everyone would tell about what had happened that day, how it went, what we did.

This type of satisfaction was most deeply undermined when the family climate deteriorated or did not meet expectations for improvement. Satisfactory family relations were an important stablizer for early adolescents.

Older adolescents focused on peer relations, and then school issues, as the major sources of both satisfaction and problems. Their identity issues reflected the importance of social status. Many emphasized the importance of achieving satisfaction in an area in which their competence or mastery was independent of familial or social values. Paul, 19 years old, expressed a typical feeling:

> I feel I had to grow up a lot quicker. You have to decide what you want to do, where to go, and where to work, totally on your own. It doesn't depend on anyone. No one can be your model. No one has lived a life like yours. And there is no place to go back to. No connections. No home. No one holding you back. You are really free.

When interviewing more than one member of a family, it was clear that each individual's perceptions were colored by his or her degree of success in the area of greatest importance to his or her self-esteem. There was no consensus about the experience of relocation.

However, pervasive social attitudes and the isolation of mobile families often combine to devalue or deny the legitimacy of many personal sources of satisfaction that do not match standard cultural expectations. Looking back, a woman who spent two years in France described the clash between her personal and social worlds:

> I loved being able to deal with foreign currency. My language skills kept getting better and better, more vocabulary, more control of tenses. It was the first time I felt I was learning something difficult. I felt smart. My parents were very proud. They said it would help me get into college. But when we moved back to the states, suddenly things I had done overseas were irrelevant, even embarrassing. I was still proud, inside myself, but once I got back it didn't make any difference. The fact that I'd been away was a "so what" kind of thing. In less than a month I smoked dope and wore jeans. By 11th grade I was using birth control pills. I didn't even take French in school. The kids would think I was trying to show off. It's amazing how fast you drop your identity to conform!

Many mobile adolescents have long-term feelings of marginality that interfere with a positive self-concept that comes from a feeling of belonging or attachment.

"The hardest question people always ask is 'where are you from?' " said a young man in his early twenties. "I hate that. I never know what to say. I guess I don't seem to fit in anywhere. I don't feel as if I really belong anywhere."

Isolation, social pressures, and value conflicts increase stress and start a self-feeding spiral of anxiety, impaired performance, and lowered self-esteem that shows up in temporary withdrawal or developmental regression.

Sue, age 16, developed fears characteristic of a much younger child. Her mother described a problem that persisted for four months after their move.

> Sue became very alarmed if we went out at night. In spite of the fact that she had been staying all alone for years, she now panicked at the thought of being alone. She was unable to go to sleep until we showed up and was furious with us when we were late.

In extreme situations, this spiral leads to stress-related problems such as personality and behavior disorders, family violence, dependence on and abuse of alcohol and drugs, adolescent runaways, and school failure.

Stereotypes of Mobility

The American stereotype of the "nuclear family" has a powerful grip on mobile families. It is a source of many values about mental health. Because mobile families are nuclear families, many identify with this stereotype, and let it set the standard for family behavior and relationships. Thus, they believe that self-sufficiency is a sign of maturity and good health; conflict is a sign of illness.

The danger of the nuclear-family model, especially as used by mobile families, is that it ignores crisis, personal growth, and life-cycle changes. It is a static model, created but untouched by patterns of social and geographic mobility. It ignores the stress of loss, separation, and isolation, and denies major conflicts of commitment, values, and goals within and outside the family unit. It cannot prepare people for the disorienting elation/depression of relocation and the physical and psychological difficulties that challenge personal standards of normality.

The nuclear-family image does not provide options for sorting out and balancing personal priorities and needs. It contradicts the need for flexibility and creativity in meeting new situations. A young woman who describes her family as "typical as apple pie" thus revealed her dissatisfaction with the stereotype:

> By moving around so much we saw my mother's total dependence on my father and the company. It was a big drawback to her. I think it motivated us to become independent, not to depend on any man. I think a lot of us who grew up like that, moving so much, saw what it did to our mothers and we decided to put off marriage as long as possible.

Like the stereotype of the "broken family," the mobile family is trapped by narrow expectations that discount individual differences. The myth promotes a sense of loss, shame, and deviance in those who feel outside the mainstream pattern. A young man thus describes the loss of legitimacy his family experienced:

> We were forced to admit that we were no longer a typical American family. It was hard to accept. We were now outsiders, different, foreigners. We had to give up the cozy idea that our lives would go on as before.

The feeling of shame keeps people from asking for help. They view therapy or counseling as a sign of serious problems, rather then a temporary support. Admitting difficulties to an outsider cracks the veneer of the strong family image.

Gross categories of success or failure are erroneously applied, especially in recognizing the obvious initial difficulties of relocating families. Such dichotomous thinking in terms of success or failure obscures the true complexity and diversity of the transition process. Individuals and families have their own rates and their own styles of adjustment. Initial difficulties do not doom an individual to persistent adjustment problems.

An all-or-nothing pattern of thinking, common in high-risk situations, leads to unrealistic expectations for performance. A young woman describes how she was a victim of this attitude:

> I expected to move in, unpack, and immediately jump into the swing of things. But that wasn't how it worked out. And the whole scary thing was that my father kept telling us that if we didn't make it right from the start, chances were that we wouldn't make it at all.

The glowing examples of perfection and martyrdom and horror stories of humiliation and breakdown that are popular conversation pieces in highly mobile and expatriate communities intensify fears about exposing both personal difficulties and personal accomplishments. The pressure of managing a total reorganization of everyday life and the need to appear independent and in control makes stereotyped roles seductive even though they do not fit real life. Thus needs are frustrated, and thus, as the full range of feelings are unexpressed or ignored, the feelings of loss and alienation grow. One man concluded: "Moving means throwing away things you love a lot."

Because the full spectrum of the transition experience is obscure, the relatives, friends, and acquaintances of relocated individuals tend to trivialize or gloss over problems with a "buck-up" attitude. They advise mothers and children to keep up a smiling disposition to avoid burdening the father whose psychological resources must be devoted to his job. Symptoms of distress are labeled a syndrome or a phase "that everyone goes through." Although the knowledge that everyone goes through it may offer some small comfort or reassurance, it does not encourage individuals to look for ways to relieve the pressure, or to learn to manage the situation more effectively.

Afraid of the stigma of failure, many people avoid professional help. Moreover, psychiatrists, counselors, and social workers add to these fears by their clinical focus on patients and their problems, that is, his inability to adjust, her incapacity to cope. In addition, the literature on relocation is full of unsympathetic references to "problem wives," troubled teenagers, weak marriages, and sick families. By identifying "movers and nonmovers" or individual and family pathologies, employee-screening techniques attempt to avoid expensive "performance failures." All this intensifies the pressure for family members to appear in control and happy.

The hidden or unexpected stresses of these social and psychological pressures increase isolation and alienation. They impede satisfactory adaptation, and negatively bias attitudes about mobility. They weaken the family's ability to respond to the needs of its members, especially the adolescents, whose positive self-concept depends on a supportive family climate.

Relocation: An Opportunity for Growth

This investigator observed that those adolescents able to view relocation as self-enhancing had a supportive family whose psychological resources encouraged them to understand and accept their experience, and to learn ways to respond to frustration that reversed the downward stress spiral. Moving evokes the painful emotions of loss, separation, and loneliness. Adolescents must learn constructive ways to express these feelings. Parents that help their children to share feelings in appropriate and acceptable ways provide important guidelines that are useful throughout a person's life. Two brothers, Alan and Arnold, discussed how they were given this kind of valuable aid for dealing with their intense feelings.

> My family was really into celebrating events with ceremonial dinners, toasts, formal goodbyes and declarations, a time to be sad and emotional. I think that really helped. We all cried together.

> My mother always encouraged us to have a hobby, something we could carry with us, some solitary activity that would get us through the lonely times . . . things we could do on our own. It took the pressure off of us. We didn't feel so bad when it took a while to find friends.

This investigator asserts that those who attribute a satisfactory adjustment to certain personality traits, corporate relocation policies, or the right school or neighborhood, overemphasize factors that seem unchangeable and/or out of the individual's control. In contrast, the evidence suggests that mobile adolescents have many adaptive options for actively managing the stresses of relocation. They have or can learn many adaptive skills. They have many opportunities for ego growth as well as failure. Though a pessimistic picture can be painted of the mobile adolescent's developmental prospects, relocation can also foster adolescent development. An eighteen-year-old woman who enjoyed four moves within five years said:

> I never would have known how to cope and face new people and new situations. Now I have a lot of confidence that I could move to a new town, that I could get along in a new place. I think that is important for the future.

The confrontation of the unknown and the unexpected evokes personal resources rarely exercised to such an extent in familiar settings. The potential of these challenges to enhance ego strength is often overlooked in the throes of transition, but is clear in retrospect.

When asked how mobility had influenced their lives, those individuals providing retrospective views mentioned the following: many close friends, more professional colleagues abroad of diverse cultural backgrounds, more acquaintances in other cities and other countries. Most people had enjoyed the chance to compare and contrast different ways of life, and considered their experience a big influence on their career aspirations. The marketable skills or background they had acquired included languages, cultural interest and knowledge, educational advantages, experience abroad or in major American cities that gave them a sense of worldliness, confidence, and flexibility. Most people felt that they traveled more often and kept up more with current and foreign events than others who had not had the same kind of life.

Others reported a change in direction or in self-consciousness that improved the quality of their lives:

> Speaking two languages made me feel really different, as if I knew more about life. If I had not moved away I would have been in the same school for fourteen years.
>
> It was an expansive thing. I became aware of what America was and what it meant to others and to me. Moving all over the place I became aware of how different I was from others, and certainly from most Americans. I couldn't speak for anyone else but me or let anyone else speak for me.

Many people who were generally positive about the relocation experience could also specify values that they had adopted as a result of their unique perspective.

> I learned to hate certain types of American tourists—loud, shallow bullies, sticking out like a sore thumb in groups, not interested in new things, disappointed with everything and letting everyone know it. Maybe it is partly wrong to look down on tourists, not to want to be associated with them. But it pissed me off when tourists came in in their own bubble, influencing the kind of experience the other people had, like stepping on something that's already growing.

Psychological Preparation: An Educational Intervention

Unfortunately, most employers consider "psychological concerns" to be a private matter, and are reluctant to suggest that individuals or families may need "that kind of advice." One supervisor of international personnel for a large corporation said, "people who need that kind of thing are screened out of relocation assignments."

Professionals such as teachers or counselors with an over-view of mobility can compensate for this attitude. They can meet the needs of mobile adolescents and their parents by using educational vehicles to demonstrate the value of psychological tools that promote supportive family climates. They can prepare adolescents and families to manage the stresses of relocation, to anticipate its effects, and to recognize the growth-enhancing possibilities of new roles, relationships, and skills.

The educational model is useful because it is prevention-oriented. It recognizes and builds on the high motivation of people eager to learn how to better manage a new challenge, life task, or stage of development. It reduces the threat and eclipses the stigma of problem-oriented intervention.

I hope that this discussion will stimulate parents and professionals to find ways of offering the appropriate supports and services to mobile adolescents and each other. The purpose of this paper is not to eliminate the inevitable pains and frustrations of adolescence, but to enhance the mobile adolescent's potential for developing a healthy self-concept in spite of the problems and challenges that mobility adds to an already stressful stage of growth.

Arenas of Adolescent Development

The disruption and discontinuities of relocation challenge adolescents in three major arenas that shape their self-concept: family relations, social status, and school achievement. This section discusses some of the ways that mobility exacerbates the developmental process.

Family Relations: Meeting Needs for Security

The parents' ability to provide guidance and support to the adolescent depends on how well their attitudes, values, and expectations are suited to the adolescent's ego development. To nurture the adolescent's self-esteem, they must find a delicate balance of needs. There must be a subtle shifting of controls to allow for the adolescent's desire to exercise independence. Adolescents must break habits of dependency on their parents, and transfer loyalties and attachments to peers.

When a family moves, the need to establish a sense of security in the face of uncertainties, to relieve fatigue and tension in spite of stress and disorientation, and to maintain self-respect in spite of frustrations and isolation conflicts with its ability to encourage any family member's autonomy, separation, and experimentation, especially the adolescent's.

The adolescent's developmental achievements are undermined when increased needs for security decrease the parents' willingness to recognize previous levels of autonomy. In many families it reintroduces battles that had been fought in earlier years. Family members disagree about the limits and controls appropriate in the new environment. The parents' protectiveness conflicts with their son's appetite for exploration and his need to make friends:

> We were constantly worried about John's safety. It wasn't that we didn't trust him. It was just that he didn't take our warnings as seriously as we think he should have. He was too open and eager to make friends. He wanted to explore everywhere, right away. He complained that we were trying to scare him. Our fights started to sound like the fights we had had when he was thirteen.

The need to discover and assert their individuality leads many adolescents to test the level of control that parents deem appropriate. In new environments parents and adolescents differ about real and imagined dangers just as they would in familiar settings, but the conflicts are intensified. One father describes the parents' increased sense of helplessness:

> Our biggest fights were always about getting home on time. One night Sandy was late getting back from visiting a friend. There was a hopeless sort of feeling—no way we could really find out where they were. Finally, at one A.M., they showed up at the door. It turned out that there wasn't any terrible crisis from their point of view. They couldn't understand, wouldn't understand, why we were hysterical. After that there was a general cracking down, absolutely having to know where everyone was at all times. It was important being in a strange city. More important than if we had been at home. A special feeling of concern: especially about their general safety. But they didn't understand what all the fuss was about.

Relocation intensifies family conflicts and emotional responses. New and stressful situations produce many psychological and physical reactions that cause unexpected changes in moods, perceptions, and reaction patterns. Parents and teachers who are unaware of patterns of adolescent development may be hostile or indifferent to an adolescent's special vulnerabilities and the behavior they typically use to defend themselves.

Early adolescents go through rapid and uneven physical, emotional, and intellectual changes. Relocation can intensify their disorientation, or prolong episodes of heightened emotionality. Adolescents are thus more susceptible to sudden attacks of excitement and fatigue, to minor irritations and disappointments, to anger and severe depression. Sudden tears, sweats, or a pounding heart, extremely common during adjustment to a new setting, can cause them chronic worry about their physical and mental health. Strong, uncontrollable feelings make them feel childish, or as if they are coming apart when they most desire to demonstrate emotional control or self-mastery. The unpredictability and intensity of these reactions undermines the self-confidence they need in order to view their life in positive terms.

The desire to demonstrate emotional control and the threat of seeming weird or crazy can make adolescents hide or deny their fears and feelings. Some adolescents withdraw to minimize family tension at the expense of increasing their own alienation. They may hide illness out of fear or guilt. Or they avoid peers, to escape the risk of exposing their emotional turmoil.

Temporary withdrawal is an adaptive mechanism, but it can delay important social learning. In extreme cases, adolescents who cut themselves off from social interaction are deprived of the chance to develop the experience and skills they need for learning norms of self-disclosure and reciprocity, the protocols of their new peers. Delayed in separating from the family, the adolescent is limited to certain role behaviors appropriate to adult–child relations. With peers, the adolescent practices social competence as an equal, plays out various roles, and tests behaviors useful in the adult world.

For the adolescent, tests of fitness, strength, and maturity are attempts at mastery, and also tests of normality. Adolescent experiments with special diets, drugs, vitamins, and exercises are some ways that they try to control or regulate their feelings and behavior.

Family attitudes that idealize self-sufficiency and independence increase the adolescent's tendency to control or hide ambivalence negative feelings, or fears of failure and inadequacy. This further alienates adolescents, who then carry the burden of these scary feelings alone. Giving them the chance to clarify fears about their physical and psychological well-being makes it more likely that adolescents will find the reassurance and support they need. Parents, teachers, or friends who understand the need to explain and empathize can offer insight into what is happening. Those who don't understand become confused, rejecting, angry, worried, and guilty. Family climate and friendships deteriorate as misunderstandings and resentments escalate.

Because moving causes so many problems for teenagers, and seems to threaten so many of their developmental goals, parents feel guilty, and can be manipulated by that guilt:

> We were worried about the effect of moving on Jane. We were eager to make it easier for her. We let her make long-distance calls to her best friend back home. We drove her anywhere she wanted, bought her clothes, gave her more money than we were comfortable giving. We didn't pressure her about school. After four months she was still miserable and pushing us around like servants.

Adolescents, unlike children, are very powerful family members. Power struggles and role reversals allow adolescents to exploit parental guilt, anxiety, and friction.

Relocation intensifies and creates tensions between parents. Adolescents without peer attachments are acutely sensitive to parental tension. They become anxious, worried, critical, and hostile.

Some adolescents stage a crisis to capture parental concern. They stay out overnight, run away, fail in school, steal or vandalize, get pregnant. A family in isolation is extremely vulnerable to these crises. They do not have the emotional and social supports that would help them keep their balance and perspective. The stakes are higher. Legal trouble may result in the loss of the father's job. Whether or not adolescents are aware of the implications of their acts, when their escapades have such serious outcomes the whole family suffers guilt, shame, and bewilderment.

When an adolescent develops exaggerated feelings of loyalty to one parent, the alliance drives the parents further apart. Especially when anger and resentment destroy parental communication and cooperation, adolescents assume powers inappropriate to their age and position.

Sidney describes how she, her father, and her siblings ganged up on their mother, who became the scapegoat for family difficulties.

> We were horrible brats. Mother was slow to pick up the language and we ridiculed her accent and mistakes. We kidded her about her fear of driving and her fear of going out. She was an easy target. . . . We hated ourselves for it later but then we couldn't stop. It was too easy.

Stan describes how he joined his mother in emphasizing their sacrifices and blaming discomforts and problems of relocation on his father.

For a while Dad couldn't come home without getting all our aches and pains blamed on him. We made him feel like a rat.

Other role imbalances occur when a lonely parent demands the adolescent's companionship. Especially when fathers work long hours, take long or frequent trips, and the family is new in the community, adolescents have conflicts of loyalty to their mother and to the need to pursue their own social life.

As adolescent interests and values are increasingly shaped by peer-group attachments, conformity to peers is a threat to parental authority. The urgency to conform in sometimes trivial aspects of appearance and behavior is a sign of the adolescent's intense desire to be accepted by peers. Drastic changes in behavior, language, and appearance upset parents, already sensitive to changes in family patterns and relationships. When adolescents rebel against parental wishes and values, they typically choose drug use, sex, running away, dangerous escapades in which they place themselves at risk of personal harm, and poor performance in school.

Although sex is an important area in which adolescents forge their individuality, parents are especially sensitive to provocative threats and forms of dress. Their own sexual relationship is strained because of the fatigue, friction, and frustration of relocation. When their sexual tension is high, they are easily upset by the adolescents' sexual behavior. Even benign adolescent romances can ignite intense battles over privacy, curfews, dress, and behavior.

Under stress, the adolescent's apparent impulsiveness, irresponsibility, and intense emotional outbursts may reinforce parental beliefs that adolescents are not mature enough to make decisions about school, sex, friends, and how to spend their time. For example, Becky recalls her intense mixed feelings as moving day arrived:

When I first heard we were going overseas I thought I was very special. I got out of gym class to study Italian. But at the end of the school year I was hysterical. I didn't want to lose my friends. I felt my position in the group would be lost. I felt ripped away. I cried and tried to make myself sick so they would have to leave me behind.

Peer Attachments and Newcomer Anxieties

Loss of peer attachments causes intense anxiety and conflict when families move. The loss of important social bases and the ambiguous and marginal status of the newcomer mean there is an extended period when the adolescent does not have reference points that help define appropriate behavior. When families lack the skill of "saying goodbye," unresolved feelings of sadness and loss can contribute to the newcomer's disorientation during and after the move.

In addition to broken friendships, the adolescents' social status is the most easily identified loss of relocation. Their social prestige and status depend not only on belonging to a group, but on winning approval and acceptance from the group that matches specific social needs and aspirations. Unfortunately, most teenagers find out that the social skills, activities, and values that mean acceptance and status in one social system do not easily transfer to another school or community. The prestige value of grades, spending money, and privileges varies within schools and subgroups. Acquiring an interest or proficiency in the activities that have high prestige value in the new social group takes time, and a rate of involvement that can conflict with family or academic priorities. Furthermore, peer-group values may discount academic achievement.

The social significance of appearance explains why early adolescents are extremely sensitive to and preoccupied with their physical features. Early or fast developers gain status and privileges that are delayed for late bloomers. Height, weight, shape, and skin condition are a constant source of concern. Fears of change and differences lead to conformity in speech, makeup, dancing, hairstyle, and dress. The sudden loss of personal standards for appearance and behavior further undermines the self-confidence and sense of self that adolescents require in order to negotiate status in the new setting.

Academic Status and School Selection

In addition to social status, an adolescent's level and area of school achievement influence feelings of competency, social-group acceptance and position, parental approval and respect, level of aspiration, college options, and career goals.

Parents are extremely concerned about choosing the school that will maintain or improve academic status. A typical parent says:

> We were looking for a feeling of familiarity and continuity. The schools had to be like what they were used to if they were going to make the most of the year.

Parents investigate performance standards, social climate, curricula, teacher expectations, extracurricular activities. They want to match adolescent needs, temperaments, interests, and abilities to maintain and improve performance levels. Some parents want their adolescent to fit into a standardized program to minimize disruptions. Other parents value schools that promote individuality. Among mobile families, there is little consensus on the type of school that is most likely to satisfy a particular individual or family. One mother said:

> We didn't want to put the kids into local schools that would pin them down when we wanted to take them on vacation. Private schools seemed to be more flexible. They were used to treating kids like individuals.

Evaluating schools is difficult. If parents have a choice, it must be made quickly, without enough time to check the school as thoroughly as they would like to. It is hard to compare schools and neighborhoods with various assortments of services, resources, activities, and interest groups. When schools lack several features, there is not always time to investigate whether community facilities are available.

In addition, identifying adolescent educational needs is frustrating; their needs, interests, and abilities are in flux. The esteem-values of academic achievement, social status, and personal accomplishment are variables that depend on a constellation of personal and family values. Priorities are hard to pin down. Parents and adolescents have different priorities. Decisions that seem right in May can be totally unacceptable in September.

When local facilities are inadequate or unacceptable, boarding schools are usually the answer. The more mobile the family, the more likely they would be to choose a boarding school as a way of giving their adolescent a continuous educational experience. Margaret describes a typical response to boarding school and the impact it had on her family:

> The biggest problem about being away in boarding school with your parents moving around so much is that we never knew where to call. And on vacations we went "home" to a totally strange city where we had no friends. We were stuck in an apartment with nothing to do. Other kids would go home to towns where they knew people, but not us. And the first three days home on vacation were always bad. It took time to adjust to living with each other. We weren't used to a personal form of authority. In school, it was all equal. There were certain rules that we had to follow. Once we were home, our parents began to impose all their little rules. What to wear, when to be in, our language, how we fixed our hair. The good part was that we never had problems with boyfriends and parents never had to hassle us about homework. We escaped a lot of conflicts.

School Adjustment: Sorting Out the Problems

Expectations for school adjustment run high and disillusionment is swift when school transition is difficult. Adolescent anxieties cause chronic complaining, unhappiness, performance blocks, teacher–student conflicts, and student evaluations lower than parent or student expect. The more responsibility parents take in choosing the school, the harder it is for them to avoid guilt and self-doubt when adolescents have trouble.

It is rare to find a headmaster like the headmaster of the private international boarding school who flew to North Africa to reassure thirty-seven sets of American parents about first-term grades. More typically, parents do not find teachers or administrators reassuring or helpful.

It is difficult to clarify the source of student problems. When it is difficult for parents to communicate with adolescents about school, relocation creates additional blocks to effective cooperation. Problems are obscured by everyone's tendency to attribute changes in behavior to obvious events such as relocation when, under stable conditions, the same behavior would be perceived as independent of the environment or related to manageable factors. Problems in school seem to confirm parental fears about the effects of relocation on adolescent development.

Because achievement is highly valued, adolescents may be guilty about performance problems. Parental criticism intended to motivate adolescents cripples their performance, especially if it is already blocked by unrealistically high expectations. Other performance problems can be caused by self-criticism, and conflicts about seeking approval from teachers, peers, and parents. Early adolescents are particularly vulnerable to outside criticism because their increasing abstract thinking ability makes them idealists. They have strong feelings about how people should behave, and what is right and wrong. They see people as either heroes and idols or villains and enemies. They tend to be reformers and critics, usually targeting the people closest to them, parents and siblings. Although this critical ability helps adolescents differentiate their own and parental values while they experiment with new attitudes and behaviors, parents poorly tolerate the onslaught of criticism when they are going through their own periods of self-doubt. Thus, at a time when adolescents need parental encouragement and understanding, school problems can set off a cycle of recrimination, guilt, alienation, and misunderstandings. This undermines their ability to perform. They often find approval and acceptance from other angry young people to make up for the lack of approval from parents. This, in turn, can lower their self-esteem, and provoke parents to criticize their choice of companions.

The adolescents' reluctance or fear of asking for help also makes it difficult for parents (and teachers) to help them with school problems. Although adolescents may need it, they may resent parental intervention, interest, or even sympathetic concern. Without recognizing the true weight of external pressures, adolescents internalize failure. This is reinforced by attitudes that make teachers attribute personal and social anxieties to personality problems. The older the adolescent, the more he or she is expected to be in control. School counseling services are usually problem-oriented. Intelligent, well-adjusted students are expected to take care of themselves. The burden of adjustment is on the individual.

Furthermore, parents of adolescents are not used to an active role in high-school affairs. Schools rarely encourage parent involvement, and consider parents "overprotective, pushy, and troublemakers" when they initiate contacts, especially on behalf of older adolescents. And adolescents usually agree that parents should stay out of their world.

"At first it never occured to me to question the moves. It was just something we always did. But I have very strong feelings about school . . . that was my life," writes a woman who says that her parents never interfered with her course selections, study habits, or extracurricular involvements. They cooperated when she requested to change schools in the middle of tenth grade, and "didn't bat an eyelash when I dropped out to go to a full-time dance school."

More often, school choices, course selections, school grades, and evaluations are the focus of parent–child power struggles.

In communities where parents derive vicarious status from their child's academic accomplishments, adolescents can embarrass their parents with poor grades and other school problems. Poor grades reflect rebellion against parental values, especially when other avenues of achieving developmental goals of separation are blocked by the dependencies created by relocation. Poor grades also show indifference to future goals, or that peer relations or extracurricular activities have a higher priority than grades. The adolescent's primary needs are for acceptance and personal accomplishment.

Planning for the Future: An Educational Imperative

When an adolescent ignores planning for the future, parents fear that relocation contributes to predicaments that a young woman describes below:

> I never considered higher education in Europe but always worried about getting accepted back in the States. I never did any research on colleges. I picked an eastern school because that was where my best friend was going. But I hated it. I gained twenty-five pounds the first year. I visited another friend in Colorado, liked it, and transferred. But a year later I was going somewhere else. Nowhere felt like home.

This statement raises an important question: Why does she expect college to feel like home?

Her problem reveals a "side effect" of mobility in many older adolescents once they leave their families to live on their own. She moved willingly, enjoyed the adventure of it, made friends easily, and could depend on good relationships with her parents. But this woman is now an adolescent lost in young adulthood. She lacks the focused interests, developed talents, and internal value system that set patterns of future behavior and prepare her to make decisions on her own. Her mobility was not a "problem," but the functional value of her present-tense orientation, typical of early adolescents, expired when she "grew up."

She has none of the satisfactions of an independent identity. Her needs for belonging, self-mastery, and achievement are unbalanced, because mobility made belonging a priority for so long. Her alienation masquerades as autonomy, but her feelings of dissatisfaction and marginality reveal that her self-image is not yet fixed. Friends cannot substitute for purpose. She has an empty, "homeless" feeling that plagues young adults when they have no goals to help organize and give purpose to new observations and encounters. She is inexperienced in making commitments, even to herself.

This young woman's education, traditional and academically excellent, has been inadequate. She does not know what she wants to do with her life. She has not learned to articulate her skills, interests, and goals. She treats each step in her life as a clean break with the past. Past events and relationships have a vague emotional power, but she does not know how to use them as reliable guides to satisfying future decisions. She does not value her personal career, nor does she recognize it as a legitimate reference. Frustration leads to avoidance, escape, and regressive behavior instead of experimentation, learning, growth, reflection, and planning.

Learning to Manage Transitions: Suggestions for Parents and Teachers

People who view transitions such as relocation as a set of tasks avoid many uncertainties and psychological traps. Those who learn to translate stress-producing anxieties into specific threats to their well-being can anticipate, plan, and participate in the transitional process. They can work to overcome identifiable obstacles, and achieve definite goals. They become agents in promoting their own adaptation to the new environment; thus, they reinforce their sense of control, competence, and security.

Parents and teachers can best support adolescents by learning to recognize the subtleties, ambiguities, and conflicts of adolescent autonomy, to tolerate the contradictory swings in mood, interests, and impulse control, to interpret the behavioral clues that reflect adolescent developmental needs for security, belonging, esteem, recognition, and praise.

The stereotypes that shape perceptions and expectations of adolescents depict wild behavior, impulsiveness, and lack of responsibility. The stress of mobility inflates stereotypes, and can discount first-hand knowledge. Thus, adults are more likely to respond to problems with control and punishment, to value discipline more than growth. They fear emotionality and resist necessary changes in family patterns, rituals, and roles.

Responsible adults must learn to avoid the Scylla and Charybdis of adolescent development. Adolescence is a process as well as a stage of growth. The transitional process can be delayed if parents enforce the same structure and props over time and interfere with the adolescent's continual attempts to test limits. Development is distorted if adults emotionally abandon the adolescent too soon and too quickly, failing to recognize different individual needs. Parents must be sensitive to the signs of stress that feed a cycle of frustration and alienation that undermines adjustment.

The adolescent process of self-definition is a continuous task with enormous implications. The danger of intense or prolonged transitional crisis is continual stress; emotional and physical resources are depleted, performance falters, confidence falls, risktaking slows, learning stops, and the ability "to adjust" fizzles. It is unrealistic to expect adolescents to manage the stress of change alone. They need a supportive family, school, and community.

Social bases such as the schools, churches, community centers, libraries, and adult-education programs have the structure and the legitimacy to help adolescents and their families to overcome their isolation and to meet their needs in a new community. They provide opportunities for socialization; they channel tension, excitement, and frustration into acceptable outlets. They can maintain libraries of special-interest materials and offer programs to help adolescents to universalize their physical and emotional changes, to learn about social patterns, new roles, relationships, and skills, to plan educational agendas and career goals. They can create support groups and hot lines, and provide counseling for parents, adolescents, and families. They must reach out to the adolescent with information and resources to help reduce the stress of unexamined and unproductive family tension.

Educational programs and services can support the successful adjustment of adolescents by focusing on the most important factor in their lives, their family. A supportive family climate helps adolescents to consolidate changes by teaching ways to maintain equilibrium and establish a sense of purposeful direction. By translating the behavior patterns of supportive families into teachable concepts, professionals can help parents improve the quality of life for themselves and their children. Teachers, health-care professionals, and others who interact with adolescents also can benefit from training that identifies specific tasks, learnable habits of thought and behavior that people use in order to cope successfully with stress and change.

For example, patterns of communication that clarify new expectations and commitments are essential for creating or maintaining a supportive family climate. These patterns include: the habit of verbalizing perceptions, feelings, and expectations to compensate for the loss of predictable family routines and traditions; the habit of discussing personal fears, fantasies, and associations to provide a broader personal context for understanding how changes affect family members and to provide clues to the intensity and style of their responses; the habit of anticipating a new situation and discussing a variety of emotional responses to increase flexibility and a sense of preparation. These behaviors help adolescents to understand, accept, and learn from their own responses. It provides them with a feeling of self-confidence and growth.

Problematic family relationships cannot provide the basic trust and security that help adolescents safely explore the options that lead to true self-definition and autonomous functioning. Without security, their behavior patterns tend to be defensive and repetitive; innovation and risktaking are blocked. They lose the optimism and flexibility necessary for managing transitions, for orienting themselves in a world of rapid unpredictable change, for comprehending a new milieu, and for maintaining an internal equilibrium at different stages of growth when different needs have ascendency.

With support and preparation, parents of adolescents can be innoculated against unnecessary disorientation and guilt; their effectiveness as parents is enhanced. They can better help adolescents avoid the unnecessary stress and frustration that undermine their ego growth. Moreover, the needs discussed in terms of adolescent development apply to the parents' own developmental and transitional needs, and thus help them to achieve security in the face of uncertainty and change.

Additional Reading Suggestions

Bower, E. M. American children and families in overseas communities. *American Journal of Orthopsychiatry,* 1967, *37,* 787–796.

Coelho, G. V., Hamburg, D. A., & Adams, J. E. (Eds.). *Coping and adaptation.* New York: Basic Books, 1974. Recommended for more specific information about stress, stress-producing situations and conditions, stress-induced behavior, and stress management.

David, H. P., & Elkind, D. Family adaptation overseas. *Mental Hygiene,* 1966, *50,* 12.

Duhl, L. J. (Ed.). *The urban condition: People and policy in the metropolis.* New York: Basic Books, 1963. See chapters: Grieving for a lost home, by Fried, Population mobility in the American middle class, by Gutman, & Effects of the move from city to suburb, by Gans.

Fensterheim, H., & Baer, J. *Don't say yes when you want to say no: How assertiveness training can change your life.* New York: McKay, 1975. Discusses people with deficient social networks and offers specific advice on how to overcome isolation and achieve a satisfying social network. Breaks the idea of making friends and developing intimacy into learnable concepts.

Giammattei, H., & Slaughter, K. *Help your family make a better move.* Garden City, N.Y. Doubleday, 1970. Advice from experienced mothers.

Gordon, R. E., Gordon, K. K., & Gunther, M. *The split-level trap.* New York: Geis Associates, 1961. Chapter 1 describes the sensitizers, the pressurizers, and the precipitators—the kinds of stress that the authors have found attack mobile people.

Hall, E. T. *The silent language.* Garden City, N.Y. Doubleday, 1959.

Hall, E. T. *The hidden dimension.* Garden City, N.Y. Doubleday, 1966.

Hall, E. T. *Beyond culture.* Garden City, N.Y. Doubleday Anchor, 1976. These three books by Hall describe those elements of culture and communication that are below awareness, and the importance of context for meaning.

Harris, P. R., & Harris, D. L. Preventing cross-culture shock. *The Bridge: A Journal of Cross Cultural Affairs,* Winter, 1976–1977.

Hopkins, R. *I've had it: A practical guide to moving abroad.* New York: Holt, Rinehart, & Winston, 1972. A good discussion of moving and ways of coping. Discusses culture shock and adjustment. Suggests sources for more information.

Kenniston, K.: The Carnegie Council on Children. *All our children.* New York: Harcourt Brace Javanovich, 1977. Contains a good discussion on the myth of the self-sufficient family, its history, and the social costs of perpetuating it.

Noer, D. M. *Multinational people management: A guide for organizations and employees.* Washington, D.C.: Bureau of National Affairs, 1975. Discusses the people dimension of multinational business. Offers an interview worksheet for relocation candidates that helps pinpoint areas of concern. It is especially useful for families evaluating their priorities and strengths.

Packard, V. *A nation of strangers.* New York: McKay, 1972. See page 206ff. Describes the kinds of relationships that are necessary for maintaining self-esteem and well-being.

Ruina, E. *Moving: A common sense guide to relocating your family.* New York: Funk & Wagnalls, 1970. A thorough discussion of moving, with especially useful advice on selecting a school and community. Recommends making self-inventories.

Sommer, R. *Personal space: The behavior basis of design.* Englewood Cliffs, N.J. Prentice-Hall, 1969. Useful for interesting information about how people are affected by different spaces and contexts.

Transplanting the executive family: how to minimize the shocks. *Business Week,* November 15, 1976.

Werkman, S. Hazards of rearing children in foreign countries. *American Journal of Psychiatry*, 1972, *128*, 992–997.

Werkman, S. *Bringing up children overseas*. New York: Basic Books, 1977. Includes a special chapter on adolescents.

West, E., with J. P. Carter and members of the staff of the Boston Children's Medical Center. *Keeping your family healthy overseas*. New York: Delacorte, 1971. Addressed to mothers because they feel that the family's experience is based on the mother's ability to manage and adjust. She is the focus of all frustrations and the source of alleviation. Useful discussion of culture shock, p. 65.

11

Coming Home: Adjustment of Americans to the United States after Living Abroad

Sidney L. Werkman

The task of readapting to the United States after living overseas is, for many, the most difficult hurdle in the cycle of international life. People who have lived overseas emphatically report that it is far less stressful to leave the United States and find a place in a new country than it is to experience the unexpected jolt of coming back home. As a 20-year-old woman recalled: "People pushed and shoved you in the New York subways; they treated you as if you simply don't exist. I hated everyone and everything I saw here and had to tell myself over and over again: 'Whoa, this is your country; it is what you are part of.'"

Very little attention has been paid both in research studies and in the literature of the behavioral sciences to the issues involved in return and readjustment to the United States of people who have lived overseas (Borus, 1973a, b; Bower, 1967; Cleveland, Mangone, & Adams, 1960).

Yet clinical experience and the anecdotal reports of returnees—in the United States we do not even have a sanctioned term such as "colonials" to describe them—indicate that the problems of fitting into the United States once again can be serious and, at times, long-lasting. Both clinical and research interests have focused my professional

Sidney L. Werkman • School of Medicine, University of Colorado, Denver, Colorado 80262.

concern in this area (Werkman, 1972, 1975) which promises to become an increasingly important one for students of transitional phenomena.

This chapter, based on my professional experience, will describe the nature of psychological stresses brought to bear upon the traveler who returns to the United States. Certain characteristic reaction patterns in these stresses will then be delineated. From a knowledge of characteristic stresses and patterns of reaction a group of methods for dealing effectively with the transition to the United States will be developed.

Subjects: The population from which these observations were made consists of four groups: (1) adolescents and adults interviewed by the author during consultation trips to international schools overseas; (2) extensive tape-recorded interviews with 30 university students (average age, 21) who had lived overseas at least one year and were attending the University of Colorado at Boulder; (3) patients from the author's clinical practice whose problems began in relationship to living overseas; and (4) a research sample of 172 adolescents living overseas compared with a control group of 163 adolescents who had never lived overseas. The research sample, matched for age, sex, and socioeconomic background, consisted of young people who had not sought psychiatric help and were overtly adapting effectively.

Approximately 1,700,000 Americans live overseas, of whom 230,000 are children attending international schools. In 1970 military personnel and their families were the largest category, 1,375,000, followed by 236,000 people in private-sector activity, and 110,000 government employees. Of the civilian adults, the major categories were, in decreasing order: religious workers, engineers, teachers, scientists, and technicians. The nonmilitary population, with whom the author has had the greatest experience, is predominately a highly selected group of professionals and college-educated administrative personnel, largely with intact families, who probably represent a more than usually stable, psychologically resilient, competent group of people.

As the prevalence and incidence of psychiatric disorders in the United States cannot be stated accurately, it is even more difficult to estimate the number of such disturbances overseas. However, as most people living overseas are of at least middle-class socioeconomic status, we would expect fewer clinical psychiatric disorders to be present than are represented in a continental United States population. Indeed, this generalization appears to be accurate (Kenny, 1967; Smith, 1966). It must be recognized, however, that the stresses of overseas living may well be important predisposing factors in the development of later depressions, character difficulties, and subtle but important aspects of a person's interpersonal relationships and thought processes. Obviously, it is impossible to generalize accurately from such a diverse group of people. However, some recurring, seemingly fundamental characteristics common to Americans who have been overseas have emerged, regardless of the range of differences in life experience that occur throughout the world with its enormous variety of cultural and work patterns.

To best understand the psychological styles and problems that develop when Americans return home, it is necessary to consider issues related to the entire cycle of leaving the United States, settling in overseas, and then being uprooted a second time in order to reenter the United States. Many of the problems that come into focus upon return to the United States are composed of elements that developed in one or another stage of the complex cycle of living overseas.

Leaving

Common to all overseas Americans is the experience of leaving the United States at least once for a prolonged period of time. At first glance this may not appear to be a significant stress. However, anyone who moves overseas must relinquish ties with relatives and friends, give up the many sociocultural supports present in the United States, and attempt to find substitutes for these crucial social-system elements in a foreign country. The act of leaving involves issues of separation, repudiation, and loss that often have important consequences for later adaptation.

Goals: Motivations for a move overseas are indeed various. An overseas assignment is a normal part of careers in international business, the foreign service, or the military, and people who pursue such careers may not even question the need for such travel. For others, moving overseas contains a large measure of uncertainty combined with a search for meaning and significance in life. Families may move to bolster an unhappy marriage, to give children a fresh start in a new culture, or to escape painful living situations. Obviously, the consequences of such diverse motivations may turn out to be adaptive or pathological. However, some element of dissatisfaction must play a part in the decision to choose the complicated, demanding peripatetic existence of living overseas.

The majority of families who go overseas for the first time are young families. For them, leaving the United States is not a particularly difficult hurdle, as they are in the process of seeking their roots in society and tend to be open to new experiences. Older families who have already made life-defining commitments may encounter more serious opposition to the need to invest in new ties. As one father whose family had lived overseas for a considerable period of time described it: "It is easier to travel when your children are small; then it gets more and more difficult. Move by move, I felt more and more resistance to change from everybody in the family, until the time came when they just refused to be bounced around anymore."

Certain characteristic styles of behavior and thought may be developed in people who live overseas; indeed, such styles may well be decisive factors in the choice of an overseas career. Many people overseas seem to thrive on novelty and short-term relationships. Conversely, they may become uncomfortable at the prospect of longer-term, intimate and dependent involvements. They like to complete a job and then start over somewhere else. A Foreign Service officer put it this way: "I really thrive on getting a new pack of cards every four years; you're not stuck with anything, and when you finish a job you just close your desk on it and go off to a new one." The more disturbing side of such a view of life was described by a teenager who said: "We're pretty good at adapting to other people's manners, but part of you gets lost when you do that too much. You get to look at everything as an outsider. Nothing is good or bad anymore." This important group of character issues will be discussed in detail later in the chapter.

Settling In

Fathers of overseas families tend to be highly selected members of their business or government organizations, and often must travel, frequently on special assignments. They become highly visible representatives of the United States in the city and country in which they live.

Because of the father's heightened sense of role, a greater child-rearing burden is placed on mothers who, themselves, are undergoing a cultural transition. One overseas-reared adult remembered his life in Hong Kong this way: "I don't have any recollection of my father when we lived in Asia. He doesn't fit in. My mother was around a good deal, and I divided my time between her and the maid."

Settling in overseas may place unusual strains on the relationship between a husband, heavily involved in a demanding career as a representative of a company or the United States government, and a wife who is often left without the support of relatives or close friends in an alien culture in which she has little work to do and little opportunity to pursue a career outside her home. Such strains may contribute to drinking problems, depression, or other incapacitating symptoms that powerfully affect her ability to nurture her children. Because of the high visibility of American families overseas, many psychological difficulties are hidden from the view of others, only to become apparent when the family returns to the United States. The feeling of being constantly on display may prevent a family from attempting to initiate changes that would help them settle in satisfactorily. An executive described the pitfalls of his showcase existence in this way:

> Here in Afghanistan we all watch each other too closely. Call it a fishbowl syndrome. People won't admit they have problems with drinking or with their children or when they feel their jobs are on the line. Most people want to hide their problems, and in a way they are right. There is just no place to talk about the things that bother you, much less the little things that make you feel good. You always have the sense that somebody is going to judge you, somebody that doesn't know you very well. People just want to get through the year or through their tour and get out. It's a feeling that they just don't want anyone else to know what their problems are.

Many overseas families employ servants and nursemaids in their household, whereas the possibility of having such help would be rare for them in the United States. The introduction of a new and significant member to the household may enrich or destroy the nuclear family in ways that have no precedents in ordinary family life in the United States.

A major geographic change necessitates an adaptation to new culture and language, new friends, new neighborhoods and schools, as well as novel social and recreational activities. As Americans overseas tend to be on one- to four-year tours, it is difficult for them to put down deep roots in any community. Instead, they must become adept at developing short-term friendships and comfortable in fitting into differing international sports, celebrations, styles, and inconveniences. And while doing all this, they must keep in mind the necessity of eventually returning to the culture of the United States.

Too much change may result in the well-known syndrome of culture shock, as described by a young woman who finally had to be returned to the United States. She said:

> When I got to North Africa I suddenly lost all my perspectives. It was a pretty hard blow, because I had never been knocked down in my life before. Everything friends would talk about would make me really homesick. I had no concept of the dirt and filth, of people urinating in the streets or the way they treat animals here. The food was terrible, and they didn't repair the toilet I had to use for three months.

An inability to settle in overseas, even for some who do not need to be evacuated, may breed feelings of defeat and pessimism that continue to plague them on their return to the United States.

If a unique and satisfying way of life has been discovered overseas, the United States may seem pallid by comparison. A disappointed returnee put it this way:

> In London there was a touch of something different. There is one pub to every fifty hamburger joints in the United States, but that one pub is better than the fifty put together. Something about living there gives a special flavor to things. You make an occasion of even just walking down to the Green Park in London. Yet, though we have so much available in America, it just doesn't seem worth the effort to get away from the TV to see it.

Having learned a fresh and thrilling way of experiencing life, such a person finds it exceedingly difficult to settle down in the United States once again.

Special Competences

People living overseas accumulate a group of special competences that help them adapt to new situations successfully. They learn to use the Paris Metro skillfully, to interact tactfully with a wide variety of people, to converse about restaurants, museums, monuments, and political parties throughout the world. Mastery of these bits and pieces of knowledge makes life infinitely easier overseas, and contributes to an aura of distinctiveness. Many variations on the foregoing themes can be described, all of which contribute to make up a special consciousness in people who have lived abroad. Special competences are of little use when a person returns to his original home. As one returnee put it:

> You have a language, French, that is a part of you, and you cannot share it. You learn about painting and architecture. In Paris you really learn to look at things, and they are worth looking at. The people back home don't care about how I have lived, who I have met, what I have done, what I am now. They ask questions, but they don't care about the answers.

Return to the United States

Separation and Loss

Returnees leave a significant part of themselves behind when they give up a foreign way of life. They must learn to live without cherished friends and family customs at the same time that they attempt to find secure places in American culture. Unfinished tasks, unfulfilled dreams must be dropped or forgotten. The need to abandon intense friendships and cultural supports frequently results in disturbing feelings characteristic of a grieving process. Though most returning Americans seem to make a good surface adjustment to this country, that adjustment may, at times, cover over a host of barely contained feelings of uncertainty, alienation, anger, and disappointment. The following reports describe some of the underlying feelings:

> I felt out of everything when I came back. I didn't know about the music, what to wear, or how to get into the tight cliques that have formed from people who have been together all their lives.

> I had lots of friends and played on the school soccer team overseas. When I got back to the United States nobody noticed me in school and nobody went out of his way to be nice to me. My marks slipped and I was miserable for two years.

> My junior high school graduating class in Saudi Arabia had just 15 other kids. This high school has 2,000 kids and it is unbelievable. You even have to get a pass to go to the john. The school is filled with cliques, the kids who do dope, the cheerleaders, the sports kids. I just couldn't get in with any of them. I didn't like them and they didn't like me. I felt I was more mature than the other kids, and the things they thought were important seemed trivial to me. "What am I going to wear to school today?" "Who am I going to walk home with?" Those are just not big things in my life. I was afraid of these kids because, even though I felt mature, they knew a lot more about living in America than I did.

Values may be difficult to reconcile. A teenage girl put it this way:

> I was miserable the first few months, and my parents never knew what I was going through. I was over-developed physically, but I never really realized it until I got into school here in the United States. That and boys expecting you to go to bed with them right away were just too much for me. It made me feel so odd. I just wasn't aware of it ahead of time, that people really do go through these kinds of problems.

For some, the return to the United States is seen as merely one move among many. A Foreign Service officer had this to say about the issue:

> Once the habit of moving gets into your blood, you always itch for the next challenge. I love the contrast in work from one country to another. When I go back to New York I can see my friends doing the same thing they have always done and it depresses me.

Men who had been involved in multinational negotiations overseas may feel let down when their work no longer affects global politics. Women must make career choices that have been postponed during an overseas tour; they must take on again the roles of chauffeur, cook, handyman, and sole caretaker of the children. Children begin again in their quest for friends and a place in the social sun.

A Life of Fantasy

Living away from the United States frees a person from participation in family and community problems. He cannot visit aging parents or comfort lonely aunts, nor can he participate in church committees and local political campaigns. His lifelines to current issues in this country, *Newsweek* and the *International Herald Tribune*, are no substitute for daily participation in what is happening here. Issues that grip people in the United States lose their urgency, and the person overseas frequently develops a new group of interests based on an entirely different premise, that of being an observer and guest. As a visitor, his attention tends to be drawn to the timeless, proud expressions of a host country's culture—art, music, architecture, theatre, holidays—rather than the mundane, daily ones. He loses contact with the anchoring points of daily life both in the United States and overseas.

Such conditions foster the development of a rich fantasy life, that may flow over into the creation of a fantasy life difficult for others to comprehend. An adolescent girl told me that, when she was in Paris, she had lived out a vision of herself as a turn-of-the-century beauty, strolling through museums and stopping along the Champs Elysees for a cup of coffee. No one questioned the role she was playing out, least of all herself.

On return to the United States little support can be found to nurture international fantasies, and, in addition, the returnee has lost track of the events current in this country. Bafflement and frustration may ensue on all sides.

Communication of Experience

Much of our experience is primarily nonverbal. It is difficult to translate into words certain of our touch, taste, smell, or sight perceptions, even though these perceptions may exert a potent influence on one's consciousness and self-definition. A businessman lamented:

> You simply can't describe the feel of the hot wind on your skin in Sicily or the noise and commotion of traffic in Rome. It just can't be reproduced in conversation. When I try to tell people what it was like, it probably sounds like I just want them to envy me. But it's not that. I just want them to know what I felt, who I am.

This large component of experience, nonverbal and unshared, creates a painful barrier to comfortable communication, and the person returned from overseas, isolated from the world around him, may find that he becomes a prey to all kinds of distorted perceptions and disturbing fantasies.

Though we possess a time-honored tradition of farewell parties to send people overseas, and a considerable literature of adventure and self-discovery to guide them, we are endowed with very little ritual and writing to help them on their return. Thomas Wolfe's *You Can't Go Home Again* emphatically warns the traveler not to do what he must do, and Malcolm Cowley's *Exile's Return* praises the timeless, dreamlike pleasures of life in Europe, rather than any creative satisfactions to be discovered in the United States. We are in great need of a literature that will interpret America to returning Americans and, at the same time, explain them to those who have remained at home. Similarly, we need to devise societally recognized events that will reintroduce travelers to the people of their home country, and guide them to a recognized place in their community.

Symptomatic Problems of the Returnee

Many returnees describe feelings of discomfort and vague dissatisfaction with their lives, though they cannot pinpoint the basis of their difficulty. They are able to adjust to the United States, but are not comfortable with that adjustment. Most of the problems I have encountered fit into this category of vague adjustment reactions, rather than any more traditional psychiatric diagnosis. Long-lasting feelings of being restless, out of place, rootless are typically recalled, even by those who are overtly well-adjusted to their return from overseas.

Nostalgia for a Lost Way of Life

A teenager will often remember his time overseas as one of having fallen in love, not only with a girl friend, a boyfriend, or a culture, but as a time in which he engaged in an exciting love affair with a foreign way of life. Such a love affair is recalled as an ideal time, the disturbing, boring or unsatisfying moments forgotten, in which he gave all of himself to an experience. A wise teacher recalled the syndrome in this way:

> American teenagers who come to our school develop strong attachments and sometimes intense crushes with Italian youngsters. They become inseparable, for each sees in the other, at a time when both are bursting with vitality, an opportunity to fulfill all of life's wishes. The breaking off of these friendships when the American goes home can be devastating for the visitor and the Italian child alike. Both suffer the effects of separation for such a long time.

Young people who have had such an experience may find themselves unable to become involved with a new school and new friends.

They make intolerable demands upon teachers and friends, demands that cannot be met no matter how hard new acquaintances try, because the teenager is yearning for an ideal person, an ideal time, that probably never did exist. These feelings of loss and disappointment develop into a kind of nostalgia for a great and perfect past. Sometimes this nostalgia takes on the proportions of a true disease, a yearning and grieving that seemingly has no end. A high school student in the United States wrote to his cherished friends in Tunisia about his sense of longing as follows:

> Vienna, Virginia is about as middle class as it can be. Our school is gigantic, made up of commuters and freaks, with no sense of spirit and a lot of people just trying to get it over with. It is tough to rise above the general sense of apathy. I set here with your yearbook and am so envious of all of you there in Tunisia and think of how lucky you are to be isolated from the mass education we have here in the homeland.

This idealization of memory occurs because of the mind's wish to ignore unpleasant reality—the discomforts and uncertainties of the past—and to recall only the blissful fulfillments. A preoccupation with nostalgia succeeds in helping one forget the frustrations of the present and the efforts necessary to engage in a new life.

A Different Self-Concept

Attitudes of teenagers overseas about themselves and others were examined in a recent research study (Werkman & Johnson, 1976). Differences in attitudes and values between teenagers reared overseas and those reared exclusively in the United States were studied by comparing 172 teenagers who lived overseas with 163 teenagers matched for age, sex, and socioeconomic status who had lived exclusively in the United States. Using the Semantic Differential technique, the subjects' reactions to certain concepts were rated on the dimensions of evaluation, activity, potency, and sensitivity. The groups were separated at a statistically significant level in rating the following concepts.

Teenagers who had lived overseas rated themselves as less strong, good, or happy than those in the United States. The future was not so strong, colorful, stable, or close to them. Friends were less important, close, strong, and colorful. Loneliness was more interesting, close, stable, and comfortable for them. Restlessness was more interesting, good, and happy for them.

The results suggest that overseas teenagers are unusually searching and open about themselves, and especially capable of acknowledging potentially disturbing affects. They appear to be less secure and optimistic than adolescents who live exclusively in the United States, but in many ways more psychologically sensitive. The self-concepts of overseas teenagers appear to be less positive, and they seem to show less of a feeling of security and optimism about life in general. These results do not suggest that teenagers who have been reared overseas are less psychologically healthy than those reared in the United States, but rather that overseas experience does have a significant effect on their values and attitudes.

Roots

Because of the attitudes described, it well may be that overseas teenagers are candidates for becoming restless, possibly rootless people, who have a constant need to be on the move. Indeed, it is a general clinical impression that the majority of people who have grown up overseas do not want to settle down in one place during their adult lives. Approximately two-thirds of the people seen in the Boulder sample, as well as in my clinical and consulting practice, hope to return overseas, and expect to live geographically mobile lives. The following statements describe their views well:

> I just think of myself as belonging to the world: not any one place, not Pennsylvania, but anywhere I can do what I want to do. I doubt if I will ever settle down or if I will ever need to.

> I want to belong to a creative and openminded group of people who are interested and concerned about the world. Everywhere I stay is home because I don't need a single home, just a sense of the place where my family lives.

Recommendations for Successful Reentry

The findings from a number of studies (Borus 1973b; Hamburg & Adams, 1967; Silber, Coelho, Murphey, Hamburg, Pearlin, & Rosenberg, 1961; Werkman, 1977) have agreed that the following attitudes and strategies appear to be central to the achievement of successful transitions. The person who makes life transitions successfully seeks out advance information about the new situation to be mastered, finds ways to try out the new behaviors and attitudes required, and utilizes peer-group interactions to gain support, test out new behaviors, and learn about values needed in a new situation. He recalls successful experiences in the past when confronted with new challenges.

These conclusions were substantially confirmed in the subjects studied by the author. In addition, it should be noted that the person in transition does better if he is able to recognize and grieve over the losses involved in moving to a new situation.

The Importance of Goodbyes

In the United States it is rare for friends to part forever, but the overseas family must say final goodbyes repeatedly. Many families, finding the experience of leavetaking to be an increasingly wrenching one, slip into the unfortunate habit of evading the rituals of farewells. Though an effective short-term pain reliever, such evasion may result in a legacy of anxiety and guilty sadness. Unfinished farewells may return to haunt the memory of returnees.

Integrate the Past with the Present

A mother described her experience as follows:

> The first year home we bought an expensive barbeque grill, grilled steaks every night, and dressed in "mod" clothes. It was ridiculous. Then we began to realize that we didn't need to have a color TV or sugar-coated cereal every morning. My husband and I realized that we had learned some things overseas, and decided to keep those things we had learned for ourselves and our children.

Another put it this way:

> When you get back, you can't expect people to want to hear about your travels, but you do bring things that people need—the sense of community you notice in small towns in Greece; the humane way people deal with each other in villages; the slow pace of people who take time to be with each other.

Overseas Life Leaves Its Imprint

Recognize that living overseas has long-lasting effects on personality. Several returnees described those effects as follows:

> My experience in Germany has made me a much more reserved person, a more quiet one than I would have been if I had stayed here. I don't make quick judgments about things, but stand back and really look at them.

> Living overseas has given me a chance to look at things from a different perspective. Now, I want to know more about everything. I am more involved in searching and in finding, and know that there may not be a pure truth.

> You end up with a double concept of yourself. There is this sense that you have an extra talent, your knowledge of another language and another

> culture, that has no value except in planning your own later life. I think where people get in trouble is that they tend to come back and know they have this asset, but think that everybody else should know about it and feel "you should respect me because I've got this extra experience," but it doesn't apply. What you gain from experience abroad is going to be maintained, but it's better to tuck it away. What you have learned not only doesn't get you anywhere, but it tends to threaten or irritate people.

> It makes a person much deeper. You see that there are different kinds of people. I think here in America people are only what they are in relationship to others, instead of being what they are to themselves. The people in the school that I am in now seem to care so much what other people think about them. For example, when there are girls all crowded together in one part of the room you have to sit with your best friend or they think you are weird. But it doesn't bother me in the least.

A common theme running through these reports is a recognition of a deep sense of aloneness together with a need for individual self-definition. Returnees tend to view life in comparative terms and characterize themselves as observers rather than active participants in social experience. Their reports contain many themes of existential alienation described succinctly by Maddi (1967).

The Need for Support Systems

A parent described her family's successful efforts to find support systems in this way:

> We helped our 13-year-old get into the Boy Scouts through a talk with the boy who delivered papers. My daughter immediately looked for a place to rent a horse, and found friends around the stables. We joined a community swimming pool where our children made lots of friends. Our kids didn't have any real problems of reentry into the United States, but it is something you have to work at. It doesn't just happen.

A young woman spoke about what she felt she needed on returning:

> Nothing helped too much, but I wish I had been able to get in touch with friends who had shared my experience. The problem was that those friends were in a different part of the country. The only friends I have in high school have never been abroad and can't understand these things. They couldn't understand this cultural shock. It seemed so fake to them that I was having trouble readjusting to my own culture. I needed to talk to someone and tell them about it, so I wrote letters to friends and family. They helped because they had culture shock when they came back. I have always had this yearning to get back to Europe after every tour, but this time I almost wanted to give up my citizenship because I felt so lonely.

Emotional Cost of Adjustment

Transitions and reintegration take time, and inevitably are accompanied by pain. The various curves of adaptation to geographic mobility devised by researchers (Gullahorn & Gullahorn, 1966) all emphasize that periods of loneliness and discomfort alternate with periods of effective coping. They comprise part of the cost of adjustment.

Programs

People moving from one culture to another need advance information that can be offered through seminars and discussion groups before they move. They need guides and mentors in their new homes. The use of peer counselors (Hamburg, 1974) has been recommended as a useful way of helping students effect transitions successfully.

Schools, colleges, businesses, and government agencies would do well to set up transition groups to aid in the integration of returning Americans. However, such programs should last more than a weekend, and include more than a welcoming ceremony. As one returned college student observed: "It would have been helpful to have someone show me how to locate friends and find my way around here, someone who could have taken care of me during the first month back. I mean someone just to watch out for me."

The task of readapting to the United States demands the best efforts of people who have been away, and those present to welcome them. Some of the pitfalls of this transition have been described in this chapter. Useful strategies for dealing effectively with the stresses of geographic mobility have been summarized. The challenge that this transition be recognized as a serious rite of passage for many people remains to be taken up in the future.

References

Borus, J. F. Reentry. I: Adjustment issues facing the Vietnam returnee. *Archives of General Psychiatry*, 1973, *28*, 501–506.(a)

Borus, J. F. Reentry. III: Facilitating healthy readjustment in Vietnam veterans. *Psychiatry*, 1973, *36*, 428–439.(b)

Bower, E. M. American children and families in overseas communities. *American Journal of Orthopsychiatry*, 1967, *37*, 787–796.

Cleveland, H., Mangone, G. J., & Adams, J. G. *The overseas Americans*. New York: McGraw-Hill, 1960.

Gullahorn, J. E., & Gullahorn, J. T. American students abroad: Professional versus personal development. *The Annals of the American Academy,* 1966, *368,* 43–59.

Hamburg, B. Coping in early adolescence. In G. Caplan (Ed.), *American handbook of psychiatry,* vol. II. New York: Basic Books, 1974.

Hamburg, D., & Adams, J. E. A perspective on coping behavior. *Archives of General Psychiatry,* 1967, *17,* 277–284.

Kenny, J. A. The child in the military community. *Journal of the American Academy of Child Psychiatry,* 1967, *6,* 51–63.

Maddi, S. R. The existential neurosis. *Journal of Abnormal Psychology,* 1967, *72,* 311–325.

Silber, E., Coelho, G. V., Murphey, E. B., Hamburg, D., Pearlin, L. I., & Rosenberg, M. Competent adolescent coping with college decisions. *Archives of General Psychiatry,* 1961, *5,* 517–527.

Smith, M. B. Explorations in competence: A study of Peace Corps teachers in Ghana. *American Psychologist,* 1966, *21,* 555–566.

Werkman, S. L. Hazards of rearing children in foreign countries. *American Journal of Psychiatry,* 1972, *128,* 992–997.

Werkman, S. L. Over here and back there: American adolescents overseas. *Foreign Service Journal,* 1975, *52,* 13–16.

Werkman, S. L. *Bringing up children overseas: A guide for families.* New York: Basic Books, 1977.

Werkman, S. L., & Johnson, F. *The Effect of Geographic Mobility on Adolescent Character Structure.* Paper presented at annual meeting of the American Society for Adolescent Psychiatry, Miami, Florida, May 1976.

12

Running Away in America: The History and the Hope

James S. Gordon

Introduction

Runaway young people have always been regarded with ambivalence. Their desire for escape and adventure, their search for change, and their challenge to accepted norms have excited the imaginations and elicited the sympathy of a nation which values independence and admires youthful courage. On the other hand, their premature departure from American homes has been regarded as a continuing subversion of the families which we are, often desperately, concerned with preserving; and their presence in the community and on the street has been seen as an offense to decency and, often enough, a threat to the social and economic order. Although these young people have been glamorized in fictional presentations, they have, in fact, been treated rather badly by our society: originally regarded as deviants to be corrected, they have more recently been seen as confused and misguided children who must be returned whence they have strayed. Sometimes they have been the object of a concern not unmixed with fear, contempt, incomprehension, and condescension; sometimes they have simply been fair game for economic and sexual exploitation.

During the past 10 years, a persistently high incidence of runaway young people has been accompanied by a new perspective on their flight. Instead of stigmatizing them as immoral, deviant, or psycho-

James S. Gordon • Center for Studies of Child and Family Mental Health, National Institute of Mental Health, Rockville, Maryland 20857.

pathological—or, indeed, romanticizing their rebellion—a number of my colleagues and I have come to see their departure as a sign of familial turmoil, to find in it a critique of a society which affords many of its young people few useful roles and little hope for the future. In the context of a new kind of residential facility—the runaway house—we have tried to help young people to use their departure as a catalyst to individual and family change, to provide a microsocial setting in which some of the inadequacies of contemporary adolescent life may be redressed.

After providing an historical context for understanding running away, I will trace the origins of this new paradigm, and sketch its elaboration in contemporary runaway houses.

Running Away in America: A History

Concern about the vagaries of young people and their susceptibility to evil influences preoccupied the colonists even before they landed on North American shores, and provided one of the many reasons for them to leave Europe. In his "General Observations on the Reasons for Migration to the New World," John Winthrop, the first governor of the Massachusetts Bay Colony, noted that in England "most children, even the best of wits and the fairest hopes, are perverted." He fervently hoped that in "the fruitful and convenient" New World, families would be more cohesive and the social order more binding (Bremner, Barnard, Hareven, & Mennel, 1970, pp. 18–19).

In the tight theocratic communities of Colonial New England, patriarchal families were the primary social and economic unit as well as the model for authority. By the time they were seven or eight years old, children were productive participants in family life, assistants in their parents' fields or kitchens, or apprentices in nearby households. Adolescents who tried to live outside of their own or their neighbors' families were regarded as a loss to the family's economy, as well as defectors from its morality.

Like single older people, orphans, and bastards, these young people were quickly placed in family settings. The justification was biblical ("God settleth the solitary in families"—Psalms 68:6), but the arrangement also had its political and economic advantages: the community was spared the danger of a potentially seditious force and the labor of these young people became available to the families that took them in.

This view of the young person as a potential economic asset, and of running away as a social and economic disruption as well as an offense against God, continued through the 17th and much of the 18th century. In the late 18th and early 19th centuries an accelerated rate of immigration, the importation of large numbers of young servants, and the nation's gradual secularization, industrialization, and urbanization combined to decrease the economic utility of American children and to increase the numbers of those who did not live with their parents. Large numbers of young people ran away from rural areas, where they had been supplanted as laborers by stronger and no more expensive immigrants, and flocked to the cities. Some found work in newly opened factories; others, along with the children of impoverished Irish and German immigrants, wandered the streets.

In some smaller communities these young people were classified with paupers and other indigents, and auctioned off at "vendue" to whomever could keep them with the least expense. In large cities increasing numbers of them were confined with the poor, the mad, and the chronically ill in alms houses (Bremner *et al.*, 1970, pp. 262–281).

By the first half of the 19th century these homeless young people had come to be regarded as a special and serious social problem: the "class," according to reformer Charles Loring Brace (1880), "of a large city most dangerous to its property, its morals and its political life" (p. 11). Believing that they should be confined in "the best of all asylums," the farmer's bouse, Brace transported thousands of young runaways and street people to the western territories. There, in rural settings, their labor was profitable to farmers, who in turn instructed them in the virtues of honesty, family life, and hard work.

Meanwhile, others who were concerned with the runaway young began to create large institutions for their care. The Orphans Society of Philadelphia, for example, was founded in 1814 to "rescue from ignorance, idleness and vice unprotected and helpless children and to provide for them such support and instruction as may eventually render them valuable members of the community" (Bremner *et al.*, 1970, p. 653). At the same time "schools of reform" and "houses of refuge" were created for juvenile offenders who were judged criminal or vagrant. In 1837 the doctrine of *"parens patriae"* gave full legal standing to the institutionalization of children not living at home or otherwise considered difficult (*Ibid.*, pp. 691–693).

By the late 19th century the combination of their ever earlier sexual maturity and their diminishing economic utility—and the resulting prolonged dependency on their parents—provided the material basis for an anxious segregation of the young. (See, e.g., Laslett, 1971; McKeown & Record, 1962.) Gradually, the belief that particular adolescents—among them runaways—needed to be reformed, began to yield to the view that adolescence was itself a particular, and particularly treacherous, stage of life. Laws prohibiting child labor, enforcing compulsory education, and creating a separate juvenile justice system (Handlin & Handlin, 1971) provided a social structure for containing and protecting the young. The developing fields of psychiatry, psychology, and psychoanalysis offered tools for the understanding and treating of the more recalcitrant members of this newly designated group of "adolescents."

The chief ideologue of this process in the United States was G. Stanley Hall, author of the mammoth text, *Adolescence*. Although many of his theoretical contributions have since been repudiated, and though anthropological data, such as that gathered by Margaret Mead (1928, 1930) in the 1920s, contradicts it, Hall's view of adolescence as a stage of development characterized by continual crisis has persisted. For the last 75 years writers on adolescence have continued to make the effect (the difficulty of being a young person in twentieth-century America) into the cause (adolescence is a time of great stress).

At its best, a psychological perspective has helped parents and those charged with care of the young to understand adolescents in the light of their feelings and motives, as well as their behavior. It has provided a basis for understanding running away as a response to a familial, social, and economic situation young people can neither understand nor change, and for approaching the individual runaway with the compassion due a victim as well as the firmness required by a delinquent.

In the first decades of the twentieth century it was clear to investigators that social and economic factors were preeminent in turning ordinary young people into runaways. In the first extensive study of runaways, Armstrong (1932) indicated that the enormous pressures of urban poverty and the absence of adequate employment were combining to force large numbers of immigrant children from their families. She noted that 65% of the homeless young people who came to the attention of the New York City courts were the children of foreign-born parents (7% were foreign-born themselves), and that 55% of them came from homes that were "broken," usually by parental death, marginal economic and social status, and overcrowding.

During the Depression, pervasive social and economic pressures forced unprecedented numbers of the children of all Americans from homes that were simply unable to sustain them. Both Minehan (1934), who met runaway young people on the road, and Outland (1938), who studied them in federal camps, emphasized the importance of poverty and familial disintegration in producing runaways. Minehan noted that only 35% of his sample of 1,500 "boy tramps" came from two-parent homes, and Outland learned that 36% of the young people in camps had left—or been sent away—for explicitly economic reasons.

In the decade after the Second World War in increasingly crowded urban centers the ties that bound poor and black families and communities began to fray. Meanwhile the economy's demands for mobility helped fragment the white families, which fled first to one suburb and then to another. Divorce rates, incidence of single-parent families, and numbers on the welfare roles all rose.

The lives and behavior of teenagers were of course shaped by these and other social and economic changes. The lurching disruptiveness of a society rapidly expanding its geographic and economic limits had been replaced by the constant pressures of a society that demanded ever greater degrees of technical specialization and geographical mobility, and ever higher levels of consumption. While a continually shifting job market was forcing their families toward increasing mobility, young people were being wrenched from the continuity of a multigenerational family and community. In rigidly age-segregated schools they were being asked to set aside more and more years to prepare for a life of work which was increasingly further removed from their experience at home—or indeed at school; to plan for a future that seemed to change continually and capriciously.

During these years their increasing social isolation, their disappearing economic role, and the burgeoning influence of an individually-oriented medical perspective combined to obscure the social and economic factors which shaped the lives of adolescents, and pushed some among them from their homes. A few investigators, like Shellow, Schamp, Liebow, and Unger (1967) underlined the endemic nature of family problems and the similarities between most of those who ran and those who remained. The vast majority of the physicians and behavioral and social scientists who began to examine runaways emphasized their deviance. They regarded the runaways they met in jails and mental hospitals as particularly unruly members of a generally disturbed age group, as subjects to be tested for evidence of psychopathology and/or criminality.

Some clinicians and researchers, like Riemer (1940), Jenkins (1969, 1971), Jenkins & Boyer (1968), and Foster (1962) emphasized behavioral factors common to runaways and "other delinquents," and others, including Leventhal (1963, 1964) and Robins and O'Neill (1959) focused on the individual psychopathology which running away was presumed to reflect. In their 30-year followup study of child guidance patients, these latter authors suggested that running away was indeed a "predictor" of both delinquency and psychopathology. They noted, among other findings, that runaways had "an adult incarceration rate that was fourfold that of other patients" and that they were one of the groups "most likely to show psychotic signs as adults."

By 1968, running away was not only a crime in more than half of our States (Beaser, 1975), but an officially sanctioned psychiatric diagnostic category, the "Runaway Reaction of Adolescence (1968)." Large numbers of young people who had done nothing more than leave their homes were being confined for long periods of time in penal and mental institutions.

Although the vocabulary had become "scientific" rather than religious, moral, or economic, the stigmatization of earlier descriptions and the insensitivity of earlier treatments had remained. No longer a slipped gear in the economic machinery, a public shame, or a nuisance, runaways became, with the passage of time, a species of patient in need of diagnosis, treatment, and cure. Young people who could be were to be reintegrated into their families; those who could not were to be removed to institutions where their behavior might be reformed, their thought and mood disorders set straight, and their futures shaped by drugs and/or behavior modification along conventionally acceptable, essentially white, middle-class lines.

By the 1960s their shared isolation from the concerns and lives of adults—and the tendency of adults to label and stigmatize their particular stage of development—had helped to make many of the young skeptical of the dominant values of American society. The Civil Rights movement inspired some to see their own powerlessness as a mirror of black peoples', and to think about "youth rights" as well as civil rights. Soon the contradictions between the American ideals of truthfulness, peace, democracy, and self-determination and the American actions in Indochina would alienate large numbers of young people who had been only marginally touched by the civil rights struggle: revolted by the televised slaughter of the Vietnamese, and terrified by the hypocrisy of its justification, many of them came to fear that the powerful weapons of the American military establishment might some day be turned on them (see e.g., Gordon, 1972). In this climate, disputes about politics and sex, drugs and grooming, tended to escalate to the bitter and implacable confrontations. In their wake large numbers of middle-class, as well as poor, young people left—or were told to leave—their homes.

Young people—from puritan rebels to 19th-century street people and 1930s hobos—had always hoped to find a better, or at least a less dismal and confining, life on their own; in the city or on the road they looked for comrades to keep them company, to strengthen them in their quest. Only in the 1960s, however, did large numbers of young people begin consciously to regard running away as a political protest, and their fellowship as the basis of a "culture" and a movement. While psychiatrists were discovering a new behavior disorder and debating their long-term "prognosis," young runaways, and their advocates, publicly declared that their departure—voluntary or forced—was a legitimate rebellion against a restrictive family and a dangerously oppressive society.

By the mid-1960s a small number of runaways had begun to gather with the Beatniks and their Hippie descendants, with civil rights and anti-war activists, in the centers of what soon came to be called the counterculture. In the Haight–Ashbury section of San Francisco, in Manhattan's East Village, Washington, D. C.'s DuPont Circle, and in college communities like Ann Arbor, Madison, and Cambridge, they created new styles of dress and music, politics and art, interpersonal relations, and intoxication—amalgams of past and present, technological innovation, economic necessity, and imaginative fantasy. The relaxed and sensual way in which they lived together, their opposition to materialism and competitiveness, to hypocrisy and war, and not least the intensity of media attention, soon drew tens of thousands of other young people after them.

Local groups formed to respond to the immediate needs of the thousands of homeless and penniless young people who flocked to their communities. Building on the interests and talents of natural helpers, drawing on the skills and energy of the young people who came for help, they swiftly constructed a network of human services. In San Francisco, The Diggers, borrowing their name from 16th-century English egalitarians, improvised daily bread and soup for thousands of Haight–Ashbury residents. Switchboard directed telephone callers to crash pads, free clothes, and legal services. The Haight–Ashbury Free Clinic, staffed by street people and local physicians, dealt with the ailments of a young and transient population which was experimenting with its limits of physical and mental endurance.

In these programs homeless young people found food and shelter, sympathetic medical attention, and a caring community in which they could help as well as be helped. Teenagers with venereal disease were treated without smirks or moralism, and those on bum trips were gentled down in quiet rooms, not jabbed with mind-numbing doses of tranquilizers. In contrast to the doctors, social workers, and counselors, the schools and hospitals of the larger society, these counter-institutions and those who worked in them were responsive to the expressed needs, and respectful of the desires, of the young runaways.

Runaway houses, in Haight–Ashbury and Washington, D. C., in New York City and Boston, were among the first and most important of the alternative human services. In these houses, amidst a welter of mattresses and graffiti, shabby overstuffed furniture, ringing phones, psychedelic posters, and endless pots of rice and beans, runaways found both a refuge and a redefinition of their situation. Older people who wore the same kinds of clothes and listened to the same kind of music helped hopeful but uncertain and frightened young people to see running away as part of a process of personal growth and social struggle, a symptom of decaying families and a society in turmoil. Living and working together in a runaway house, runaways and their counselors forged a cross-generational alliance of older and younger brothers and sisters.

The Vietnam war and the movement which grew to oppose it, the huge urban counterculture and the economic boom which helped sustain it, all dissipated in the 1970s. The numbers of runaways remained high. Each year between three-quarters of a million and a million people—the same percentage as in the late 1960s—are continuing to leave their homes (Ambrosino, 1971). Few of them are hoping to find a movement or a counterculture to shape their disillusionment to social change or communal satisfaction. Many of them—30% among the predominantly black youth that run to the Washington, D. C. Runaway House, and half of the middle-class white teenagers who come to the Huntington Youth Service Bureau—report that they are leaving at least partly because they have been physically abused by their parents or guardians. Others simply feel angry, depressed, and isolated at home. They are bewildered and dismayed by the lack of a social and economic role, offended by being labeled as "the family problem," unhappy at school, and suspicious of the future and its promises. Although these young people are still regarded as runaways, many of them feel that they have been "pushed out" or "thrown away" by their parents and their society.

This massive, continued, and often violent separation of the young from their families may well signal a profound change, both in the meaning of running away and in the familial and social situations from which the young are fleeing (Epps, 1976; Gordon, 1975a; Gordon & Houghton, 1977). In earlier eras, runaways tended to come from families or sectors of society made perilously vulnerable by poverty, death, or the cultural, social, or economic dislocation attendant on immigration, rapid industrialization, or economic catastrophe. The National Statistical Survey (Youth Development Bureau, *DHEW*, 1976) indicates that urban fragmentation, poverty, and broken homes are still significant causes of running away: the children of the unemployed run twice as often as those whose parents work; runaways are more likely to come from one-parent families; young people who live in rural areas still leave their homes only half as often as their urban or suburban peers. On the other hand, recent statistics also reveal that broken families, poverty, social and cultural dislocation, and family violence are pervasive facts of American life: almost 17% of *all* children live below the official poverty level, and as many more are, in fact, poor; 40% of all marriages end in divorce; 15% of new births are illegitimate; parents even of intact families spend less and less time with their children; all adults—and their children—move from city to city and house to house at an ever accelerating rate; and, as many as two million children, many of them adolescents, may each year be abused by wealthy as well as by poor parents (see Bronfenbrenner, 1976). In the 1970s, the familial situations that force young people from their homes are endemic among all classes and races.

Redefining Running Away

In 1968, workers in half a dozen runaway houses struggled on borrowed money and donations to meet the needs of the casualties of a cultural phenomenon which, they assumed, would soon subside. Within a year or two, it had become clear that running away was increasing in every part of the country, and that young people who stayed near their homes faced many of the same problems as those who fled to the big cities.

Concerned citizens in medium-sized cities, middle-class suburbs, urban ghettos, and rural areas began to open programs that drew their inspiration from those in the Haight–Ashbury and the Lower East Side, and their particular style and substance from life in Prince George's County, Maryland, or Burlington, Vermont. Some were started by young college graduates who hoped to bring the spirit of the antiwar and civil rights movement to their own communities, to bring the politics of human liberation down to a personal scale. But, increasingly, these projects were sponsored by establishment organizations. Local YM/and YWCAs, Salvation Army groups, and churches recognized their own failure in understanding and meeting the needs of the disaffected young. Runaway houses seemed the "only viable option" for young people who would otherwise be exploited on the street or treated as criminals in the juvenile justice system. They saw the house as a bridge (a name adopted by a number of them), between their group of adults and young people, and a place to learn from as well as help the young. They hoped that it would also be a link between runaways and their families. By 1972, more than thirty houses were providing short-term lodging and food, and individual, group, and family counseling for runaways.

At about this time the federal government became actively interested in runaways and runaway houses. Previously, federal concern had been confined to establishing "transient camps" during the Depression, and keeping yearly statistics on the number of runaways arrested. Now, Senator Birch Bayh's Juvenile Delinquency Subcommittee began to hold hearings on the "Runaway Youth problem."[1] Although some of the experts spoke of the connections between delinquency and running away, by far the greater part of the testimony combined an understanding of the psychological and familial problems of young people with an appreciation of their displacement from the economic and social order. Witness after witness described the disaffection of the young from traditional social services as well as from their families and schools. Many concluded by recommending that the federal government fund more runaway houses.

In 1974, the National Institute of Mental Health recognized that runaway houses were more effectively meeting the social and emotional needs of young people than the traditional mental-health services that they ordinarily funded. NIMH made 1.6 million dollars available to augment the services and training programs of some thirty-two runaway centers (Gordon & Houghton, 1977), to study how effective runaway services really were. Later that year, after the bodies of 27 young people (presumed to be runaways) were found in Houston, Congress passed the Runaway Youth Act. Five million dollars was appropriated to fund runaway centers, to gather statistics on the numbers of runaways and the reasons they left home, and to establish a nationwide "hot line" for runaways and their families.

Since that time, the number of runaway houses and the kinds and extent of services that each of them offers have steadily increased. Currently there are some 200 runaway houses, 150 of which are funded through the 11-million-dollar allotment of the Runaway Youth Act. Last year these homes provided housing and comprehensive crisis-oriented services for 50,000 runaways, and nonresidential services to a total of perhaps a quarter of a million young people and their families.[2]

Many of these programs for young people have managed over the years to combine the responsiveness and respectfulness of the first runaway houses and a careful attention to the particular needs of runaways with a growing understanding of the larger and longer-term

[1] United States Senate, United States Congress, *Runaway Youth*, Hearings before the Subcommittee to Investigate Juvenile Delinquency of the Committee on the Judiciary, 92nd Congress, First Session, Legislative Hearings on S–2829, the Runaway Youth Act, January 13–14, 1972.

[2] Annual Report on Activities Conducted to Implement the Runaway Youth Act, Youth Development Bureau, DHEW, 1977.

personal and social issues that confront young people and their families. The way that they work with young people to do this has helped to redefine the nature of running away, to transform it from a stigmatized and pathological act to a potential catalyst for individual and family growth, from a threat to society to an opportunity for community change.

The Context of Running Away

The most important fact about runaway houses, now as eleven years ago, is their physical existence. Most offer a centralized shelter; a few operate out of a drop-in center, and house young people with neighboring families. They are, above all, a refuge—from the street and its exploitation, from intergenerational conflict, pathological stigmatization, and legal strictures.

In earlier times, and in contemporary detention centers and mental hospitals, runaways who came to the attention of the authorities were summarily confined as deviants, criminals, or mental patients, as if their flight, and not the conditions they left behind or the feelings that forced them to leave, was the problem. In contrast, young people who come to runaway houses are welcomed as guests in a household, and are free to leave when they wish. The rules of these households are created not to reform them or to modify their behavior, but simply to ensure the house's survival and the comfort of all those who live and work there. The workers in the houses are primarily counselors and friends, not warders and judges.

In this context, young people who have been desperately running for weeks or months are able to relax and consider their situation; runaways who have previously avoided adults—and their coercive power and punitive judgment—begin to ask for advice. Knowing that they are not confined, they do stay; feeling they are trusted and respected, they begin to trust and respect. Though some few young people continue to disobey the rules that have been established to ensure the house's survival, many of those who had been recalcitrant at home find it easy to live within limits which seem neither capricious nor arbitrary, and to take part in preserving a setting which has so obviously been created to meet their needs.

The Meaning of Running Away

Historically, running away has been seen by adults in power as a defection from the family and the social order, a crime against the community, and a sign of mental ilness. The perspective of the young people who run has been ignored, and their right to define their situation denied. This denial is, if anything, more obvious in individual families (Gordon, 1975a, b). There is "no reason" so many parents say, no reason for her to leave home. When "she" tries to tell them "the reason" they ignore her or shout her down, denying that *their* child may actually have chosen to leave. Still, it seems, blame must be placed. The child, they say—desperately trying to deal with, to define, and therefore to establish control over the situation—must be "bad" (delinquent) or "sick" (mentally ill). Either that, or it is someone, or something, else's fault—evil friends, hippies, drugs, sex.

Law-enforcement and mental-health agencies tend to perpetuate, not remedy, this process of isolation and labeling. If a psychologist or a probation officer declares a child to be sick or delinquent, or "in need of supervision," and insists on testing or confining him, these actions and attributions outweigh any references to family problems or social and environmental influences. The young person remains the labeled patient; his or her opinions and options must inevitably be qualified by this dependent and deprecated status.

In the context of a situation where they feel comfortable, in the company of people who are willing to credit their perspective, young people can begin to disentangle themselves from others' definitions of them, to explore the reasons why they really did leave home. For some, it is simply a matter of escaping from unbearable, humiliating, physical punishment or sexual abuse. For many more, running away feels like a desperate assertion of selfhood. Many young people no longer can be or wish to be the "good" (often sexless, well behaved, or hypocritical) child their parents seem to insist on. Others are furious that their attempts at independence seem always to be defined as a species of behavior or thought disorder. In running away these young people are escaping as much from familial definitions as they are from physical control. It is these definitions that they describe and experience as murderous or prisonlike. Again and again young runaways repeat the same phases: "I couldn't be myself," "They were killing me," "They kept pushing me into a corner."

From their first hours in a runaway house, young people are encouraged to see that running away is neither pathological nor heroic, but a temporarily necessary and positive act. Instead of reformulating their situation in the language of social work or psychology, counselors urge young people to describe in their own words what concerns them. They are encouraged to look critically at the situations from which they have come, and the way they have behaved; to see their absence as a way of calling attention to family problems that have been too long ignored; to use their counselors as advocates and aides in improving their situation; to find ways to get along, with or without their family, at school or work that conform to their own, not their parents' or teachers', or indeed the runaway-house counselors', ideas about them or their future. In daily groups with other runaways they find that even their most unhappy experiences and desperate insights may be of use to others who are experiencing similar problems, as well as to themselves. In time many of them will find that leaving home has been, paradoxically, a necessary prelude to returning there.

A study (Howell, Emmons, & Frank, 1973) of young people in one program suggests that in the context of a runaway house this process of redefinition is successful. Though they had experienced "major difficulties during their run," 66% of the young people who stayed at Project Place in Boston "believed in retrospect that running away had been a positive growing experience for them." My own work at the Washington, D. C. Runaway House and elsewhere (Gordon,1975a, b, 1978; Gordon & Houghton, 1977) confirms these statistics. Their time at the runaway house is the first opportunity that many young people have to think and act for themselves. Some who had come to believe that they were hopelessly stupid, inadequate, or impulsive have patiently worked out solutions to complicated personal and family problems. Others, habitual runaways and diagnosed "schizophrenics," have discovered that in the context of a respectful setting they can behave sanely and responsibly.

Running Away and the Family

Running away is a communication to the rest of the family as well as an act of self-assertion. It is impossible for parents—even if they deny the importance and meaning of the behavior—not to know that their child is missing. Whether they accuse the young person of betrayal, belabor themselves with guilt, or are secretly pleased, they feel a loss and an uncertainty. The balance in the struggle between parent and child has shifted. If they wish to continue their contact with their children, the parents must pay attention to, even if they do not yet respect, their children's point of view and their wishes. Runaway house counselors use this moment as a lever to urge the family toward confrontation and change.

While the parents are wondering why the young people left, the young people in a runaway house are looking critically at their situation and, with their counselors, exploring their options for the future. Most of them quickly see the need for meeting with their family: obviously they cannot return home if things are unchanged; nor, given their legal status and earning capacity as minors, can they survive on their own without the support of parental resources or the protection of parental permission: even foster placement is dependent on their parents' signatures. Runaways who are sure that their counselors will help them present their point of view at a family session are often eager to set up a meeting. It is not uncommon, after a few days or a week, to see young people who have always hated and feared "shrinks" urging their parents to come to family therapy so "we can communicate better, and maybe work things out."

Sometimes, even in the first session with a family, counselors are able to help the young person articulate the content of the protest that has been expressed in running away, to help the parents and other siblings to hear its meaning. Sometimes the family arrives at a mutual understanding which facilitates practical compromise and a swift return home. More often, the counselors must begin by trying to create a safe place for the family to be together in all its mystified contrariness. Slowly they try to help family members to find a common language of understanding in which habitual, often incoherent quarrels can become mutually intelligable.

Sometimes formal counseling lasts for only one session, understanding for just a moment. Over the years, those of us who work with the families of runaways have learned to value that moment as an example of the possibility of communication and closeness, one that may later be referred to and enlarged upon. Sometimes there is only the sharpening of conflict. Here the session provides a safe place for disagreements, the opportunity to clarify them. The family discovers that impasses may be broken, that choices are possible, and that differences do not necessarily spell disaster.

In time, many runaways realize that the pressures which have been brought to bear on them are not unlike those their parents feel. They are able to see that their families either are or feel socially marginal, that many of their parents lack both intimate friends and close ties with an extended family, that they have tried desperately to shape their children's lives to fit ideals and ideas that haunted their own childhoods, to make them behave in accordance with the demands of a society by which they themselves feel oppressed. In time, it becomes clear to the young people that their parents' angry and confused imprecations are reflections of their own bewilderment and betrayal. Sometimes they realize that their own flight from home and the struggles which led up to it are far less catastrophic and far more remediable than their parents' alienation.

Long-Term Needs and Long-Range Perspectives

Early in their evolution a number of runaway houses discovered that some runaways were able neither to return home nor to live on their own. Over the last seven years, counselors in more than forty programs have created long-term alternatives to institutions for these young people. The very existence of such facilities simplifies the work that runaway houses do with all young people and their families. It makes unnecessary the extremes of "home on our (parents') terms" or "you have to be locked up." Neither runaways nor their parents nor indeed the counselors who work with them have to feel compelled to make decisions immediately, or "settle things once and for all." For the small group of young people who eventually need them, these group and individual foster homes offer the same kind of respectful and responsive living situation that they have grown to appreciate at the runaway house.

As they have become aware of other needs, runaway houses have been quick to improvise other services. As they became more sensitive to the particular problems of female runaways—41% of all those who leave home, but 60% of those who seek shelter and counseling at runaway houses—a number of programs began to offer special programs for young women, "girls' groups," where they have the opportunity to explore together the conflict between the pride and hope that the women's movement has helped them to feel and the pressures toward conformity and passivity which continue to pervade our society; to discuss their feelings about their sexuality and its implications for their relationships with parents, boyfriends, and girl friends. More recently, runaway houses have created specialized counseling programs and residences for rape victims (as many as two-thirds of the young women in some urban houses), for young prostitutes, and for young people—males as well as females—who are or feel that they may be homosexual.

Similarly, runaway centers in large cities have become acutely aware of the needs of the minority young people who live around them. With the abolition of many of the "Great Society" programs, the deepening of the recession and decline in employment, and the increasing fragmentation of their families, more and more of these young people have had to come out of the ghettos to seek help elsewhere. Urban runaway programs which once housed no more than ten to fifteen percent of minority youth are now working with a population that is overwhelmingly black or Hispanic, with a group of young people whose handicaps—material, educational, and vocational—are enormous. These houses have hired a proportion of minority counselors to match the numbers of young people, and have made specific efforts to address their cultural identities and economic needs.

As they have become more firmly rooted and more widely respected in their communities, runaway centers have become advocates and resources for all young people. According to the annual report of the

Youth Development Bureau (National Statistical Survey, 1976), more than 25% of the young people who use the counseling facilities or runaway centers are still living at home.

Both those who come to stay at the house and those who simply drop in there may come to regard their experience as a touchstone. At the house they are allowed to be themselves; their rights and wishes are respected; their counselors acknowledge and insist on their responsibility for their own lives. At home again, under stress, they draw strength from knowing that the house is there. They are not really trapped. They can always call or write or drop by at the runaway house. Knowing that they can leave. They are free to choose to stay; remembering, feeling, their own strength, they are less likely to be overwhelmed in the struggle to work things out.

In the last several years, a number of runaway houses have tried to institutionalize their responsiveness to young people's needs, to create an ongoing living and working community which young people can continue to be a part of. All of the houses have included young people on their boards of directors, and many have created peer-counseling programs that help young people to use their experience as runaways as a basis for helping others. In fact, some of the most successful programs, including Aunt Martha's in Park Forest, Illinois, and the Youth Emergency Services in St. Louis, have included people under 18 at every level of work—volunteer and paid—from individual, family, and group counseling to administration and policy setting.

In recent years, this concern for reversing the social and economic passivity of young people has been the basis for a spate of new programs. At a time when as many as 60% of the young people in some inner city communities can find no work, in an era when few young people know or trust adults other than their parents, when most teenagers are bewildered and uncertain about their future, they are trying to provide a bridge to adulthood for the young people who are outside as well as those who are within the runaway-house program. Some have solicited federal money to train young people to work as counselors, maintenance people, administrators, office help, etc., in their own and similar programs. Others have tried to extend the feeling of community and the intimate personal learning that pervades their own projects to shopkeepers, crafts people, and local community businesses where they place young people as apprentices.

Conclusion

Runaway houses cannot, of course, reverse the difficult economic and social conditions that affect families and propel young people from their homes, or single-handedly alter the contemporary situation of adolescents. They can, however, offer the increasing number of young people who leave home a time and a place for themselves, a chance to take a critical, and often compassionate, look at the families with which they have been hopelessly struggling, and an opportunity to make a transition to an adulthood that is at best uncertain, in the company of some older people who care. For those of us who are not so young, they also offer some hints about the kind of meaningful community that may be necessary to sustain the young and, indeed, ourselves.

References

Ambrosino, L. *Runaways*. Boston: Beacon Press, 1971.

Armstrong, C. *660 runaway boys*. Boston: Badger, 1932.

Beaser, H. *The legal status of runaway children*. Final report for a study conducted for the Office of Youth Development, DHEW, by the Educational Systems Corporation, 1975.

Brace, C. L. *The dangerous classes of New York*. New York: Wynkoop & Hallenbeck, 1880.

Bremner, R. H., Barnard, J., Hareven, T. K., & Mennel, R. M. *Children and youth in America: A documentary history*, vol. I, 1600–1865. Cambridge: Harvard University Press, 1970.

Bronfenbrenner, U. The disturbing changes in the American family. *Search 4*, Fall, 1976, 4–10.

Epps, H. C. *Why Did You Split: A Typology of Adolescent Runaways*. Unpublished doctoral dissertation, University of Michigan, 1976.

Foster, R. M. Interpsychic and environmental factors in running away from home. *American Journal of Orthopsychiatry*, 1962, *32*, 486–491.

Gordon, J. S. The Vietnamization of our children. *The Washingtonian*, November 1972, *8*(2), 78–81.

Gordon J. S. The Washington D. C. runaway house. *Journal of Community Psychology*, 1975, *3*(1), 68–80. (a)

Gordon, J. S. Working with runaways and their families: How the SAJA Community does it. *Family Process*, 1975, *14*(2) 235–262. (b)

Gordon, J. S. The runaway center as community mental health center. *American Journal of Psychiatry*, 1978, *135*, 932–935.

Gordon, J. S., & Houghton, J. *Final Report of the National Institute of Mental Health Runaway Youth Program*. January 1977, Center for Studies of Child and Family Mental Health.

Hall, G. S. *Adolescence: Its psychology and its relation to physiology, anthropology, sociology, sex, crime, religion, and education*, vols. I & II. New York: Appleton, 1904.

Handlin, O., & Handlin, M. *Facing life: Youth in the family in american history*. Boston: Atlantic, 1971.

Howell, M. C., Emmons, E. B., & Frank, D. A. Reminiscences of runaway adolescents. *American Journal of Orthopsychiatry*, 1973, *43*, 840–853.

Jenkins, R. L. Classification of behavior problems of children. *American Journal of Psychiatry*, 1969, *125*, 1032–1039.

Jenkins, R. L. The runaway reaction. *American Journal of Psychiatry*, 1971, *128*, 168–173.

Jenkins, R. L., & Boyer, A. Types of delinquent behavior and background factors. *International Journal of Social Psychiatry,* 1968, *14,* 65–76.
Laslett, P. Age of menarche in Europe since the 18th century. *Journal of Interdisciplinary History,* 1971, *2,* 14–20.
Leventhal, T. Control problems in runaway children. *Archives of General Psychiatry,* 1963, *9,* 122–126.
Leventhal, T. Inner control deficiencies in runaway children. *Archives of General Psychiatry,* 1964, *11,* 170–176.
McKeown, T., & Record, R. G. Reasons for the decline of mortality in England and Wales during the nineteenth century. *Population Studies,* 1962, *16,* 94–122.
Mead, M. *Coming of age in Samoa.* New York: Morrow, 1928.
Mead, M. *Growing up in New Guinea.* New York: Morrow, 1930.
Minehan, T. *Boy and girl tramps of America.* New York: Grosset & Dunlap, 1934.
Outland, G. E. The home situation as a direct cause of boy transiency. *Journal of Juvenile Research,* 1938, *22,* 33–43.
Riemer, M. Runaway children. *American Journal of Orthopsychiatry,* 1940, *10,* 522–528.
Robins, L. N., & O'Neill, P. Prognosis for runaway children. *American Journal of Orthopsychiatry,* 1959, *29,* 752–761.
Runaway reaction of adolescence. *Diagnostic and statistical manual of mental disorders* (2nd ed.). Washington, D. C.: American Psychiatric Association, 1968.
Shellow, R., Schamp, J., Liebow, E., & Unger, E. Surburban runaways of the 1960s. *Monographs of the Society for Research in Child Development,* 1967, *32* (3).

IV

Stressful Situations of Foreign Students: Challenges of Cross-Cultural Education

Introduction

An important aspect of modernization is the planned movement of students across cultures for advanced training abroad. Especially important is the program for international exchange of persons, which involves a transfer of knowledge and skills developed by graduate students and senior scholars who spend several years in educational institutions abroad. Although a temporary sojourn abroad for young students is probably the most benign and protected form of uprooting, it is nevertheless a stressful transition. There are several reasons for using the cross-cultural educational experience as a "natural experiment" for examining coping behavior in stressful situations. First, the student makes a voluntary decision to participate in the program of education abroad. Second, the environments in which the students are relocated have academic traditions that are fairly comparable or at least predefined. Third, since relocation is usually limited in duration, the progressive stages of adaptation can easily be studied. Fourth, the counselors and administrators who have followed the progress of students on their campuses are able to provide insights regarding the conditions most favorable for adaptation. Thus, foreign students provide a unique population of individuals for the study of the impact of a new culture on human adaptation and the development of coping behavior.

Klineberg, in Chapter 13, examines the phases of this adaptation process from the initial selection procedures in the home country, to the student's decisions in planning his sojourn abroad, through the decisions that lead to permanent residence in the host culture, or to the more common case of return to, and employment in, the home country. Drawing upon a general review of international research on foreign students in France and in other countries, Klineberg focuses on the potential stressful issues at various stages of the sojourn abroad, and considers means for mitigating them. Foreign student selection procedures that are designed to screen candidates planning to study abroad recognize competency in technical skills or academic work, but pay little attention to personality characteristics and degree of maturity. Klineberg's recommendations include increased attention to preparatory activities such as language training, provision of information about the university to be attended, and an introduction to the cultural norms and customary social behavior of residents in the host country. Orientation programs carried out by sensitive cross-cultural counselors can help to minimize the severely stressful consequences of the culture shock that students, especially those from developing countries, often confront during the early stages of their sojourn in a highly technological mobile western society.

Cross-cultural counseling can facilitate the alleviation of potentially stressful situations. Klineberg emphasizes that counselors should pay attention to areas in which a student is likely to experience a blow to his self-esteem, especially in situations where the student may be exposed to condescension, exploitation, or lack of cultural sensitivity on the part of his host. Since coping is not merely an individual enterprise, Klineberg reiterates the theme of many other authors, that the resident population must also be sensitized to the presence of the uprooted student, and must learn to appreciate some of the cultural and social resources that the foreign student brings to the community. The problems which the student faces, and his means of coping with them, will be greatly influenced by the reaction and the reception of the host community in the foreign culture.

A major task of adaptation for foreign students in a new culture is that of recognizing the diverse roles in which one is interacting with his university and community members. In Chapter 14, Pedersen suggests that an identity crisis may be accentuated as the foreign student must learn to handle multiple new roles. His role as a student and as young adult may be complicated by his being perceived as a cultural ambassador who needs to explain and justify the policies of his country. Each of these roles—the foreigner, the university student, the young adult, and the cultural ambassador—has implications for the way he interacts with his hosts. The student must learn to be able to differentiate and yet integrate these conflicting role assignments. He may experience conflict in managing simultaneously the representative roles that are assigned to him, and those of scholar and student which he considers most important in his education abroad.

A range of variables may influence the degree of stress felt by the individual. In reviewing the extensive foreign-student literature, Pedersen hypothesizes that the greater the cultural contrast in the way of life and social mores between the home and the host countries, the more complex and stressful are likely to be the demands on the individual's adaptive resources. Furthermore, adaptation is easier if the individual's role conflict is minimized, and if prior expectations prove to be more in accordance with the reality situation. Students often suffer the shock of status change, and usually status loss, which is particularly stressful. Pedersen also emphasizes the importance of the role of conationals in providing support, advice, and reassurance. Co-nationals who share a common cultural background often provide the most valued assistance to the individual in the course of his or her adaptation. Notwithstanding the vast literature on foreign students since World War II, there is limited knowledge about the processes of coping and adaptation of students in stressful situations of cross-cultural education.

Pedersen recommends that comparative studies be designed on such topics as the selection of foreign students, the effects of foreign study on the home-country populations, and the impact of foreign-educated leaders on home universities.

In Chapter 15, Spaulding and Coelho review the cross-cultural educational activities of several government branches, and the participation of the United States in international organizations such as UNESCO. They call attention to the valuable contribution of foreign study both to the student's host country and to his home country. Spaulding and Coelho make several recommendations. First, there should be transnational communication on a continuing basis between former students and their teachers. Second, the counseling and advising of potential foreign students should not be under the propaganda arm of the host government. Third, government agencies should collaborate in recruiting foreign students, with specific attention to the possibility of eventual employment in the home country. Fourth, legislative authority and support must be advocated in the United States in order to promote the activities of international organizations and institutions which deal with programs of student exchange. Finally, Spaulding and Coelho argue that the social impact of the uprooting of students is largely determined by educational agencies and cultural institutions which establish priorities in policy. The role of national policy is therefore central in the developing of the organizational infrastructure and institutional supports that are necessary for an effective transfer of training and technology from the host to the home country.

13

Stressful Experiences of Foreign Students at Various Stages of Sojourn: Counseling and Policy Implications

Otto Klineberg

My purpose in this paper is to look at the foreign sojourn of students as a process which goes through various stages, from selection, which determines who is to go abroad, to the return to the country of origin. The approach resembles that of a case history, which includes an account of the antecedents of a particular experience, a description of the experience itself, and an indication of its consequences in the life of the persons concerned. The process may even be described as analogous to a life story, since a full understanding of selection will involve attention to what led up to the desire as well as the opportunity to go abroad, and the impact after the return home may in some cases last through the rest of the person's life. The stages to be discussed are clearly not entirely separable or independent, since what happens at the outset may be closely related to what follows; but the chronological approach should be helpful in seeing the foreign sojourn as an integrated experience within a time perspective. This would appear to be all the more necessary since the bulk of the research literature in this area is limited to the study of foreign students at one particular point of time. In what follows, each successive stage will be discussed in terms of its relevance to the stressful experiences with which it may be associated, the problems which it raises, its implications for coun-

Otto Klineberg • Ecole des Hautes Etudes en Sciences Sociales, Paris, France.

seling and policy, and, whenever applicable, the kinds of research which may still be necessary before definitive conclusions can be drawn.

The Problem of Selection

Some years ago (1964) a study in the United States by the Center of Information in America, published in *Vital Issues,* estimated that about one-third of the foreign students had been carefully screened, one-third only partially, and one-third not screened at all with regard to their chances for satisfactory academic and personal adjustment. Even this relatively modest estimate has been challenged as being too optimistic regarding the proportion of students who have received even a moderate amount of screening.

The problem is complicated by the fact that selection occurs in so many different ways that any conclusion regarding its nature and its importance will necessarily be limited. When scholarships are granted by governments, or even by intergovernmental organizations, presumably some screening occurs, but the process may be complicated by the need to arrive at an acceptable geographical distribution among the foreign students, and political relationships (as with former colonies, for example) may play an important part; the possible role of nepotism, or the granting of fellowships to students related to persons in power, can certainly not be ruled out. In a study now in progress we interviewed foreign students in France, and our results showed selection to be a varied and rather haphazard process. Many students come because of general admiration for French culture; some because they are working in some area of French literature or history; some are eager to work under a particular professor of whom they have heard; some because scholarships were available, which they accepted even though they would have preferred to go elsewhere; a few because their parents sent them. Even if we assume that at least some degree of selection occurs before scholarships are granted, we are faced with the fact that many foreign students come on their own, with money from their parents or in the hope that they can earn enough to support themselves; these students are in a sense self-selected, and presumably highly motivated. I am not aware of any studies which have systematically compared the experiences of those who go abroad with and without scholarships, respectively. It may be that the difference in motivation may be balanced by the fact that those who "select themselves" may have more serious financial difficulties which interfere with their adaptation.

In one sense, however, all foreign students are self-selected. It is only rarely that they are forced by their parents or their governments to go abroad. With the rarest of exceptions, they all go willingly, and in many cases enthusiastically. Some may prefer to stay home, but go because they feel that a foreign degree or diploma will help them in their future careers; some may wish to escape from problems at home; some may be anxious and worried about what they will find; it seems clear, however, that the large majority welcome the prospect. Why?

I hope I may be forgiven if I illustrate this query with an account of a personal experience. During 1927 I was finishing my Ph.D. thesis at Columbia University; I already had a taste for travel, but until then it had been limited to Canada (where I was born) and the United States. I applied for a fellowship which would permit me to carry out an investigation in Europe. My interest in the research topic was genuine; so was my desire to go abroad. While I was waiting for an answer to my fellowship application, I received an offer of an instructorship at a university, at a time when university appointments were rare and very difficult to obtain. All my friends and colleagues, without exception, urged me to accept the appointment; my definite preference, which was realized, was to go to Europe. Again I ask: Why?

The problem of selection may be expressed in the form of two simple—perhaps oversimplified—questions. Who wishes to go abroad for study? Who is qualified to do so? The first question, which to my knowledge has never adequately been approached, is pertinent to the issue of stress and adaptation; the motives for going may play an important part in the success of the sojourn that follows. Better knowledge of the range of personality characteristics related to the desire for travel abroad should certainly help in the process of counseling. Some of the motives which may operate seem positive and promising; curiosity, the desire to learn more about the rest of the world, the hope of contributing to better relations across national boundaries, awareness of the international character of learning, need to obtain training not available at home; a wish to overcome the insularity resulting from experience limited to one region and to one society. Others may strike the counselor as less encouraging; the desire to escape from an unpleasant personal situation, the feeling that the grass is always greener on the other side of the river, the belief that quality is associated with distance from home. It is not always easy to identify motives, but it seems reasonable to conclude that when that is possible, it should result in more intelligent guidance that can be offered to the prospective traveler.

This issue is clearly related to the second question, as to who is qualified to go. It is my impression, based on considerable personal experience, that, when selection is taken seriously, it is usually limited to the issue of competence in the particular discipline involved, with little attention to more personal attributes; for example, adaptability to new situations and flexibility in solving unexpected problems, attitudes toward people of different ethnic and national background, freedom from excessive identification with the patterns of behavior and values characteristic of one's own society, a high degree of independence, an openness to new varieties of personal relationships. These and related factors are not easy to identify, but failure to pay attention to them may result in the wrong kind of selection.

There is considerable discussion in the literature as to the age or level of educational attainment at which a student might be encouraged to go abroad for study; the consensus appears to be that a considerable degree of maturity is desirable. To take one example, Aich (1962) is strongly of the opinion, on the basis of his study of Asian and African students in Germany, that it would be preferable to admit only those students who have already terminated their studies at home and wish to obtain specialized training abroad, or those who are already embarked on a career such as teaching or research, and feel the need for supplementary experience or instruction. Our interviews in France also indicated that a number of our subjects themselves felt that they should have waited another year or two before leaving home. We should perhaps make an exception in the case of the Junior Year Abroad, in which the program includes the kind of supervision by a representative of the home university which helps to ensure that the young student receives the guidance he requires. Even in this case, however, the degree of relative maturity reached by the student is by no means irrelevant to the process of successful selection.

One further point which requires attention in this connection refers to what happens when the foreign student is accompanied by his or her spouse. This issue arises more frequently at a relatively mature level (in the case of professors or research workers, for example), but a number of students are married at the time of their foreign sojourn, and problems occasionally develop as a consequence. There has been some discussion in the literature (Torre, 1963) of cases in which someone is sent abroad on a mission and is on the whole fairly successful, except that the spouse (in these cases almost invariably the wife) is unhappy abroad, makes life miserable for her husband as a consequence, and reduces the effectiveness of the sojourn. This has resulted, in at least some cases, in drawing attention to the need for interviewing both spouses, since the lack of adaptability on the part of either one may have a deleterious effect on the mission as a whole. Another possibility, fortunately rather rare, is that the marriage may suffer as a consequence. I know from my own experience of two cases where the sojourn abroad was followed by divorce, and in our present study one person reported that he himself was pleased with his stay abroad, but added "watch our for your marriage." Although such cases are not frequent, the unfortunate outcome should strengthen the still rare tendency to take into consideration the characteristics of both spouses before making the final selection. This may be a delicate procedure occasionally, but it should be undertaken wbenever possible.

The issue of selection leads directly to a consideration of the second major stage in our case history, namely, the preparation of the students for their foreign sojourn.

Preparation for the Foreign Sojourn

Once the student who wishes to go abroad has been offered a scholarship, or has decided to go on his own, the problem arises as to how he may best prepare himself in order that the sojourn may be successful. Preparation has a number of aspects, of which perhaps the most obvious is the facility of the foreign student in the language of the country to which he is about to go.

Language Competence

In spite of the clear importance of adequate preparation in the language of the host country before entering the new university, experience indicates that this requirement is not receiving the attention it deserves. In our study of the problems faced by foreign students in seven countries (Klineberg, 1976), we found that our colleagues everywhere laid great stress on the difficulties created by the lack of adequate advance preparation in this respect. This may constitute a problem even in the case of students who come from countries where, for example, French or English is the language of the universities, and who now are studying in French- or English-speaking areas; this is, of course, the situation in many of the former colonies, from which a very high proportion of foreign students both in France and in Great Britain originate. They already speak the language of the host country, but usually not so well; after all, it is their second language.

For students from other countries, the language handicap may be considerable. This was brought home to us very strongly in a series of interviews conducted in France as part of a larger study now in progress. Of 28 foreign students, 12 had little or no knowledge of

French on arrival; they could not follow what was presented in their courses, and were incapable of entering into any kind of rapport with fellow students. The lack of adequate preparation in this respect constituted a truly stressful situation at the beginning of the foreign sojourn. By the time of our third interview, toward the end of the academic year, most of these students were able to handle French reasonably well, but by that time a lot of valuable time had been lost. Not surprisingly, a number of them felt that one year was not adequate for study abroad, since such a substantial proportion of their time had to be devoted to the adaptation process, mainly because it took so long before they felt at home in French. Many of them urged that greater attention be paid to language preparation before departure.

In a number of countries there are attempts made to overcome the difficulties encountered by foreign students because of inadequacies in language. The *Alliance Française*, for example, offers a large variety of courses in French at all levels, and many students from abroad speak of these courses with great enthusiasm; the fact remains that those who arrive badly prepared in this respect may frequently have to devote the whole of their first year to filling this lacuna. In Japan, foreign students receive language instruction which starts as soon as they arrive, since few of them speak the language beforehand; in some universities, the language department arranges a number of intensive courses designed specifically for foreigners. It is reported, however, that even after two or three years many still have difficulty in following lectures with reasonable ease. This constitutes a very real, and so far unresolved, problem. West German universities also give special courses for foreign students, as does the Goethe Institute, with branches in various parts of the country. In the United Kingdom, on the other hand, many universities have a careful screening process which includes the testing of linguistic facility, both oral and written, and elimination of those who do not reach the minimum standard regarded as essential.

The language problem is clearly one of the most difficult in the whole area of international university exchanges, and there is no easy solution. Insofar as policy implications are concerned, it may be helpful to keep the following considerations in mind.

It would appear to be useful to make a distinction with regard to language requirements when the foreign sojourn is expected to be relatively brief (not longer than one academic year), as compared with the situation in which a more extensive stay abroad is planned. In the former case, unless the student is really familiar with the "new" language, he will have to devote so much of his time to learning it after his arrival that a substantial portion of the foreign sojourn may be wasted. In that case, he should either demonstrate an adequate linguistic facility at the time of selection, or he should be expected to devote a period of time before departure to an intensive learning experience designed to overcome this handicap before he seeks admission to the foreign university. Such requirements can be imposed in advance in connection with the Junior Year Abroad programs, or through the bilateral commissions responsible for the Fulbright fellowships; this cannot be done in the case of students who "select themselves." The same requirements may be applied when the student requests admission to the new university; here we may anticipate so much variation among countries as well as universities as to make any generalized statement inapplicable at this time, but it seems reasonable to urge them to adopt such criteria for the relatively brief (one academic year or less) sojourn.

When the project involves remaining for a number of years, it is less "costly" to accept the foreign student even if he must at the outset devote considerable effort to becoming truly competent in the new language. Here, too, valuable time will be lost, but proportionately much less than in the case of a briefer sojourn. Since it is so much easier to learn the language in a social environment in which it is widely spoken, it would be a pity to discourage those students who plan to stay long enough to overcome their handicap and still have time to profit from their foreign sojourn.

It would be very helpful if psychologists could develop a simple test of linguistic capacity. This would, theoretically, enable us to distinguish between those who, though they know little of the required language, could learn it reasonably quickly, as contrasted with those for whom such a task represents almost insuperable difficulties. There is probably little hope that such a test would be satisfactory, unless it took into account the general attitudes of the student toward the new language and the people who speak it; such attitudes apparently play a significant role in determining what is usually described as linguistic competence or capacity.

Advance Knowledge of the New University

Some difficulties might be avoided if the foreign student knew a little more in advance about what to expect at the new university—what his relations with the professors will be, what they will expect of him, the nature of his contacts with fellow students, and so forth. These issues will be raised again when we deal with the actual sojourn abroad, but they are also relevant to the process of adequate preparation. Such advance knowledge may be helpful in reducing some of the strains and stresses which frequently develop.

One striking example arises out of our interviews with foreign students in France referred to above. There was a great deal of complaint, especially in the early stages of the sojourn, about the inaccessibility of French professors. Students who had difficulties in fulfilling the tasks required of them, and who would have liked to talk things over with the professor, found it almost impossible to do so. Anyone with any degree of contact with French universities could have told them so in advance, and thus have prepared them for this frequently frustrating experience. At the time of the uprising of the French students in 1968 (usually referred to as *"les événements de mai"*), one of the major criticisms levelled was at the *mandarinat*, the tendency of French professors to consider themselves as mandarins, seated high above and remote from their audience, too busy with their own projects to have time for personal involvement with students. The 1968 events were supposed to have changed this situation, but with several honorable exceptions the changes have been minor. The late Kurt Lewin was once asked what differences he noted between American and European universities. His answer was that the main difference was that the doors to the offices of American professors were always open. When I first came to teach in Paris in 1962, I asked questions about my courses, and then: "What about office hours?" "Office hours?" came the reply; "you have no office!" I should add that professors in France do become accessible to advanced students working for their doctorates, but the large majority of students find such contact extremely difficult. In the reverse direction, Bochner (1972) points out that the informality characteristic of most American universities may be confusing and disturbing to foreign students accustomed to a more "respectful" attitude toward the professors. Bochner also mentions that Asian students in Australian universities had difficulty in satisfying their professors' insistence on developing their own critical approach, rather than relying on authorities.

These differences in pattern will undoubtedly continue to create problems, but their severity might be reduced if foreign students knew a little more about what to expect. One suggestion that has been made, and which should prove useful, is to organize Alumni or Alumnae associations made up of nationals who have studied in institutions to which the students plan to go, and with whom such students could meet before their departure. Cultural officers attached to the relevant embassies could also help.

Advance Knowledge of the New Culture

The issue here relates to whether the foreign student has learned enough about the culture of the new society to adapt to it successfully. What was said above applies here as well, but extended to the society as a whole, and not just to the university. Associations of former students as well as cultural attachés might play a similar useful role. What has been called "culture shock" would not be entirely avoided, but it could be reduced by imparting knowledge beforehand. The foreign students we interviewed in France frequently complained of their lack of adequate preparation in this respect. Many of them, for example, found French students and their families almost as inaccessible as their professors, insofar as establishing friendly contact was concerned. They became used to this in time, and concluded that "the French are like that," but again some unhappiness might have been avoided if they had been warned of this in advance.

This is one of our most complex issues. How do we inform students about the host society which they are about to enter? What kinds of information should be transmitted? Where is it available? How can it be taught? What is involved runs the whole gamut from table manners to the most intimate personal relations, and it is difficult if not impossible to anticipate the kinds of problems which the student will encounter. One of my Dutch friends told me of his disillusionment when he discovered that Americans who called him by his first name, and thereby (by his own standards) accepted him into the close intimacy of friendship, turned out to be no more than casual though "friendly" acquaintances. Everyone who knows Americans could have told him that the use of first names means very little, but how should he have learned this in advance? Books on the American national character would probably omit it as too trivial to mention, and courses on American life and culture are unlikely to include it. Yet it played an important part in the impression that this Dutch professor carried away. I would suggest that a new kind of text might be written, based upon interviews with students who have been abroad, and including an account of the difficulties, great and small, which contributed to their "culture shock." This could be useful not only in terms of specific content, but also as indicating the *kinds* of problems which the foreign student would be likely to encounter during his stay abroad.

The issue of previous academic, as distinct from what might be called personal, preparation, is also of obvious importance in this connection, but will be discussed in the next section, dealing with admission to the new university as the first stage in the actual foreign sojourn.

The Sojourn Abroad

What has been said above may serve as an introduction to this section. The difficulties due to inadequate preparation, language inadequacy, and lack of knowledge of what to expect in the university and in the community generally, contribute to the stresses and strains of the foreign sojourn, and continue to exert their influence throughout its course, particularly in the early stages. We turn now to the implications of specific experiences which are encountered after the foreign student begins his residence abroad.

First Contacts at the University

In addition to the problems mentioned above—the lack of access to the professors, little intimacy with other students—what is perhaps the first stressful situation encountered relates to the place in the university to which the foreign student is assigned on arrival. Several of the foreign students whom we interviewed in France expressed their disappointment at the amount of "credit" that they were given for work completed before their arrival, and felt that the degrees or diplomas obtained at their own universities had been appreciably undervalued. This raises the general question of "equivalences" in credits, an issue which affects both the admission and the placement of foreign students. In both these respects there are wide variations not only between countries, but also among various universities within the same country, especially where not all universities are under a uniform, usually governmental, administration. The colleges of Oxford and Cambridge, for example, are autonomous in their admissions policy. In Japan, Waseda University has adopted a different admissions procedure in the case of foreign applicants, since it was found that, largely because of their deficiences in the Japanese language, they could not be expected to pass the same examinations as were applied to local students. The German Federal Republic, on the other hand, adopted a tough admissions policy in 1963, insisting that foreign students had to satisfy the same requirements as were demanded of German students.

The French have adopted a system which might be expected to simplify the adaptation of foreign students to the French university system. There is, for example, a long list of "correspondences" between the French degrees or diplomas and those obtained from universities abroad, and this list is continually being extended by the educational authorities. There is in addition a list of "equivalences," determined each year, which can be exchanged for their French equivalents. This arrangement applies particularly, but not exclusively, to former French colonies. Recently a similar equivalence has been set up with certain German universities. The French system as a whole has considerable flexibility in this respect.

The question arises as to whether this degree of flexibility may not create additional problems. Foreign students do frequently complain that they should have been placed at a higher level than the university administration has seen fit to grant them, and, rightly or wrongly, they believe that other students at their level have been treated more leniently. There can be no doubt that this issue may result in great unhappiness for the student who wishes, or is obliged, to pursue his studies in a foreign country. German students, on the other hand, only learn on their return what "credits" they have earned abroad. At the policy level, it would seem highly desirable to work out an extensive series of equivalences among the universities of different countries, preferably in such a manner that the student knows before beginning his foreign sojourn just what he may expect. He can then make his decision as to whether—and where—to go with clearer knowledge of where he will be placed when he arrives. If this could be done, there would be a definite reduction in the unhappiness experienced by many foreign students. This, too, is a complicated task, but a beginning has been made, and it should be possible to extend the process more widely.

Academic Success or Failure

It is not easy to establish satisfactory criteria by which to judge the success of a foreign sojourn. There is at least one criterion, however, which is relatively easy to apply, and which most of those concerned with international student exchanges would find relevant and logically applicable. Has the student been successful in satisfying academic requirements? Has he passed his examinations? Did he obtain the degree or diploma which he sought, and for which (at least in part) he undertook his foreign sojourn? If he has been disappointed in this respect, he can certainly be expected to return home with the unhappy emotional burden of failure, and with a negative attitude toward his whole foreign experience. All the problems mentioned above—selection, language facility, preparation in advance—may contribute to failure. The kind of training previously received at the home college or university will also obviously have major importance in this connection.

The question then arises: do foreign students show an unduly high incidence of failure? The data available are inconclusive. The study by Aich (1962) in Germany indicated that about 80% of Asian and African students failed in their intermediate examinations, and 40% in their finals. These findings, and particularly their extension to foreign students in general, are rendered doubtful in view of the 1971 report by the German Statistical Office, which found very little difference in this respect between German and foreign students. In the study by Eide (1970) of students from Egypt, Iran, and India attending universities in West Germany, the United Kingdom, and the United States, it was found that a substantial majority of these students obtained a degree from the host country before returning home. Bochner (1972) also noted a small proportion of failures of foreign students in Australia. On the other hand, Kapuŕ (1972) found a greater frequency of failure among foreign than among British students at the University of Edinburgh, but only among males; the foreign female students did as well as the natives. Other British universities, however, such as the University of Sussex (Klineberg, 1976), have reported that their foreign students are just as successful as others, partly because of the rigorous admission procedures, but also as the result of the tutorial system which assures to all students easy access to someone who can give them advice, and help them overcome the difficulties which they encounter.

We know too little about the extent and nature of failures, not only in the case of foreign students in general, but also as regards specific national samples studying in specific foreign countries. There can be no doubt that examinations and dissertations create stress, but as yet we cannot be sure that such stress occurs more frequently among foreigners. Failure may, however, be more traumatic in their case, since it may be considered as a form of treason to the home country which has sent them abroad, and to all those who contributed to making the foreign sojourn possible.

There is a further problem which, as I have indicated elsewhere (Klineberg, 1976), is rarely mentioned but very troublesome, that is, whether greater leniency is shown to foreign students, particularly from the developing countries, on the ground that in view of their previous handicaps they should not be judged so harshly as others. This tendency is difficult to avoid; there are in fact many professors who adopt such a position deliberately. The motives involved many be generous, but there is danger that a *"magna cum laude"* or a *"mention très bien"* so obtained may arouse undue expectations in the persons concerned, as well as in their associates, and that the consequent disillusionment may create serious stresses later.

Adaptation to the New Environment

The relevant literature is rich in references to the problems related to the difficulty of adapting to a new social situation; the culture shock which can be responsible for uneasiness, uncertainty, conflict and misunderstanding. Reference has been made above to the difficulties which may arise on arrival through lack of information as to what to expect, and to the complex issue of how to prepare the foreign student for what he will encounter. What is not always realized is the reciprocal nature of this phenomenon, related both to the foreign students' attitude toward the host population, and to the manner in which they are perceived and treated by those who constitute the new social environment. Objective factors obviously play an important role; the students may be harassed, for example, by financial problems, difficulties in finding adequate housing, differences in food habits, or distance from family and old friends. The relationships with people, however, appear to play a dominant role.

In a number of the studies conducted on student exchanges, it is reported that satisfaction with the foreign experience is intimately related to whether the student feels that he has made friends in the host country. This is clearly a function of attitudes on both sides. The student is more likely to make friends if he starts out with positive feelings toward his hosts; they, on the other hand, must be prepared to accept him. Eide (1970) indicates that the students in her study reported a considerable degree of contact with the host population, with frequent mention of the development of friendly relations. Other investigators are less optimistic. N'Diaye (1962) found that African students in France speak of the superficiality of contacts with French students in the majority of cases, and of the fact that most of their newly formed friendships were with other Africans, or occasionally with other foreigners.

My own study with Ben Brika (1972) indicated that students from the Third World attending universities in Austria, France, and the Netherlands reported many contacts, and a few instances of what appeared to be true friendship, but there were frequent feelings of isolation and loneliness. Both in Austria and in France a substantial majority express dissatisfaction with regard to human relations, but the variations even among students of the same nationality in the same country of sojourn are striking. Some were satisfied, some dissatisfied, and some were, or claimed to be, completely disinterested in making contacts with the local population. The desire for more contact is expressed with some frequency in the analysis by Tajfel and Dawson (1965) of essays written by African, Asian, and West Indian students in the United Kingdom; their book has the significant title, *Disappointed Guests*. Our current study of foreign students in France indicates that a majority of them felt depressed and frustrated by their inability to make French friends; their relationships were almost entirely with their own nationals and other foreigners.

This whole issue is complicated, as suggested above, by the difficulty of knowing whether to assign the responsibility to the hosts, the visitors, or to both. If to both, it would be useful to know in what proportions the responsibility is divided, and also how these proportions vary with the characteristics and attitudes of the hosts and the visitors, respectively. The research with which I am familiar has emphasized mainly the reactions of the visitors, and has tended to neglect those of the host population. It would be interesting to know, for example, whether French students (and others) are more or less favorable to the presence of foreigners at their universities than are students (and others) in the United Kingdom or the United States; how university communities within the same country differ in this respect; what measures are taken, if any, to make the foreign students feel a little more at home, and what indications there are of the success of such measures. We know, for example, how difficult it is to create an atmosphere of friendliness through university-sponsored activities without giving the impression of artificiality or condescension.

Policy implications are difficult to identify in the absence of this information, but the stress resulting from loneliness can be so devastating that, at the very least, we must urge that more attention be paid to this issue. A start is being made in France, thanks in large part to the appointment of foreign student advisers who, at least in some cases, are giving advice to foreigners as to the kinds of social activities in which they might care to participate. It is ironic to note this development in France, stimulated by the American experience, at precisely the moment that many American universities are eliminating such advisers from their staff.

The Problem of Prejudice

Mention was made above of the study by Tajfel and Dawson (1965) which indicated that with great frequency the nonwhite foreign students in the United Kingdom were "disappointed guests," and that a common cause of their disappointment was the racial or ethnic prejudice which they encountered. There are similar reports by Davis, Hanson, and Burnor (1961) and by Veroff (1963) of the frequency with which black students in the United States mentioned discrimination in various situations. The latter indicates that such nonacademic problems clouded the perspective of these students on their American sojourn. In my study with Ben Brika in 1972 we discovered that about half the students from the Third World in all four countries (France, Austria, the Netherlands, and Yugoslavia) complained of discrimination due to their "race" or ethnic origin. We found this result all the more remarkable in view of the fact that these four samples varied so much in their ethnic composition. The foreign students we interviewed more intensively in France as part of our current international investigation, particularly those of Arab and black African origin, also attributed some of their problems to the alleged French prejudice against foreigners.

In many cases, the negative reaction is produced not so much by the personal experiences of the students themselves as by their perception or judgment of the racial situation in general. In the study by Useem and Useem (1955) of Indian students who had been in the United Kingdom or the United States, it was found that fewer than one-fourth had been the subjects of discrimination themselves, but more than three-fourths knew of discrimination against others, and were upset by it. Similar results are reported in a number of more recent investigations, and we observed the same phenomenon in our own studies of foreign students in France.

This raises the important issue as to the extent to which discrimination may be allied with what psychologists refer to as "pluralistic ignorance," in which all the members of a particular group express one view, while being certain that all the others in the group would disagree. (The members of a college fraternity, for example, all state their willingness to accept blacks or Jews, while attributing to all other members an unwillingness to do so.) Without of course denying the reality of discrimination, which may affect housing and jobs as well as personal contacts in general, it would be important to determine how much of it is real, and how much imagined.

This whole issue is complicated by the question of national status, that is to say, the status accorded by the host country to the national groups represented by the foreign students. Morris (1960) regards this factor as playing a major part in the attitudes of foreign students to the United States, and as explaining the greater hostility of those who feel that Americans ascribe a lower status to their country than they themselves regard as justified. Another related phenomenon is that of "status shock"; students from India, for example, who occupy a high social position in their own country, now find themselves identified with the mass of Indian students and regarded as similar to all the others, their higher status completely unrecognized. These and other related situations may create the impression that discrimination exists where none was intended, but the impact on the individual concerned may be no less unpleasant as a consequence.

The policy implications of discrimination, real or imaginary, relate to the whole complex issue of ethnic and international relations, and it is hardly to be expected that any solutions can be offered at this point. There are some actions, however, which might be taken, and, even though they touch only specific aspects of the situation and not the problem as a whole, they should be helpful. They are presented here as examples, without any pretense that the list is complete. It seems logical that a much more complete preliminary briefing of nonwhite students before their foreign sojourn, with regard to the existing pattern of intergroup relations in the country to which they are going, would at least reduce the element of surprise and shock, and contribute to a more realistic appraisal of the situation actually encountered. Indian students of high status might be informed in advance as to what to expect. A major source of irritation might be removed if university authorities helped the foreign students obtain housing without running the risk of applying where they are not wanted; the same applies to finding work. Social events organized by the university or by religious organizations might be improved by some expert advice as to how to make sure of eliminating what is often seen as condescension (mentioned above), or the patronizing exploitation of what is foreign and exotic. More information needs to be given to the host population as to what to expect, and to the foreign students as to what to avoid (for example, preparing food with unfamiliar and therefore frequently unpleasant odors in areas frequented by others). In certain universities, particularly the smaller ones, it should be possible to bring to the attention of the local students what they can do, naturally and unobtrusively, to make the foreigners feel more at home, and to counteract their tendency to see discrimination even where it does not occur.

The Mental Health of Foreign Students

Reference was made above to the feelings of isolation, depression, and homesickness occasionally reported by foreign students. More seriously, the question has arisen as to whether cases of psychological maladjustment occur more frequently among foreign than among native students. This is the view presented by Alexander, Workneh, Klein, and Miller (1976), who estimate, on the basis of their own research on Asian and African students in the United States, that psychological problems occur among foreign students more frequently than is usually assumed. They state that their "research has shown that the vast majority of non-Western or Third World students . . . feel vulnerable and at risk during much of their time in the United States" (p. 83). They conclude that foreign students are a high-risk group; that they seek psychological help only after all other resources have been exhausted, and are reluctant to ask for psychological help because this would constitute a loss of status; and that psychosomatic and emotional problems follow rather similar patterns in all groups of foreign students. It is suggested that statistics based on the frequency of appeals for help would probably underestimate the occurrence of psychological problems among foreign students. Earlier, Banham (1958) spoke of the disturbing incidence of mental breakdown among Nigerian students in the United Kingdom, usually because of failure in the course of study which they had chosen.

Psychological problems do arise with foreign students, as has already been indicated, and there is some likelihood that their frequency is greater than among natives; but to speak of the "vast majority" as being affected is, in my judgment, an exaggeration. (This point will be discussed more fully below.) The fact remains that, whenever it does occur, some provision has to be made to give the student the care that he requires; and it seems most natural to expect the university counseling services to fulfill that function. Serious questions have been raised, however, as to their capacity to do so, largely because of the cultural differences between the counselor and the foreign students whom he may be called upon to help.

The whole problem of counseling across cultures is discussed in an important recent publication edited by Pedersen, Lonner, and Draguns (1976). They point out that cultural sensitivity on the part of the counselor has recently been considered to be an ethical imperative. In his chapter on the field of intercultural counseling, Pedersen remarks:

> The American Psychological Association sponsored a conference on patterns and levels of professional training at Vail, Colorado, in July 1963. One of the recommendations of that conference was that the counseling of persons of culturally diverse backgrounds by persons who are not trained or competent to work with such groups should be regarded as unethical. . . . It is apparent that cultural sensitivity and awareness will play an increasingly important role in the training of counselors. (pp. 35–36)

Pedersen believes that the trained counselor is not usually prepared to deal with members of groups whose values, attitudes, and general lifestyles differ from his own.

In another chapter in the same volume, "Racial and Ethnic Barriers in Counseling," Vontress gives a number of specific examples of the difficulties that may arise. Clients from some cultures show less openness, and are unable to speak freely of their own troubles to a comparative stranger; this is true of American Indians, who usually communicate so little as to render the counseling process largely ineffective. Japanese-Americans may hesitate to express their feelings in the presence of individuals of higher status. In a later chapter Sundberg refers to the great difference between Indians and Americans "in regard to decision making—a very important aspect of counseling young people" (p. 164). A number of years ago, Gardner Murphy (1953) reported that in India a greater change in attitudes occurred when students were addressed by an authority figure than when they arrived at a group decision; this finding has since been confirmed by subsequent research. Sundberg (1976) points out that there are other Indian cultural factors which affect the process of counseling interaction. The concept of privacy and openness is not the same as in the United States; certain family matters are not mentioned to outsiders such as counselors; nondirective counseling may be interpreted as meaning rejection or lack of interest in the Indian client. Sundberg adds that all of these factors, together with differences in customs, make the transfer of counseling to Indians a serious problem. After raising the question as to whether perhaps too much stress may have been placed on cultural factors, he writes: "Most of us in counseling are left between the two dangers of being overly concerned with cultural differences and being not concerned enough" (p. 141).

It should be possible to find a satisfactory position somewhere in the middle of the road. There are similarities as well as differences between cultural groups; anthropologists have helped us to understand cultural variations and their consequences, but they have also drawn our attention to the common human. It is probable, however, that most counselors are more likely to proceed as if the techniques that work at home may also be applied elsewhere; as a consequence, it is in the area of variations that they are in need of further training. The policy implications of this discussion are therefore, in my judgment, clear. The counselor, within or outside the university, who is called upon to help a foreign student resolve his psychological problems must know as much as possible about the cultural background of his client. In view of the composite character of the foreign-student population, this requirement is difficult if not impossible to realize fully. At the very least, however, there should be an awareness of its importance as part of the training of a counselor, who may then be in a better position to obtain the additional information he needs. The alternative is failure.

The U-Curve

One of the most striking research findings related to the whole course of the foreign sojourn is that students show a pattern of reactions described as a U–curve. It has been described by Lesser and Peter (1957) as follows. When the students first arrive, they go through a spectator stage, in which they are happily engaged in an exciting adventure and enjoying the new experience. They then become involved, have to face many problems, and suffer disillusionment and even depression. Finally there is a third phase, in which the students learn to face and solve their problems; and the curve of satisfaction rises as they succeed in the process and prepare for their departure.

Information about the U–curve should be therapeutically useful in the case of many foreign students. If they are depressed as a result of their difficulties, the knowledge that this is a common if not inevitable stage which others also have to traverse before the curve of satisfaction moves upward may help them to view their troubles with more optimism. Counselors may, as a consequence, find the material on the U–curve useful with clients during the "trough" of unhappiness. To the extent that this curve is typical, on the other hand, caution must be observed in interpreting results of research conducted at a particular moment, since the students questioned or interviewed may be at different stages in the process of solving their problems. The issue is complicated by the fact that the research shows no real consensus as to the length of time it takes to emerge from the trough. Some studies report such emergence within a year; others indicate that three or four years may be necessary.

There are other complications. In our 1972 study, Ben Brika and I found that a very substantial proportion of Third World students (more than half) studying in France and Austria reported that they were worried and anxious *before* their departure. There was no initial period of euphoria and excitement when they arrived; no first stage of happy adjustment at the beginning of the foreign sojourn. In the investigation now in progress, to which reference was made above, a majority of the foreign students interviewed on three separate occasions in France felt lost, lonely, and depressed at the beginning of their sojourn, but were usually relatively happy by the end of the year. The curve may clearly take forms other than that predicted by the letter U. In a few cases, moreover, students were happy with their first impressions, and experienced no depression. This minority at least indicates that the curve of the total experience varies to such an extent as to make any generalizations exceedingly hazardous.

Insofar as policy implications are concerned, it has already been indicated that foreign students may be helped through their depressive stage (when that does occur) if they learn that their experience is by no means exceptional in this regard. The counselor, on the other hand, should be aware of the tremendous variations that the U–curve undergoes from individual to individual, for example, in the length of time during which each stage lasts, the intensity with which it is experienced, and even the presence or absence of one or another of the stages. The U–curve is not inevitable as a description of the course of the total sojourn, and it is just as much a mistake to take it too seriously as to ignore its frequent occurrence.

The Return Home

Problems may also arise when the foreign sojourn is over and the students return to their home country. It has been suggested, in particular by Gullahorn and Gullahorn (1963), that most students experience a second U–curve when they return; there is first euphoria on being back with family and old friends, and with the new degree or diploma which will surely open doors to a prestigious career; then disappointment and depression at the inability to find the job to which one now feels entitled, and the discovery that others may not ascribe much importance to one's having studied abroad (they may even consider it a handicap in some cases); and, finally, emergence from the depression when a post becomes available. Even then, as Useem and Useem (1955) have indicated in the case of Indians who have studied in the United States and the United Kingdom, the post obtained will often be below the level which the training abroad would have justified. There may also be problems of personal adjustment, and nostalgia for what one has left behind, to add to the difficulties on the return home.

Here, too, there will be tremendous individual variations, and as far as counseling is concerned there is little that I can add to what was said above. There are, however, policy implications at another level which have not always received the consideration they deserve. When governments or foundations give scholarships, for example to students from the Third World to study abroad, how much attention should be paid to the availability at home of employment in the disciplines in which they receive their training? Should such scholarships be restricted to the technical capacities of which the country is badly in need for purposes of development, or should a gifted scholar be permitted to study Chaucer at Oxford or Descartes in Paris? These questions are exceedingly difficult, and a case might be made for a variety of answers. There is no doubt that technical training is in great demand in the developing countries, but will not the new universities need specialists in comparative literature or the history of philosophy? And if not today, what about tomorrow? Perhaps some compromise can be found, but if the training obtained is completely irrelevant to openings back home, the student is faced with the choice between inevitable frustration on the one hand, and remaining abroad on the other.

The Brain Drain

Most foreign students do return home, but a fairly substantial minority, varying in extent in accordance with where they come from and where they receive their training, constitute what is usually called the brain drain. This is often considered to represent a failure in the program, since it is precisely the countries that need trained personnel the most that are deprived of the gains which had been anticipated. As has been pointed out, many motives may enter into the decision not to return, but it is clear that it is not always financial reward that plays the dominant role. It should also be kept in mind that if the training is in a complex field (for example, atomic physics or cancer research) for which no facilities exist in the home country, it is precisely the return which should be considered "wastage" or failure, and the decision to remain abroad may then result in a greater contribution to knowledge and the satisfaction of human needs. The positive side of the brain drain should receive careful consideration.

It has been said that there are today in the United Kingdom so many doctors and nurses of Indian, Pakistani, and West Indian origin that the British Medical Service would find it difficult to function without their presence. There are Germans and Poles teaching in Scotland, British and French professors in the former African colonies, Indians on the staff of universities in Canada and the United States, Americans and others of various nationalities at French universities. These are all presumably making a definite contribution, and even though some of them come from industrially developed countries, a substantial number from the Third World is included in this group.

One of the hoped-for consequences of international university exchanges is that they will contribute to friendlier relations between the peoples involved. This will hardly solve the problems of war and peace, but it may contribute to the development of an atmosphere in which contact and cooperation are rendered more likely than conflict. To the extent that this is true, the presence of foreign nationals may have a positive impact. Seeing people from the Third World in positions of high status in medicine, teaching, and the arts may help to change unfavorable stereotypes and aid the process of mediation between cultures. How successfully this role is fulfilled is difficult to determine, but at the policy level this should at least make us hesitate before considering every example of the brain drain as a sign of failure.

A Concluding Comment

Throughout this presentation, the emphasis has been placed on problems to be solved and difficulties to be overcome. This is not surprising, in view of the fact that the aim, at least in part, was to bring to the attention of counselors and other concerned individuals and institutions some directions in which they might possibly help. After all, when things go well, no great outside aid is needed. This emphasis may, however, have given to people unfamiliar with the research on this topic the impression that failure is more frequent than success, and problems more common than feelings of satisfaction and enjoyment of the foreign experience. Fortunately, this is not the case. The majority of foreign students interviewed in our own and earlier research terminate their sojourn with a definite conviction that it was worthwhile, that they learned a great deal, that they enjoyed themselves, and that their future career had been helped in the process. They also usually find themselves feeling more friendly to their hosts (see Coelho, 1962; Flack 1976).

Not always, however. There is a minority for whom the negative aspects of the experience outweigh the positive. It is difficult to determine the size of that minority, since it varies from one study to another, but there is no doubt that it exists. The efforts of many investigators, including our own, are directed to reducing the size of that minority, on the ground that a single failure is one too many if it could possibly have been avoided. The purpose of this chapter has been to draw attention to the possible sources of failure through the nature of the selection process, the preparation for the foreign sojourn, the academic experience, relations with the host population, issues of racism and prejudice, mental-health aspects, and the return home. Successes are, however, more frequent than failures, and nothing that has been said in this chapter should reduce our general confidence in the conclusion that international student exchanges should be encouraged.

References

Aich, P. *Farbige unter Weissen*. Cologne: Kiepenheuer & Witsch, 1962.

Alexander, A. A., Workneh, F., Klein, M. H., & Miller, M. H. Psychotherapy and the foreign student. In P. Pedersen, W. J. Lonner, & J. C. Draguns (Eds.), *Counseling across cultures*. Honolulu: The University Press of Hawaii, 1976.

Banham, M. The Nigerian student in Britain. *Universities Quarterly*, 1958, *XII*, 363–366.

Bochner, S. Problems in culture learning. In S. Bochner & P. Wicks (Eds.), *Overseas students in Australia*. Sydney: The New South Wales University Press, 1972.

Center of Information in America, 1964 (see Klineberg, 1976).

Coelho, G. V. Personal growth and educational development through working and studying abroad. *Journal of Social Issues*, 1962, *18*, 55–67.

Davis, J., Hanson, H., & Burnor, D. *The African student: his achievements and his problems*. New York: Institute of International Education, 1961.

Eide, I. (Ed.). *Students as links between cultures*. Oslo: Universitets Forlaget, 1970.

Flack, M. J. Results and effects of study abroad. *The Annals of the American Academy of Political and Social Science*, 1976, *424*, 107–117.

German Statistical Office, 1971 (see Klineberg, 1976).

Gullahorn, J. T. & Gullahorn, J. E. An extension of the U–curve hypothesis. *Journal of Social Issues*, 1963, *XIX*, 33–47.

Kapur, R. L. Student wastage at Edinburgh University. *Edinburgh University Quarterly*, 1972, *Summer*, 353–377.

Klineberg, O. *International educational exchange: An assessment of its nature and its prospects*. Paris: Mouton, 1976.

Klineberg, O., & Ben Brika, J. *Etudiants du tiers-monde en Europe*. Paris: Mouton, 1972.

Lesser, S. O., & Peter, H. W. Training foreign nationals in the United States. In R. Likert & S. P. Hayes, Jr. (Eds.), *Some applications of behavioural research*. Paris: UNESCO, 1957.

Morris, R. T. *The two-way mirror: National status in foreign students' adjustment*. Minneapolis: University of Minnesota Press, 1960.

Murphy, G. *In the Minds of Men*. New York: Basic Books, 1953.

N'Diaye, J. P. *Enquête sur les étudiants noirs en France*. Paris: Réalités Africaines, 1962.

Pedersen, P. The field of intercultural counseling. In P. Pedersen, W. J. Lonner, & J. C. Draguns (Eds.), *Counseling across cultures*. Honolulu: The University Press of Hawaii, 1976.

Sundberg, N. Toward research evaluating intercultural counseling. In P. Pedersen, N. J. Lonner, & J. C. Draguns (Eds.), *Counseling across cultures*. Honolulu: The University Press of Hawaii, 1976.

Tajfel, H., & Dawson, J. L. (Eds.). *Disappointed Guests*. London: Oxford University Press, 1965.

Torre, M. (Ed.) *The selection of personnel for international service*. New York: World Federation for Mental Health, 1963.

Useem, J., & Useem, R. H. *The Western-Educated Man in India*. New York: Dryden Press, 1955.

Veroff, J. African students in the United States. *Journal of Social Issues*, 1963, *XIX*, 48–60.

Vontress, C. Racial and ethnic barriers in counseling. In P. Pedersen, W. J. Lonner, & J. C. Draguns (Eds.), *Counseling across cultures*. Honolulu: The University Press of Hawaii, 1976.

14

Role Learning as a Coping Strategy for Uprooted Foreign Students

Paul B. Pedersen

Foreign students are uprooted both in their temporary adjustment to the host country, which is unfamiliar to them, and, following the completion of their studies, once again back home, which has changed in their absence. In some cases, the foreign students experience a greater adjustment in returning to their once-familiar home country than in their original host country. The problems of cultural adjustment for foreign students are as great as or greater than those experienced by permanent immigrants or tourists and business personnel located more temporarily in a foreign culture.

The circumstances encountered by foreign students suddenly and simultaneously impose a variety of competing and sometimes contradictory roles which must be learned. When the requirements of those roles are realistically perceived and effectively learned, the experience is likely to be "successful"; but when the roles are not accommodated, the resulting identity diffusion and role conflict may affect the students' emotional well-being, and present serious obstacles to the achieving of their educational objectives. This chapter explores the negative consequences of foreign students' and their contacts' either *underemphasizing* cultural differences, by assuming that all people perceive and are perceived similarly, or *overemphasizing* cultural differences, by isolating foreign students into stereotyped boxes. Although foreign students come from widely diverse backgrounds, they are expected to

Paul B. Pedersen • Culture Learning Institute, East-West Center, Honolulu, Hawaii 96844.

"adjust" to a narrowly defined set of behaviors that requires them to learn their proper role very rapidly. Failure to learn their new role will result in confusion about their own identity and conflicts with those around them. Role learning then becomes a necessary coping strategy, whether the foreign student decides at the conclusion of his studies to return home or to emigrate.

Table I identifies the regions sending students to study in the United States in the last several years, there being a rapid rate of increase here, especially in numbers of students sent from the non-Western cultures.

Although there is a general increase in numbers of students from each geographic area, the most dramatic increases are from African and Asian countries, particularly from the oil-rich Middle East. The flow of foreign students is directly related to the national and international political situation in the student's home country. Decreasing enrollments of United States nationals in higher education also makes the universities more receptive to tuition-paying foreign nationals, especially those who do not require financial aid. Despite the increasing numbers of foreign students, universities have cut back on specialized services and scholarships for foreign students, thus intensifying the adjustment problems.

The Diversity of Foreign Student Roles

The skills of adapting to cultural diversity can become resources of great strength, and invaluable assets in helping persons to learn from one another and about themselves. Intercultural encounters are likely to highlight otherwise hidden conflicts in the person's own behavior at a rate many times faster than culturally homogeneous contacts (Hall, 1976). We all learn certain roles in order to function adaptively in our own cultural milieu. One's self-esteem and self-image is validated by significant others who provide emotional and social support in culturally patterned ways. Moving to a foreign culture suddenly deprives the student of these support systems. A normal response to the withdrawing of support is anxiety, ranging from irritation and mild annoyance to the panic of extreme pain, and the feelings of disorientation that accompany being lost. Every decision now requires a deliberate effort, and concentrated energy. Each of us has experienced a mild form of the same phenomenon when we have communicated with persons different from ourselves (Higbee, 1969; Singer, 1977).

Take the example of a male foreign student early in his period of

Table I. Distribution of nonimmigrant students in the U.S.[1]

	1973–1974	1974–1975
Eastern Africa	2,746	4,040
Middle Africa	311	410
Northern Africa	1,605	2,710
Southern Africa	406	610
Western Africa	6,669	10,600
Africa, unspecified	41	—
Total—Africa	11,778	18,400
East Asia	27,216	30,720
Middle South Asia	12,345	13,890
Southeast Asia	11,211	13,850
Southwest Asia	16,965	23,910
Asia, unspecified	28	—
Total—Asia	67,765	82,370
Eastern Europe	700	900
Western Europe	10,560	12,850
Europe, unspecified	14	—
Total—Europe	11,274	13,740
Caribbean	4,830	6,500
Central America	5,450	7,270
South America	9,732	12,490
Latin America, unspecified	15	—
Total—Latin America	20,027	26,270
Australia and New Zealand	1,084	1,260
Pacific Ocean Island Areas	1,070	1,390
Oceania, unspecified	1	—
Total—Oceania	2,155	2,650
Stateless	156	150
Country Unknown	4,559	2,370
Grand total	125,116	154,580

[1] Institute of International Education, *Open doors 1975*. New York: IIE, 1976.

studies who wants to "fit in" to the university community. All his American friends go out on dates regularly, but he can never seem to get a date with an American female of his choice. He has come to have the opinion that his strangeness as a foreign national is the reason for his being rejected. Consequently, he has become very critical of his own culture, and has begun avoiding his own fellow countrypersons. He is feeling very lonely and isolated from everyone, and is beginning to withdraw from any contact with other students. By the time the counselor becomes aware of his problem he has stopped going to classes, and has spent the better part of a week without leaving his room except when absolutely necessary. He has become extremely lonely for "back home," where he had been quite popular, and is becoming very bitter about the way other students are treating him. He is ready to give up and return home immediately.

We can cope with our conflicting cultural roles (1) because we rank-order them in terms of the importance of each role for our own identity, (2) because most identities apply only in certain contexts and are constantly changing, and (3) because these rankings and the identities themselves are constantly changing. The roles we value most highly define our "primary" identities, which we have learned gradually since childhood, or to which we have been converted as adults (Singer, 1977).

Foreign students provide an example of a population having to learn a wide range of culturally defined and typically unfamiliar roles in a short time under conditions of considerable stress.

A wide range of institutional and individual rationales have been developed in support of the sending of students to foreign countries to study (Eide, 1970). First, there is the idea that more knowledge leads to more empathy and finally results in improved international relations and world peace. Second, there is the idea that increased knowledge will stretch the imagination about the alternative interpretations of human culture and increase our tolerance for those different interpretations with a stabilizing effect. Third, diffusion of knowledge among cultures is expected to result in a more homogeneous world, taking the best from each culture in a more refined synthesis of cultures. Fourth, intercultural contact will teach people the necessity of interdependence. Fifth, increased knowledge of others will clarify our knowledge of ourselves. Although each of these goals may result from intercultural contact, Amir (1969) reviews the evidence that they will occur only under carefully structured "favorable" conditions, and not under the less structured and more randomly encountered conditions of spontaneous intergroup contact.

The critics of international educational exchange programs argue that foreign study is not the most efficient means of "developing" world resources (Bochner, 1972). First, foreign students refuse to return home after their studies, and end up emigrating to the host country, thus contributing to the "brain drain." Second, students return home with the "wrong" skills, unsuited to their home country. Third, returned students cannot apply scarce and needed skills because the local scientific infrastructure is inadequate. Fourth, students are selected for study abroad through family influence or for political reasons, rather than because of their superior qualifications. Fifth, returned students are embittered by real or imagined injustice, which turns them into enemies of the country which educated them. Sixth, returnees use their overseas-learned skills to widen the gap between rich and poor in their home countries, inviting political instability. Although both negative and positive evaluations of foreign study are based in part on real events, neither bias is inevitably true.

Of all the roles confronting the sojourner, perhaps the most diffuse and difficult to fulfill is a stereotype of what "foreign students" are supposed to be. There is a perception of foreign students as helplessly confronting all kinds of problems (Eide, 1970), defenseless and bewildered. Johnson's (1971) research discovered that Americans expected foreign students to have many more problems than the foreign students themselves reported. The problems faced by foreign students are not so different from problems confronted by students in general (Kahne, 1976; Torrey, Van Rheenan, & Katchadourian, 1970; Walton, 1971), and the fact that they are foreign nationals should not be allowed to obscure the identity crises they share with all other students in American universities.

> Very little awareness seems to exist that the vast majority of international students on our shores are not here solely by virtue of their intellectual achievements. They are not usually the prize-winning merit scholars of all-wise governments whose concerned foresight has made this grand adventure possible. But, like their American counterparts, many are here because neither their families, nor their governments, nor perhaps they themselves, know what else to do that makes more sense. (Kahne, 1976, p. 37)

There has been a tendency to confine foreign students to a rather narrowly defined role isolated from their peers, when in fact there is probably as much difference between any two foreign students from different countries as between either of them and any American student. By making a "special case" of foreign students we run the danger of isolating them, just as we might stereotype them by not recognizing those unique problems and resources which individual foreign students do in fact present. One extreme is at least as dangerous as the other. The fact is that adjustment by international students is a lot more complicated than the four crises of arrival, engagement, acceptance, and reentry might lead us to believe. The literature on foreign students too often substitutes labels for solutions. The "foreign student" is a multiplicity of roles, and, in actuality, there is no such person in the universe.

Although foreign students come from a wide diversity of cultural backgrounds, those students from non-Western countries, whose cultures are likely to be very different from the American host culture, are increasing proportionately. The greater the cultural difference, the more complicated the foreign student's adjustment is likely to be. At the personal/behavioral level, the foreign students define their goals in a more deliberately personal/professional/familistic context. How *they* perceive their role and learn to cope accordingly is probably more significant to this contribution. However, research promoted by American sponsoring agencies assumes a foreign-policy rhetoric, rather than a personal/professional/institutional development of the students in their back-home situation, as the evaluational criterion. At the policy level, the assumptions behind international exchange programs are defined in terms of leadership, national contribution, and international understanding.

The Foreign Student's Adjustment to Role

The research on sojourner adjustment is so varied, divergent, and unrelated in its approaches that it is difficult to develop any theoretical consistency among the research results. The population of "foreign students" is itself so diverse that, even when the same issues are being researched, the findings are often contradictory. Some of the more frequent approaches to describing or explaining the adjustment process of foreign students depend on curves or stages of adjustment, cultural shock, personality typologies and traits, background and situational factors, and social interaction. The goals of foreign-student research have not been conceptually defined, and attempts at theoretical formulations have not been validated.

Much of the research on intercultural adjustment has assumed that traits, factors, or types determine the sojourner's behavior. The trait of "flexibility," for example, has been assumed to result in an ability to adapt in a great variety of cultural situations. Mischel (1968) reviews the research on personality traits, to conclude that

> With the possible exception of intelligence, generalized behavioral consistencies have not been demonstrated, and the concept of personality traits or broad response predispositions is thus untenable. (p. 146)

Although personality traits have not been of use in predicting or explaining intercultural adjustment, there are still many advocates of this "personality approach," and the search for the "ideal" sojourning personality continues (Brein & David, 1971; David, 1972). Recent behaviorally oriented research on modeling, reinforcement management, self-reinforcement, and desensitization has shown more promise. Social behaviorism has deemphasized such inner variables as traits and intrapersonal unconscious determinants of psychodynamic theories. Cognitive social-learning theories which emphasize external environmental determinants of behavior seem more promising (Guthrie, 1975). Miles (1976) has demonstrated that "boundary-spanning" activities result in role ambiguity and role diffusion for the "integrator" or boundary role person who is expected to coordinate the conflicting demands of our multiple memberships. In a similar vein, Berry (1975) concludes that individuals with a high level of differentiation will be more independent of the incongruity and conflict of cross-cultural contacts. Not only is the content of roles differentiated, but, as Stewart (1974) points out, the functions of parent/child, teacher/student, boss/employee, or man/woman roles in American culture are also much less formal and specified than in most other cultures.

Despite considerable research, we have no theoretical basis for predicting role adjustment by sojourners. The "U–curve" or "double U–curve" has been tested and expanded to suggest a series of stages that foreign students experience in developmental sequence during their sojourn, but has been found to be unreliable, particularly for students from less developed countries (Spaulding & Flack, 1975). The "two-way mirror" concept that the foreign student's image of the United States reflects the image of his or her home country commonly held by Americans has been supported in some studies (Galtung, 1965), but, like much of the research, it has not drawn out the implications of sociopsychological factors for attitudes, social adjustment, or academic success. In part, this general weakness of the research has been related to methodological weaknesses in the design of sojourn research (Brislin & Charles, 1977). Some more general research orientations have included studies on binational or "bicultural" persons, who belong simultaneously to two different societies and who maintain two identities as they relate to their respective societies from within the context of one or the other culture (Useem & Useem, 1967). Bicultural individuals have the potential to function with cognitive flexibility, and are creatively adaptive in either culture from which they draw their identity (Berry, 1975), but there is little or no research on how these individuals can best make use of their biculturality (Niyekawa-Howard, 1970). Wallace (1956) suggests a "mazeway" theory, looking at the patterned image of society and culture symbolized in the sojourner's identity or self-image and his ability to receive or transmit information appropriately in an intercultural environment.

Research on cognitive strategies in differentiation of ingroup/outgroup perceptions has drawn from a variety of theoretical perspectives. Cognitive-consistency theory, learning theory, social-judgment theory, and cognitive developmental theory suggest that intergroup perception can be adequately accounted for as natural consequences of more general cognitive processes by which human beings structure, simplify, and give meaning to their physical and social environment (Brewer, 1977). Seligman's (1975) learned helplessness model, for example, suggests how maladaptation may result from a lack of control over one's environment, or ignorance of the consequences of inappropriate behavior. Cognitive theories can provide a basis for understanding and predicting behavior in an intergroup context.

Although there is no single conceptual/theoretical basis for the understanding of the adjustment process experienced by foreign students, a number of stage-development and trait theories have been researched, with conflicting results. Other conceptual models from social psychology appear more promising for the predicting of foreign student adjustment, even though research applying those models is lacking.

Role Learning

Bochner (1972) relates the problem of facing overseas students to four main social roles the student is required to fulfill. These are the role of foreigner, the role of university student, the traditional role of young adult, and the role of ambassador for the student's home country. Difficulties arise when behaviors appropriate to these roles are not learned effectively. This places the problem of the overseas student in the larger context of any individual adapting to multiple roles in a foreign or strange culture. The difference is that in our home culture we experience a gradual adaptation to new roles, whereas the overseas student is transplanted from the home culture to a totally different culture, and needs to achieve a rapid, if not instantaneous, mastery of the requirements inherent in multiple and conflicting new roles. In addition to learning new adaptive roles, the student will also have to maintain essential roles in the home culture, such as that of eldest son concerned about his parents' welfare.

Any change will require the learning of new roles (Foa & Chemmers, 1967). The normally stressful conflicts of being a student who needs money, lodging, food, recreation, friendships with the same and the opposite sex, peers, and the problems of developmental maturation, are further complicated by the foreign student's unfamiliarity with the host culture.

> The difference between cross-national and within-national role conflict is that two groups in the one society usually know what the other side expects of them, but do not agree in their respective definitions of what is the behavior appropriate to a particular role. (Bochner, 1972, p. 69)

The student's existing repertoire of responses may be of limited use in the host culture, and may even hinder the adjustment process. The greater the cultural differences, the greater is the likelihood that barriers to communication will arise, and that misunderstandings will occur (Mishler, 1965), particularly when the student is insecure about the home country's image overseas (Banton, 1965; Bochner, 1972; Hartley & Thompson, 1967).

The foreign student is forced to assume the role which Adler (1976) describes as the "multicultural man," based on skills of constant adaptation to new value configurations as a process rather than belonging to or learning about any particular culture. The student is tenuously suspended in a role which creates a new identity in an eclectic configuration which exists on the boundary of many cultures in a constant state of becoming. The student is required to synthesize this new temporary identity from a variety of unfamiliar value systems, refining universals across cultures, sensitive to both the similarities and the differences (Walsh, 1973). "Multicultural man," like "marginal man," or "protean man" (Lifton, 1969), is always recreating an identity. There is a change in identity as roles are learned, modified, or discarded in each discontinuous situation (Adler, 1976). However, there is, developmentally speaking, a core layer of continuity in self-image that maintains early socialization patterns.

Role learning is an attempt to relate the individual foreign student's behavior to the student's perception of the environment defined by national status, self-esteem, dual group membership, role conflict, identification, cultural distance, and other concepts related to adaptation. Superficial adjustment, adaptation to some specific goals, and global satisfaction are partial measures which need to be coordinated to include the wider context, the target areas of influence, patterns of adjustment, direction of change, phases of adjustment, and other complicating environmental factors. There is a need for theory building that relates the full range of these variables to intercultural background and outcome factors, and at the same time attempts to develop and test theoretical perspectives appropriate for explaining the processes involved in cross-cultural learning. Klein's (1977) research suggests that (1) culture is important in defining role conflict and identifying elements of stress, but adaptive coping responses are similar across cultures; (2) environmental factors are more powerful than personality in determining adaptation; and (3) self-esteem and self-confidence, with positive reinforcement of social skills and learning of new skills, are predictive of adaptation. Another researcher in the same project (Yeh, 1976) cites evidence that some foreign students may, however, be predisposed to role shock, loss of self-esteem, or psychological disorder where going overseas was perceived as a way of solving their psychological difficulties at home or some other neurotic conflict, suggesting that the student's personality is one factor in the configuration of adjustment.

The foreign student is placed in the role of mediator between the host and home culture, even though the role is not formally defined as it might be in the case of interpreter, tourist guide, industrial relations conciliator, marriage counselor, ombudsman, or an elected representative of a particular ethnic community. Through the student's knowledge of both cultures, problems of communication or adaptation can be anticipated, and an appropriate intermediary can be sought (Taft, 1976).

Role Conflicts

Roles are customarily defined as those behaviors we expect of an individual appropriate to a particular social or cultural context. These roles are based on how we feel, what we see as appropriate, what we consider rational or logical, and our own priorities in the otherwise chaotic experiences of everyday life. When roles are not carefully defined and are out of harmony with familiar values, foreign students may lose confidence, and allow family, friends, or society to make decisions for them.

The acculturation process confronting foreign students disturbs the priorities of previously learned role expectations by conflicting value orientations and expectations. The framework of role theory (Sarbin & Allen, 1968) and role differentiation (Banton, 1965; Hartley & Thompson, 1967) provides a basis for theory building in predicting the intercultural adjustment of foreign students. We depend on our group identifications for a somewhat secure self-image, and our attitudes or values become the internalized role norms of the groups with which we have identified. Our behavior grows out of differentiated roles modified somewhat by individual personality differences. You can tell much about a person by knowing that person's significant reference or membership groups. Banton (1965) describes the role-differentiation process as the extent to which membership in one role is independent of membership in any of the other roles, with some roles being more independent than others. In the foreign student's multinational contacts, national membership roles will be dominant in situations (1) where the individual is in an alien context, (2) when confronted by another foreigner, (3) when "representing" the student's culture or country to a group, (4) when home-country symbols trigger practiced "nationalistic" responses, and (5) when home-country values are challenged. Consequently, the foreign student needs to (1) reduce the salience of the national role, (2) increase the convergence of cross-national perceptions in reasonable rather than irrational conclusions, and (3) reduce the social distance between the student and other nationality groups (Bochner & Meredith, 1968).

In the adaptation process of foreign students, conflict and stress are important motivators (Spradley & Phillips, 1972). Klein (1977) describes five phases or patterns of adjustment, compounded by the familiar roles of the sojourner's self-identity, the complexity and the duration of the variables. Each new role is controlled by (1) the strength of motives for change, (2) the amount of change needed, (3) the individual's skills and coping resources, (4) characteristic stress responses, and (5) reinforcements provided by the new environment (Klein, 1977).

The greater the cultural differences or distance, the greater are the difficulties in adaptation or psychological adjustment. This rather obvious observation is particularly true for students from less developed countries going to a more developed country to study.

> The sojourners who travel in the first direction may be regarded as deprived, while those who travel in the opposite direction may be regarded as privileged. In the former case, sojourners are more subject to conformity with the host culture, while in the latter case, the culture of the host country may be, more or less, aspiring to conform to the migrant's culture. (Yeh, 1976, pp. 42–43)

Yeh's research on Chinese students in American universities cites role conflicts between home and host values, and loss of self-confidence or self-esteem, as the factors mostly responsible for maladaptation.

Adjustment is easier when role conflict is minimal and expectations are realistic. Students who are able to cope with rapid changes of social roles back home are likely to adjust well as foreign students. The literature on foreign-student adjustment has assumed the many changes to be good. Given the stress of role conflicts, Yeh (1976) advises:

> If one were to characterize the ideal adaptation to facilitate academic success and return home, it would involve only the minimum changes in behavior

> and attitudes essential for students to meet goals with confidence and success. (p. 44)

Change is viewed positively in our culture, but not so in all cultures.

It is not surprising that foreign students experience conflict, as might be expected of any person when asked to assume, spontaneously, multiple and conflicting new roles while maintaining an already complicated back-home identity in a totally different culture. The most severe culture shock does not result from dealing with external matters such as differences in food, climate, language, mannerisms, and communication, but rather from status change and status loss. As it is put by Alexander, Workneh, Klein, & Miller (1976):

> Most foreign students have been academically successful at home and are often professionally well established. Suddenly they face intense academic pressures and adjustments and a painful social vulnerability as well. (p. 82)

It is not surprising then that those foreign students tend to isolate themselves from American peers, and to create a co-national subculture as their primary support system, even at the expense of intercultural contact. Those foreign students who do seek out Americans are usually culturally similar. Foreign students from Western industrialized countries tend to socialize more with Americans than students from non-Western or less industrialized countries, probably because of cultural similarity. A greater percentage of foreign students from Western countries are undergraduates, and typically more social, younger, single, etc., and also more Europeans are in social sciences and humanities, rather than physical sciences (Torrey *et al.*, 1970). The tendency of sojourners to seek out co-nationals for their most warm and intimate relationships seems to apply to some extent among American students abroad as well (Miller, Yeh, Alexander, Klein, Tseng, Workneh, & Chu, 1971). Relations with co-nationals turn out to be extremely important for the predicting of foreign-student success, and provide a very important mental-health support system (Pedersen, 1975; Torrey *et al.*, 1970). The successful non-Western student seems to resist adaptation, while maintaining, in so far as possible, a higher priority for the predeparture home-country role values (Chunnual & Marsella, 1975).

Spaulding and Flack (1975), summarizing much of the research on changes experienced by foreign students after two or three years' stay, say that basic cultural or religious attitudes, career goals, and attitudes toward the home country change very little, whereas attitudes favoring openmindedness, the value of knowledge, and greater freedom in the relationship between sexes become much more important. Foreign students who stay less than two years are even less likely to change their basic cultural or religious values. There is no consistent direction of change in attitudes, favorable or unfavorable, to the United States, and these attitudes are more likely a function of the student's individual social and academic experiences. In any case, the maintenance of traditional values appears to serve an important function in protecting the student's self-esteem, sense of worth, and successful accomplishment of academic goals.

Theories of role conflict applied to foreign students suggest that behaviors which are goal-achieving in the host country may be unfamiliar, ineffective, or destructive back home, and important back-home values may likewise be maladaptive in the host culture. If we understand the learning process to be developmental, cumulative, and to some extent linear, then it would follow that some familiar host-country coping strategies would be easier to learn than other, unfamiliar, strategies. Still other established habits will have to be unlearned as maladaptive in the host culture. The foreign student is caught in a dilemma of having to adapt to new educational objectives in the host culture without losing the elements of a home-culture identity.

> Paradoxically, the more successful an overseas student's adjustment to the host culture has been, the greater will be the problems of readjustment to his original culture after he has returned home. (Bochner, 1972, p. 76)

Coping Strategies and Emotional/Social Support Systems

It is important to conceptualize the foreign-student sojourn as one of several phases in the adaptation process from predeparture to reentry back home. Klein (1977) identifies four possible coping strategies: (1) *Instrumental adaptation,* where the student's primary focus is on professional tasks and academic goals and on maintaining contact with the home culture directly or through co-nationals; the student experiences most stress in task performance, and has minimal social readjustment on reentry. (2) *Identification adaptation,* where the primary focus is on involvement in the host culture and social relations, and academic/professional goals are secondary; the student experiences most stress in interpersonal problems, and may run the risk of alienation or maladjustment on reentry. Widely differentiated attitudes toward the host culture are found, depending on level of achieved satisfaction. (3) *"Withdrawal" adaptation* begins with initial involvement with the host culture, so that task purposes are secondary to cross-cultural involvement with the host culture, much effort is made to accommodate host values initially, tensions arise in the interpersonal context that impede adjustment, and the student shifts from host-culture relationships to increased co-national contact and increased effort is directed toward restoring national identity toward the end of the student's stay. (4) *"Resistance" adaptation* emphasizes the role of cultural ambassador as most important, emphasizing intercultural more than interpersonal similarities or differences. Here attitudes toward the home culture are largely determined by host-culture attitudes toward the home culture; there is little attitude change, and no significant shift in national identification.

The relationship between assimilation, integration, and identification suggests that if an individual identifies with the home culture, there will be favorable attitude changes toward integration, whereas if the student identifies with the host culture, assimilation is favored. Contrary to predictions based on theories of the "marginal man," those persons who are more traditionally oriented are the ones who suffer psychological marginality, rather than those who wish to move on and cannot (Berry, 1970; Sommerlad & Berry, 1973). The elements of a student's self-imagery most likely to change are those shaped by the student's nationality, profession, and/or structure of personal relations. In each of these areas, the student may adjust his self-image through coping strategies such as internalization, identification, confirmation, or resistance. The final outcome or result will be some combination of changed self-image or maintained identity. To the extent that the new environment permits and encourages change and promises a feeling of belonging, the student is likely to change (Marsella & Sanborn, 1977). Bailyn and Kelman (1962) discovered that both the internal structure of a foreign student's self-image and the externalized new behaviors are open to change. As a result of contact with the host culture, host-culture values would seem most likely to be internalized in the part of a person's self-image that is most stable in the home situation, or where experiences in the new environment are most involving and rewarding.

There is a need for more long-term longitudinal studies of "coping behavior," such as Coelho, Hamburg, and Adams (1974) reviewed, in related areas that will help sojourners establish emotional security in a foreign culture and minimize role conflict or diffusion. The criteria of a "favorable" adjustment need also to be established. We know that individuals who disassociate themselves from home-country values are more readily able to move into the American culture, but we may hesitate to say that such an adjustment is desirable. The capability for culture learning in either direction is available through the reinforcement of one role or the other. Most of the reinforcers require a radical change, however. Natural circumstances experienced by foreign students are (1) that warm, intimate international contacts are the exception rather than the rule, (2) that the "well-adjusted" foreign student is less strongly identified with home-country values, while being more self-confident and adept in communication skills, and (3) that those who remain distant from host nationals and continue to be oriented to home-culture values are more traditonal (Alexander *et al.*, 1976).

The criteria of effective coping, then, are to help foreign students adapt creatively within their *own* culture—the one they carry with them—not what used to be back home, or the new and strange host culture. The goal of counseling foreign students, then, is to help them establish their identity, and differentiate the roles being thrust on them. Most foreign students from traditional backgrounds find academic structure and expectations to be vastly different from the home-country structure where the university system, families, professors, and their government often combined to tell them what to do. Now they are much more adrift, and unsure of how to proceed and cope with subject matter from an American school system, multiple-choice tests, informal teachers' styles, financial need, and other practical aspects of the foreign student's situation.

> The loss of anchorage inherent in moving from one culture to another only heightens anxiety, self-consciousness, paranoia, vulnerability, and concern with status. (Alexander *et al.*, 1976, p. 93)

Most of the help identified by foreign students as meaningful in making a satisfactory adjustment does not come from the formal university agencies for counseling and guidance. A sampling of over 200 students visiting the Stanford University psychiatric services showed that only 10 of the students were foreign, and all but two of these were Europeans or Canadians, though foreign students comprise 10% of the student body (Torrey *et al.*, 1970). When Stanford students were asked to rank from first to ninth all the possible places they would turn to for help for anxiety or depression, 85% ranked formal medical or psychiatric services sixth or lower (Torrey, *et al.*, 1970). Pedersen (1975) likewise discovered that cocountrypersons were the most frequently sought out source of help on personal problems, even for problems where the faculty adviser, foreign-student adviser, counseling agency, or another university office would have been more helpful. We may speculate that countrypersons are most popular because they are more immediately and readily available than host nationals, particularly if the university counseling offices have been either inaccessible or unacceptable. In cases where foreign students perceived increased internationalism or cross-cultural understanding among university faculty or staff, or took part in some sort of orientation program, the number going to their faculty adviser or the International Student Adviser's Office increased considerably (Pedersen, 1975).

Migration and Side Effects of Foreign Study

The issues of role and identity and integration of personal and professional goals ultimately focus on the question whether or not the foreign student plans to return home. Much of the literature has presumed a considerable "brain drain" of foreign students from less developed countries to the more developed countries. Glaser (1978) tested the hypotheses about "brain drain" and discovered that there is a wide variation in rates of immigration by foreign students coming from different countries. Only 12% of the foreign students in the United States responding to Glaser's questionnaire said that they would stay abroad, and 12% were undecided. In contrast, Canada's more liberal immigration policy resulted in 13% of the foreign students in Canada saying that they "definitely" would stay abroad, 13% saying that they "probably" would stay abroad, and 14% being undecided (Glaser, 1978). More students and professionals from European countries stay in the United States after completing their studies than from the large majority of developing countries. This contradicts earlier research showing higher rates of emigration by foreign students from developing countries. Glaser credits the difference to the different ways of asking questions or drawing samples. Some research defined "brain drain" as any stay for work after study, rather than looking at true permanent migration.

The most commonly stated attractions for going back home after studies were family, friends, and patriotic feelings, as well as rejection of unfamiliar social structures, and sometimes racial or ethnic discrimination in the host culture. The student returning home may anticipate problems in adjusting to the work situation, and assume financial benefits of staying abroad in the host culture, although the financial attraction is only one factor in the complicated decision of whether or not to return home. The persons who decide to stay tend to be members of cultural or religious minority groups with smaller or less devoted families at home, are more cosmopolitan in their cultural tastes, have weaker ties with their government or future employer back home, were supported abroad with their own family funds or from foreign sources, have no job waiting for them back home, and have received little or no information or news from their embassies or potential employers (Gama & Pedersen, 1977; Glaser, 1973).

Table II provides the data on how demographic characteristics of foreign student attitudes predict the decision of whether or not to return home after studies. In commenting on these data, Glaser points out that the strongest influences on migration plans at all stages of careers are quality of jobs, number of jobs, and best interests of children. Other strong predictors include the opportunity to contribute to a profession, patriotism, family interests, and the spouse's preference. As the table illustrates, the predictive power of all these reasons becomes weaker after the student has returned home. Once the student is home, there are no doubt new reasons for staying that further complicate the decision process. It is interesting to note those factors which do not predict a decision to migrate. Facilities and the quality of professional relationships are not strong predictors. Age by itself does not predict migration, although those who were married before study abroad are most likely to return home. Grades are not a strong predictor when they are high, though low grades are likely to prevent the student from staying, for a variety of reasons. Poor or wealthy students are equally likely to emigrate, regardless of the family's social class. Graduate students are no more likely to emigrate than undergraduates.

An important question in the migration of foreign students relates to whether they are being forced to migrate by shortages of jobs or other constraints. In Table III, Glaser suggests the opposite to be true. Most persons prefer the migration patterns they expect to follow, or they expect to fulfill the plans they prefer. In those cases where realistic plans and tastes differ, the 161 involuntary returnees far outnumber the 72 involuntary emigrants, and the 321 who plan to return in spite of uncertain preferences greatly outnumber the 158 who prefer to return, but have uncertain plans (Glaser, 1978).

Although foreign students are influenced by and acculturated to host culture values, most have made definite plans before their arrival. In some cases, the foreign students have left the United States with a more negative opinion than when they arrived. Separation from family, friends, and homeland is the most common complaint, especially among African and Asian nationalities, along with linguistic difficulties if the student is studying in a second or third language; and discrimination is a common complaint among dark-skinned foreign students. The closer the ties with the home country through letters and personal or professional communications, particularly when related to plans for when the student will return home, the greater the likelihood of the student's returning home. There is a suggestion in

Table II. How reasons predict migration plans at different stages of the professional's career[1]

Reasons	Answers by all respondents			Answers about state of mind at time of survey by	
	During or at end of study	During or at end of work	After return	Students	Stayons
1. Working conditions:					
a. Contribution to profession	.442	.273	.445	.481	.221
b. Income	.495	.403	.485	.541	.372
c. Quality of jobs	.512	.377	.515	.542	.350
d. Number of jobs	.512	.339	.454	.568	.412
e. Housing	.271	.218	.213	.360	.172
2. Professional needs:					
a. Contacts	.117	.268	.387	.122	.132
b. Sufficient time	.364	.342	.298	.426	.384
c. Libraries	.226	.266	.387	.193	.128
d. Equipment	.105	.168	.369	.261	.117
e. Space	.282	.235	.284	.302	.267
f. Status of professionals	.331	.289	.305	.403	.354
3. Colleagues:					
a. Fellow workers for discussion	.257	.288	.378	.269	.262
b. Assistants	.228	.216	.294	.265	.186
4. Influence of others:					
a. Patriotism	.527	.543	.225	.637	.616
b. Obligations to family	.427	.426	.180	.584	.433
c. Influence of family	.414	.466	.156	.572	.515
d. Influence of friends	.393	.211	.244	.518	.331

5. Societal setting:					
a. Cultural level	.283	.316	.207	.345	.512
b. Challenge of life	.461	.357	.321	.499	.371
6. Alienation and discrimination:					
a. Feel strange	.424	.523	.226	.484	.554
b. Discrimination	.288	.481	.162	.325	.573
7. Politics:					
a. Political conditions	.430	.490	.330	.442	.263
b. Language policies	.433	.496	.470	.417	.352
8. Citizenship:					
a. Maintain existing rights	.530	.468	.255	.590	.308
b. Acquire new rights	.674	.579	.320	.713	.538
9. Interests of spouse and children:					
a. Spouse feelings	.520	.362	.434	.623	.247
b. Education of children	.554	.585	.504	.586	.534
c. Careers of children	.578	.631	.442	.621	.404
d. Marriage of children	.412	.361	.171	.503	.549
(Number of respondents)	(4958)	(752)	(1819)	(2767)	(384)

[1] From Glaser, 1978. Complete wording of the items is in Table VII-1. The numbers are Goodman's and Kruskal's gamma. A full explanation of this table is in the text. Contents of each column . . . First column: all respondents from all surveys, based on answers by students about their current thinking, and on the recollections by stayons and returnees about viewpoints as they were finishing studies. Second column: all stayons in the United States and France, and those returnees who worked abroad; the stayons answered about their thinking at the time of the survey and the returnees told how they thought as they were finishing work abroad. Third column: all returnees from India, Ceylon, Korea, Brazil, and Colombia. Fourth column: all students from the United States, Canada, and France. Fifth column: all stayons from the United States and France.

Table III. Plans and preferences[1]

Preferences	Plans: Return and stay	Uncertain	Emigrate	Total
Return and stay	2,833	158	72	3,063
Uncertain	321	397	100	818
Emigrate	161	105	565	831

[1] From Glaser, 1978. The table combines all persons from the surveys of students, stayons and returnees. Entries are numbers of persons and not percentages.

these conclusions that when the student is not satisfied with, or is unwilling or unable to fulfill, the role requirements back home, the new "temporary" roles in the host culture become more attractive.

Implications for Research, Counseling, and Institutional Policy

There are clear implications from the research we have discussed for persons counseling foreign students in this country. (1) In working with foreign students, recognize the dangers of either overemphasizing or underemphasizing the cultural differences between yourself and the foreign student. (2) Orientation is a continuous process requiring contact with students before they arrive, during their stay, and after they have returned home throughout the continuous adjustment of the total person. (3) Identify the specific skills most likely to be helpful for students from specific cultures, and develop ways that those skills can be learned. (4) When working with foreign students, help to clarify the alternatives and reduce the ambiguity of diffuse roles as much as possible. (5) Invite foreign students to bring in a fellow countryperson to the counseling interview for additional support to the student if appropriate. (6) Strengthen and encourage the bond between foreign student and home country as much as possible. (7) Help foreign students monitor the ways in which their values and perceptions may be changing as a result of their stay in the host country. (8) Develop a career-placement procedure through contacts with potential employers in the student's home country.

There are likewise implications for international educational-exchange policy planners and institutions in the materials we have discussed. (1) The emphasis should be on the 20 to 30 productive years of the foreign students after their return to their home country or after graduation, rather than the two or three years of their stay in the educational instituion. Success in school *may* contribute to failure back home. (2) Individual goal criteria should be separated from national policy goals in evaluating educational exchange, recognizing that these two success criteria may be in conflict at times. (3) Contact with fellow countrypersons in dorms, clubs, and living arrangements should be encouraged, rather than discouraged, as an important support system for foreign students. (4) Establish a formal educational preservice training curriculum for foreign-student advisers in one or more university programs, combining elements from psychology, international affairs, and other related fields. (5) Incorporate the resources of foreign students into the classroom in roles of teacher as well as student. Mestenhauser (1976) has developed a handbook describing specific ways this could be accomplished efficiently. (6) Increase funding proportionate to increases in enrollments from non-Western cultures in recognition of institutional obligations and anticipated cultural adjustment problems for non-Western students in our society.

Further research would sharpen the implications for both institutional policy and counseling interventions to fill in the many gaps in our knowledge about international educational exchange. (1) The extensive research literature on foreign students is widely dispersed and often of poor quality. Some coordination of these materials is needed for identifying the best research now available on international educational exchange. (2) Research on foreign students should draw from existing social-psychological theories rather than the limited theoretical formulations which have been developed specifically for foreign-student research. Role theory, for example, would provide many promising hypotheses for future research about foreign students, with implications for persons having to adapt to a variety of new roles in a short period of time. (3) Foreign students are an extremely diverse group, requiring research to account for the effect of wide-ranging cultural differences within the group. (4) Researchers need to compensate for American cultural values and how they may affect their perception of conditions facing foreign students in the statement of hypotheses and interpretation of data. (5) Research should begin to develop a theory of long-range changes resulting from international educational exchange both for the individual adjustment of foreign students returned home and the effect of foreign study on decisions made by influential alumni in foreign countries. (6) We need longitudinal studies of coping behavior that define desired outcomes and success criteria for working with foreign students and international exchange programs.

Conclusion

Research on the adaptation of foreign students to host-culture universities has been characterized by isolated, uncoordinated, and fragmentary data on specialized variables, and no clear application of results to international education-policy decisions. This chaotic situation is owing partly to the extremely complicated variables in research on or about foreign students which defy control conditions, partly the difficulty in attracting highly qualified and well-funded researchers to the topic of international education, and partly the lack of a theoretical framework that might somehow tie these diffuse research outcomes together. The search for traits or factors for predicting intercultural adjustment has been less valuable and promising than applications of social behavioral theories. Attitudinal studies based on surveys at one point in time are less useful than behavioral and follow-up activity analysis. This chapter has focused on role theory as offering one perspective for viewing the foreign student experience.

The experience of foreign students is characterized by a wide diversity of newly acquired roles that compete with familiar back-home values and with one another in a variety of situational contexts. The process of coping with diverse roles is familiar to most of us, but not to the extent required for foreign students from culturally different backgrounds. In attempting to meet the host culture's expectations as well as his or her own expectations, the foreign student is often caught in an impossible dilemma.

The research on foreign-student adjustment has begun to focus on those particular situations requiring specific adaptation skills rather than some general trait of intercultural flexibility. These skills can be taught and learned as appropriate adaptations of the foreign-student role, protecting the student's investment in the host-culture educational objective and the student's back-home identity at the same time. The foreign student must learn "boundary-spanning" skills in order to integrate a new multicultural identity with more traditional and familiar values.

One of the most important skills for dealing with the role diffusion experienced by the student's adjustment is the ability to differentiate the conflicting roles from one another and move appropriately from one role to the other. The roles are defined both by the foreign student's perception of the environment in terms of their experiences of social roles during socialization, and by the expectations by others of appropriate "foreign-student" behaviors. Dissonance between differentiated roles results in conflict at a wide range of levels, and causes many foreign students to confine their more meaningful contacts to co-nationals with whom they can revert to more familiar roles. Although the new environment is change-oriented through new information and a positive valuation of the change process, the foreign students are likely to resist change, wherever possible, as destructive of their basic identity and back-home role.

A variety of coping strategies is available, although the criterion for a "successful" outcome remains vague and needs to be multidimensional. The success must be measured according to *individual* goals of individual growth and professional advancement, *institutional* development goals and *national* goals of skilled leadership and resource enhancement. The emphasis is not on either the back-home or the new-host values, but rather on the unique and personalized system of values integrated in the foreign student's own eclectic role. Formal counseling services have typically failed in the attempt to help foreign students make the necessary adjustments, forcing the students to depend on co-nationals for most of their guidance and advice.

Finally, the foreign student does not appear to change much as a result of contact with the American host-culture university community, as many studies would suggest. A surprisingly low percentage of foreign students in the United States emigrate after completing their studies, although significantly more emigrate to Canada; and there are data to suggest that those who do choose to emigrate do not feel forced to do so by conditions back home. The demographic variables predictive of emigration emphasize a feeling of belonging, and elements of a strong identification with the host-culture role identity. To the extent that back-home identity roles are satisfying, the student is likely to reject other functional advantages of emigrating, and make the decision largely on ideological grounds.

References

Adler, P. The multicultural man. In L. Samovar & L. Porter (Eds.), *Intercultural communication* (2nd ed.). Belmont, Calif.: Wadsworth, 1976.

Alexander, A., Workneh, F., Klein, M., & Miller, M. Psychotherapy and the foreign student. In P. Pedersen, W. Lonner, & J. Draguns (Eds.), *Counseling across cultures*. Honolulu: The University of Hawaii Press, 1976.

Amir, Y. Contact hypothesis in ethnic relations. *Psychological Bulletin*, 1969, *71*, 319–342.

Bailyn, L., & Kelman, H. C. The effects of a year's experience in America on the self-image of Scandinavians: A preliminary analysis of reactions to a new environment. *Journal of Social Issues*, 1962, *18*, 30–40.

Banton, M. *Roles: An introduction to the study of social relations*. London: Tavistock, 1965.

Berry, J. Ecology, cultural adaptation and psychological differentiation: Traditional patterning and acculturative stress. In R. Brislin, S. Bochner, & W. Lonner (Eds.), *Cross-cultural perspectives on learning*. New York: Wiley (Halsted Division), 1975.

Berry, J. W. Marginality, stress and ethnic identification. *Journal of Cross-Cultural Psychology*, 1970, *1*, 239–252.

Bochner, S. Problems in culture learning. In S. Bochner & P. Wicks (Eds.), *Overseas students in Australia*. Auburn, N. S. W.: New South Wales University Press, 1972.

Bochner, S., & Meredith, G. *Role and attitude modification in multi-national living.* Paper prepared for the Conference Workshop on Psychological Problems in Changing Societies, East–West Center, University of Hawaii, June 1968.

Brein, M., & David, K. Intercultural communication and the adjustment of the sojourner. *Psychological Bulletin*, 1971, *76*, 215–230.

Brewer, M. Perceptual process in cross-cultural interaction: An overview. SIETAR presentation, Chicago, February 1977.

Brislin, R., & Charles, J. Research on cross-cultural interaction. Paper presented at the meetings of the Society for Intercultural Education, Training and Research, Chicago, February 1977.

Chunnual, N., & Marsella, A. Convergent and discriminant validation of a traditionalism–modernism attitude questionnaire for Thai exchange students. *The Journal of Social Psychology*, 1975, *96*, 21–26.

Coelho, G. V., Hamburg, D. A., & Adams, J. E. (Eds.), *Coping and adaptation.* New York: Basic Books, 1974.

David, K. Intercultural adjustment and applications of reinforcement theory to problems of "culture shock." *Trends*, 1972, *4*(3).

Eide, I. (Ed.). *Students as links between cultures.* Oslo: Universitet Forlaget, 1970.

Foa, U., & Chemmers, M. The significance of role behavior differentiation for cross-cultural interaction training. *International Journal of Psychology*, 1967, *2*, 45–57.

Galtung, I. E. The impact of study abroad: A three-by-three nation study of cross-cultural contact. *Journal of Peace Research*, 1965, *3*, 258–275.

Gama, E., & Pedersen, P. Readjustment problems of Brazilian returnees from graduate studies in the United States. *International Journal of Intercultural Relations*, 1977, *1*(4), 46–59.

Glaser, W. *The brain drain: Emigration and return.* New York: Pergamon Press, 1978.

Guthrie, G. A behavioral analysis of culture learning. In R. Brislin, S. Bochner, & W. Lonner (Eds.), *Cross-cultural perspectives on learning.* New York: Wiley (Halsted Division), 1975.

Hall, E. *Beyond culture.* Garden City, N. Y.: Doubleday, Anchor, 1976.

Hartley, E. L., & Thompson, R. Racial integration and role differentiation. *Journal of the Polynesian Society*, 1967, *76*, 427–446.

Higbee, H. Role shock: A new concept. *International Education and Cultural Exchange*, 1969, *4*, 71–81.

Institute of International Education, *Open doors 1975.* New York: Institute of International Education, 1976.

Johnson, D. Problems of foreign students. *International Educational and Cultural Exchange*, 1971, *7*, 61–68.

Kahne, M. J. Cultural differences; Whose troubles are we talking about? *International Educational and Cultural Exchange*, 1976, *11*, 36–40.

Klein, M. *Preliminary overview: Adaptation to new cultural environments.* Paper presented at a meeting of the Society for Intercultural Education, Training and Research, Chicago, February 1977.

Lifton, R. *Boundaries.* New York: Vintage Books, 1969.

Marsella, A., & Sanborn, K. *The modernization of traditional cultures: Consequences for the individual.* Working paper presented at meeting of the Society for Intercultural Education, Training and Research meeting, Chicago, February 1977.

Mestenhauser, J. *Learning with foreign students.* Minneapolis: University of Minnesota International Student Adviser's Office, 1976.

Miles, R. H. Role requirements as sources of organizational stress. *Journal of Applied Psychology*, 1976, *61*, 172–179.

Miller, M. H., Yeh, E. K., Alexander, A. A., Klein, M. H., Tseng, K. H., Workneh, F., & Chu, H. M. The cross-cultural student. *Bulletin of the Menninger Clinic*, 1971, *35*, 128–131.

Mischel, W. *Personality and assessment*. New York: Wiley, 1968.

Mishler, A. L. Personal contact in international exchanges. In H. C. Kelman (Ed.), *International behavior: A social psychological analysis*. New York: Holt, Rinehart & Winston, 1965.

Niyekawa-Howard, A. M. *Biculturality and cognitive growth: Theoretical foundations for basic and applied research*. Occasional papers of the East–West Culture Learning Institute, No. 1. Honolulu: East–West Center, July, 1970.

Pedersen, P. Personal problem solving resources used by University of Minnesota foreign students. *Topics in Culture Learning*, 1975, *3*, 55–66.

Sarbin, T., & Allen, V. Role theory. In G. Lindzey (Ed.), *The handbook of social psychology* (2nd ed., vol. 1). Reading, Mass.: Addison-Wesley, 1968.

Seligman, M. E. P. *Helplessness: On depression, development and death*. San Francisco: W. H. Freeman, 1975.

Singer, M. *Identity Issues in Intercultural Communication: Overview Statement*. Paper presented at meetings of the Society for Intercultural Education, Training and Research, Chicago, February 1977.

Sommerlad, E., & Berry, J. W. The role of ethnic identification in distinguishing between attitudes toward assimilation and integration of a minority racial group. *Human Relations*, 1970, *23*, 23–29.

Spaulding, S., & Flack, M. *The world's students in the United States: A review and evaluation of research on foreign students*. New York: Praeger, 1976.

Spradley, J. P., & Phillips, M. Culture and stress: A quantitative analysis. *American Anthropologist*, 1972, *74*, 518–529.

Stewart, E. *American cultural patterns: A cross-cultural perspective*. Pittsburgh: Regional Council of International Education, 1974.

Taft, R. *The personality of the mediating person*. Paper presented for the East–West Center project, The Mediating Person, June, 1976.

Torrey, E. F., Van Rheenan, R., & Katchadourian, H. Problems of foreign students: An overview. *Journal of the American College Health Association*, 1970, *19*, 83–86.

Useem, J., & Useem, R. H. The interfaces of a binational third culture: A study of the American community in India. *Journal of Social Issues*, 1967, *23*, 130–143.

Wallace, A. Revitalization Movements: Some theoretical considerations for their comparative study. *American Anthropologist*, 1956, *58*, 264–281.

Walsh, J. *Intercultural education in the community of man*. Honolulu: The University of Hawaii Press, 1973.

Walton, B. J. Research on foreign graduate students. *International Educational and Cultural Exchange*, 1971, *6*, 17–29.

Yeh, E. K. Psychiatric implications of cross-cultural education: The case of Chinese students. Mimeographed paper, Medical School, Department of Psychiatry, Taiwan National University, Taipei, Taiwan, 1976.

15

Research on Students from Abroad: The Neglected Policy Implications

Seth Spaulding and George V. Coelho

In the past twenty-five years, millions of foreign nationals have come to the United States to pursue undergraduate and graduate study. In the late 1970s there were well over 200,000 foreign students in the United States each year, the majority financially self-supporting. An influential minority are sponsored by their own governments or by private businesses and organizations in their own countries. Some of these receive financial support under American technical-cooperation programs or through American foundations and voluntary organizations, though increasing numbers are completely financed by their own governments. These sponsored students will return to their own countries to take up positions of leadership when they complete their overseas education, as will many of the non-sponsored students.

The long-term foreign-policy-related implications of the role of the United States in preparing human-leadership resources for other countries have not gone totally unnoticed by the intellectual and political communities. Phillip Coombs, building on his experience as Assistant Secretary of State for Educational and Cultural Affairs in the early 1960s, described (Coombs, 1964) the need to place educational and cultural activities of the United States on an equal footing with economic, political, and military elements. Charles Frankel, who held the same post in the mid-1960s, furthered the theme, (Frankel, 1966).

Seth Spaulding • University of Pittsburgh, Pittsburgh, Pennsylvania 15261. **George V. Coelho** • Office of the Assistant Director for Children and Youth, National Institute of Mental Health, Rockville, Maryland 20857.

He suggested that goals of educational and cultural exchange be reformulated as: (1) lacing together of educational systems, (2) improvement of the context of communication, (3) disciplining and extending international intellectual intercourse, (4) international educational development, and (5) furthering of educational and cultural relations as ends in themselves. He called for an upgrading of educational/cultural relations in the Department of State so that their significance is recognized; changes in policy-making structures so that exchange is not seen from distorted agency perspectives; and the formation of new, more cooperative and binding relationships between government and private educational and cultural communities. The creation of a semiautonomous foundation for educational/cultural affairs with responsibility for planning and implementing policy abroad was also recommended.

Scholars such as Herbert Kelman further reinforced these notions. He asked, in 1965, whether international cooperation, and specifically educational exchange, has some bearing on conditions for peace, and whether research in this area has political relevance (Kelman, 1965). In answering, he defined four ways in which international cooperation and exchange may affect relations between nations and hence create conditions for peace.

He suggested that exchange activities may produce: (1) increased openness among key individuals in attitudes toward other nations, (2) reduced levels of tension, (3) increased world-mindedness and commitment to internationalist theory among participants, and (4) development of networks of relations cutting across national boundaries. Kelman recommended research that would evaluate exchange programs in order to ascertain whether they have achieved their goals, and offer insights into ways of enhancing the effectiveness of future programs.

More than a Decade Later

Over a decade has passed since Coombs and Frankel issued their articulate calls for greater recognition of cultural and educational affairs in the foreign policy of the country, and since Kelman's outline of the kinds of research we need to better understand the impact of it all. Although the students keep coming, policy recognition of the potential significance of these and related activities in the international-education arena has been negligible. The Bureau of Educational and Cultural Affairs of the Department of State, until April 1978, operating under an Assistant Secretary of State, was in charge of such matters, but its traditional role in foreign-policy formulation can only be said to have been modest. Since April 1978 the Bureau has been merged with the old United States Information Agency, and is now the Educational and Cultural Affairs Directorate of the new U. S. International Communications Agency. It is too early (as of 1980) to tell whether the new Directorate will be any more effective than the old Bureau, but the new agency's presidential mandate is encouraging: "To coordinate the international information, educational, cultural and exchange programs conducted by the U. S. Government," and to serve as "a governmental focal point for private U. S. international exchange programs." As of 1980, however, the fiscal picture is not encouraging, with Fulbright fellowships, managed by the Directorate, having been reduced in number by about half within the past decade, owing to inflation and lack of increased funding (President's Commission on Foreign Language and International Studies, November 1979).

Congress has neglected international-education interests in other ways. It has consistently refused to fund the International Education Act, enacted in 1966 to encourage and support international-education programs in American institutions, programs which provide the context for effective educational-exchange programs. Congress has, to its credit, kept alive through modest funding (no more than 20% of authorization) Title VI (foreign language and area studies) of the National Defense Education Act, often over pre-Carter administration recommendations of no funding at all.

A study by the International Education Project of the American Council on Education (1975) notes:

> Whatever promise there is in recent increased support of international programs and research by the National Endowment for the Humanities, the Fund for the Improvement of Post-Secondary Education, the National Science Foundation, and a few other agencies, these dispersed and disparate programs have not given international education the kind of firm and continuing base envisaged by the International Education Act and explored under Title VI of NDEA. (p. 13)

This report notes that private foundations have been partners in this "national inattention to international education." Since 1970, their support for international cultural and educational affairs has dwindled. The report further points out that our entire culture is "derivative and eclectic," and that what "visitors have traditionally brought to America's understanding of nature, art, and itself is incalculable." We now need a "citizenry conscious of global interrelationships and capable of questioning," and we must "transcend the self-limiting cultural definitions of reality that hobble [our] intellectual and aesthetic achievements."

The even more recent (November 1979) report of the President's

Commission on Foreign Language and International Studies, chaired by James Perkins, suggests even greater urgency than the 1975 ACE report. The Commission concentrates on the "danger of imminent collapse" of internationally oriented area centers, programs on international problems, schools for advanced studies, libraries, overseas research facilities, and senior research-exchange programs, all resources linked to the effectiveness of our cultural-exchange programs and our ultimate effectiveness as a nation in dealing with other nations. The report also examines the need for greater emphasis on language and international studies at the elementary and secondary level in this country, the need for greater involvement of the public, of labor, and of business in international education, and the need for sizable funding programs to reverse the trend toward insularity in our educational systems.

In contrast to this call for greater public awareness and understanding of other countries, our relationships with the various specialized agencies of the United Nations have been deteriorating, usually by Congressional edict, in recent years. All of these relationships affect our educational- and cultural-exchange interests. The United Nations Educational, Scientific, and Cultural Organization (UNESCO), whose main function is the encouragement of the exchange of ideas among nations in educational, scientific, and cultural domains, received no American financial support from 1974 until early 1977 because of Congressional disapproval of resolutions concerning Israel which had been approved by the Organization's 1974 General Conference. These resolutions voiced a criticism of Israel for archaeological digs in Jerusalem, rejected Israel's application to be included in the European regional grouping, and urged the Organization to supervise Arab education in occupied territories. The massive work of the Organization (on a very limited budget of about half what one medium-sized American state university spends) with countries around the world in educational, scientific, and cultural fields has been considered by Congress as being less significant than two or three politically motivated resolutions. Attempts to destroy the organization that brings us these political messages, of course, will not change the opinion of the countries which voted for the resolutions. Such action simply makes it more difficult to dialog on issues about which we may disagree.

Similar problems have plagued our relationship with the International Labor Organization (ILO), which deals with problems of labor and management in countries throughout the world and helps countries set up vocational- and technical-education programs suited to their development goals. Traditionally countries have turned to American expertise, sending many students to the United States through

ILO-sponsored programs and inviting American experts to advise them through ILO-managed technical-assistance activities. Socialist countries, however, are also members of the ILO, and one powerful American labor leader (now deceased) consistently objected to their membership on political grounds. The U. S. Congress responded by refusing to pay American assessment (established by treaty obligation) to the Organization, and on November 6, 1977, the United States withdrew from the Organization, returning only in 1979. Political issues again elbowed aside educational and cultural goals, probably to the long-term detriment of American relations with both socialist countries and Third World countries. Even WHO, which has been universally respected for its global mission in the health field and for its worthwhile and effective programs of health promotion, communicable-disease control, and education and training in the developing countries, has not been immune to politicization of its scientific and public-health role.

Academia is not without its blind spot vis-à-vis education and cultural affairs as a dimension of foreign policy. Michael Flack did a study some years ago for the U. S. Advisory Commission on Educational and Cultural Affairs which showed that international organizations were almost totally neglected in textbooks dealing with international affairs (Flack, 1971). Political aspects of the UN system (principally the activities of the Security Council), a small part of the activity of that organization and its specialized agencies, received the major attention in such texts. When the United Nations Development Program or specialized agencies such as UNESCO and WHO were mentioned, one or two paragraphs in an entire text were not uncommon. A review of the items listed in any of the information-storage and retrieval services dealing with political-science studies (articles, books, dissertations) will reveal a similar paucity of interest in the role of educational and cultural activities, or the agencies and institutions which handle them, in foreign policy.

There are at the beginning of the 1980s, however, some encouraging signs. The President's Commission on Foreign Language and International Studies (November 1979), mentioned above, proposes a comprehensive program of federal, state, and local support for international education and exchange activities. The United States has rejoined the international community in supporting UNESCO and ILO. The new International Communications Agency and its Cultural Affairs Directorate have new and broad mandates in the exchange and public-information fields. And if anything positive can be said to come from the many recent crises (Iran, Afghanistan, transnational corporation payoffs, energy, etc.), it may be that Congress and the public will recognize our interdependence in a world of limited resources, and thus begin to support, at more adequate levels, programs in international education and cultural exchange.

Research on Foreign Students

The fact that well over 200,000 foreign students a year invest their own or their sponsor's funds in study in the United States can only attest to the ability of people and organizations to find ways to communicate across national boundaries despite a relative neglect of such programs by policy levels of the government. Nor has such policy neglect diminished the interest of a number of scholars in finding out more about what happens to these students when they come to the United States. In fact, a State Department-funded project (Spaulding & Flack, 1976) identified over 500 studies and reports done largely since 1967 on some aspect of what happens to foreign students in the United States. Many of these studies were funded or sponsored by one of the many government or private agencies and institutions involved in educational- and cultural-exchange activities. A number of others in this review were dissertations, many by foreign students who themselves were on a temporary sojourn in this country. A brief review of what this research tells us may be useful in further identification of neglected policy issues in educational and cultural exchange.

One's impression on reviewing these studies is that we know a great deal about what happens to foreign students in the United States and how they are handled, and yet at the same time we know little. Most studies are situation-specific: they deal with a group of readily available foreign students at one university, or with students affiliated with certain sponsoring organizations, or with the problems of managing and supervising foreign students in one or several institutions of higher education.

At best, most studies seem to show that foreign students come to the United States with certain preconceived attitudes (usually positive), and that many are disappointed or have adjustment problems after arrival, only to become adjusted or to become more pleased with their experience toward the end of their stay (Heath, 1970; Tanner, 1968). This is often described as the "U-curve" hypothesis (first described in 1955 by Lysgaard), though some researchers (Selby & Woods, 1966) report finding a "V" curve, and at least one study (Becker, 1968) suggests that there is some evidence that students from developing countries demonstrate an inverted "U." This latter study would seem to indicate that students from the Third World tend to come to the

United Stated with numerous anxieties, but that they are pleased with their reception, only to become gradually disillusioned toward the midpoint of their stay. This disillusionment accelerates as the end of the sojourn approaches. Furthermore, "psychological time," the overall percentage of sojourn elapsed, appeared to be more important than actual time in examining "U–curve" (or inverted "U–curve") phenomena. Proposed is a theory of "anticipatory adjustment," the selective adoption of attitudes on the basis of their utility in easing the individual's adjustment to imminent and drastic changes in environment. Klineberg (1979), in a study of foreign students in 11 host countries, confirms that foreign students everywhere have similar adjustment problems, though there is no consistency in terms of a "U" curve.

Numerous other studies further attempt to identify the problems of foreign students, either in terms of psychosocial adjustment or in terms of academic success (e.g., Jammaz, 1972; Selltitz, Hopson, & Cook, 1956; Win, 1971). There appears to be some correlation between both and the degree of acculturation of an individual. If an individual attempts to hold to behavior patterns of his or her home country, and if these patterns are markedly different from those in the United States, he or she will tend to have more problems in adjusting and in pursuing academic work (Rising & Copp, 1968). However, other studies show that many concerns of foreign students are similar to those of American students, thus suggesting that students have some common mental-health problems related to culture goals no matter where the students come from (Pratt, 1956).

Attempts at theory and concept building are few and far between. Ibrahim (1968) posits a theory of attitude organization and deals with congruency patterns, particularly in terms of political attitude as it relates to student background. Klein, Alexander, and Tseng (1971) report on a long-term study taking a psychiatric approach to attitude building and adjustment. Coelho (1973) discusses mental-health problems and "coping behavior" of foreign students in cultures distant from their own. Several studies suggest a "two-way mirror" hypothesis whereby a foreign student's attitude toward his hosts is determined largely by his conception of the host's feelings about his home country (e.g., Davis, 1971). Kelman (1965) differentiates between "sentimental" and "instrumental" sources of attachment to the system that have been tested and found fruitful in explaining ideological disengagement from both home and host society while leaving cultural values most vulnerable to change during the foreign-student experience. Poole (1965) suggests a theoretical framework for the examination of foreign students' images of self, host country, and home country and the hosts' images of the traveler and the foreign country.

The Spaulding/Flack (1976) review of such research noted that it is surprising that

> many theories have, since their original formulation, elicited almost no research followup. No foreign-student research has sought to test E. de Vries' theorem of optimum cultural tension and discontinuity, Nehnevajsa's "actual *versus* anticipated events" impact approach to attitudes, or to such constructs as "culture themes" (Opler), "multiple loyalties" (Guetzkow), "cognitive dissonance" (Festinger), "self-fulfilling prophecy" (Merton), "subjective culture" (Triandis), "marginal man" theory (Stonequist), "goal shock" (Walton), "cross-pressures" (Lazarsfeld), "coping mechanisms" (Coelho), the "Sapir-Whorf" hypothesis, etc. . . . This is also reflected in the unsystematic and halting use that is made of such theory as obtains in cross-cultural communications, emotion, learning, exchange (Homans), perception (Kilpatrick), patterns of thinking and logic (Pribram, Glenn, *et al.*; Cole), etc. (pp. 282–283)

Surprising also is the limited replication among the many studies of foreign students. Although literally dozens of studies of foreign-student academic and adjustment problems can be identified, most use data-collection instruments which have little in common with other similar studies. A handful of studies of foreign-student problems have built on the Michigan International Student Problem Inventory (Porter, 1962), but, more often than not, such studies use certain standardized inventories originally developed for American students, or totally new checklists with little comparability with any others that have been used.

The major gap in research on education-abroad programs, in terms of the consequences for individuals and for institutions, is the lack of *internationally planned* and executed studies designed with a *comparative* perspective. A beginning effort, using the same instrument for questioning foreign students in 11 countries, has been reported on by Klineberg (1979). The same author, however, long active in research on educational exchange, earlier (1976) outlined a number of needed comparative studies which go beyond the questioning of foreign students on their problems and attitudes:

> [*Research*] *has on the whole concentrated on students; more research is recommended at the faculty level, in terms of both the immediate and the long-term consequences of a foreign sojourn.* The study of the long-term impact, and the relation of the sojourn abroad to the whole career patterns of the participants, appear to be particularly important. Special interest has been expressed in the impact of the foreign experience on research training and later research productivity.
>
> *A comparative study of different kinds of exchange programs.* Do they work better under private auspices than through government agencies; when they are bilateral, multilateral, or "omnilateral"; when a university creates its own enclave abroad or immerses its students in the local university system?

A comparison of the experience of those who go abroad with fellowships of various kinds (from their own government, that of the host country, universities, private foundations, etc.), and those who go "on their own" or are self-selected. What differences (or similarities) are there in motivations, the nature and extent of difficulties encountered, success or failure in university work, amount of contact with the host population, etc.?

A comparative study of those individuals, whether students or faculty, who are eager to go abroad and those who are indifferent or even hostile to the idea.

A qualitative study of those students and scholars who do not return home after their foreign sojourn. Much has been said and written about the "brain drain"; it is often forgotten that what is lost by one country may represent a gain for another, and in some cases even for humanity in general. Is there any information regarding the flow of scholars from the developed to the developing countries to balance the movement in the opposite direction? Can we arrive at any assessment of the gains resulting from the free migration of scholars?

A study of opinions and attitudes toward foreign students, also carried out in terms of international comparisons. What judgments regarding exchanges are found among: (a) professors and other "knowledgeable people," (b) the local students, and (c) the local population in the university city? Does "racism" play a significant role in such judgments?

A determination of the impact on a university of the presence of foreign students and faculty.

A study of academic success and failure among foreign students. Is failure in fact more common than among the general population of students? If it is, can the causes be identified?

A study of the mental-health implications of the foreign sojourn. This would include an analysis of personality problems, with particular emphasis on the "depression" in the U–curve, as well as the occurrence of reactions severe enough to require outside help. . . . On the positive side, attention should be paid to particularly successful cases of adaptation to the new environment, leading to a "differential diagnosis" of the factors related to good or poor adaptation, respectively.

Despite the rich literature on evaluation of exchanges, much more is needed, particularly at the international, cross-cultural level. Much could be learned by asking the same or similar questions in a number of different exchange situations. Such questions should emphasize what suggestions are offered by exchange students and scholars themselves regarding the ways in which their foreign sojourn might be improved.

Implications for Action

We can reasonably say that we know something of the problems foreign students face in the United States, but we know little about how to deal with these problems, in part because of lack of theory in the field and the lack of more comprehensive studies of the possible cultural, psychological, and social factors which impinge on adjustment. Most studies are point-in-time efforts, and only a handful make any attempt at tracing students from the time they decide to come to the United States until well after they return home. There is also a dearth of policy interest in and funding for innovative efforts to manage foreign-student programs more effectively.

The few studies that have taken a "tracer" approach tend to find that there are region and country differences among foreign students that have much to do with academic and personal adjustment and with student attitudes toward the United States, both while the students are here and with their satisfaction with their training upon return home. As might be expected, those with jobs waiting at home feel that they are more able to contribute significantly, whereas those who do not have such posts reserved for them are less optimistic. Those in the social sciences tend to be the most pessimistic about making a contribution upon returning home, and those in professional fields (engineering, medicine, etc.) and in the physical sciences tend to be most optimistic.

Most follow-up studies find little continuing communication between returned trainees and their academic community in the United States (v. Orr, 1971). From a foreign-policy point of view, such continuing interaction may be as important as the students' having come to the United States for study. Yet there is little (if any) attempt by the U. S. Government to encourage such continued interaction. No funds are available for American academic personnel to maintain contact, or to visit groups of foreign students who have returned home. No federal program exists to encourage academic institutions to provide follow-up services to foreign alumni.

On the other hand, there appears to be much interest among private groups and universities in the United States in encouraging foreign students to mingle with Americans so as to get to know our way of life. Many universities have schemes to house foreign students with American families and to encourage such students to live with American students in university dormitories. At first examination, this appears to be a good way to help foreign students adjust to the United States. A number of studies have found, however, that it may be healthier to encourage "co-national communities," within which the foreign students preserve cultural traditions and support one another psychologically, culturally, socially, and academically (Cohen, 1971; Gandhi, 1970; Hegazy, 1968; Kang, 1972). Furthermore, such groups maintain a lively interest in the home country, and probably encourage students to return home. One might hypothesize that such groups, in fact, contribute to encouraging a favorable attitude toward the United States, since they alleviate many adjustment problems.

Migration of talent ("brain drain") has been the subject of high-level policy concern in recent years, largely in response to the concerns of developing countries that they are losing human capital that they need back home. Again, American policy interest in this problem has been prompted by a reaction to the political and economic concerns of other countries, rather than by a concern for appropriate educational and cultural policies as an important continuing component of foreign policy.

Migration-of-talent studies have concentrated on reasons for migration (v. Glaser, 1974). These, in turn, include "push" and "pull" factors. Availability of employment ranks high as a "pull" factor in all studies and, inversely, nonavailability ranks high as a "push" factor. Salary is not always a consideration. If a post is available in the home country and institutional infrastructure is prepared to recognize and reward the returnee, this will often attract a national of that country even though he or she could make more money in the United States.

Family ties and perceived ability to break into the establishment are also factors which influence a decision to return home. Thus, there are both structural elements, related to the level of development of the countries involved, and personal elements which influence migration of talent.

To guide informed discussion of policy issues regarding the brain-drain phenomenon, it is useful to review the major conclusions from Glaser's study, based on a questionnaire survey of 5,500 respondents. Klineberg (1976) summarizes the major conclusions as follows:

> Many of those who stay on do intend to return home eventually; Indians, for example, tend to work abroad for a few years and then go back. This "work" might reasonably be regarded as an extension of their educational training. Students in agriculture, business, and the physical sciences are most likely to return; those in architecture, the arts, education, and languages are most likely to stay. Migration appears unrelated to social class, age, length of study, grades obtained (with a slight tendency for the least able students to be more likely to return); no differences were found in this respect between graduate and undergraduate students. Those with scholarships from their government or employers tend to return, but grants from the university abroad are more frequently associated with migration. Those who come "on their own" are more likely to stay than those who obtain scholarships. The quality of the university plays no clear role. The motivations for staying on or for returning vary greatly, and include not only financial considerations but also the degree of integration within the home community, relations with family and friends, work opportunities, patriotism, political considerations, and many others. The large majority of foreign students do return home.
>
> As has already been indicated, this represents one situation in which the desires and interests of the individual may be judged to conflict with those of this country. On the other hand, the student or scholar who stays abroad

> may make an important contribution to his adopted country, and through it to the world at large. It is striking to note the number of American Nobel Laureates in sciences and medicine who have come to the United States from abroad; it is possible that if they had been forced to stay at home they might never have had the opportunity to reach the same level of achievement. The "brain drain" is not just a phenomenon of loss; it may also in many cases represent a real gain. (pp. 66–67)

The question then arises, what should be done about migration of talent? Here is where short-term policy decisions based on political considerations can be in conflict with longer-term policies more in tune with a philosophy of world order based on intellectual and scientific collaboration. Largely unanswered are such questions as whether or not migration of talent is truly harmful to countries from which such talent migrates. From a purely economic point of view, many of these migrants contribute substantially to the income of their home countries through repatriation of funds. Many of them, if they returned home, would be jobless or at best underemployed. Even in the case of medical-profession migrants (probably the largest numbers are in the medical field), most are over-qualified for the kinds of medical services which are feasible to be offered in rural areas of developing countries where doctors are needed.

In the absence of good answers to such questions, questionable short-term solutions are proposed or imposed. Stricter immigration regulations are developed which, for instance, prohibit foreign students from holding part-time employment in the United States, employment which is often helpful in their training. Home countries impose bonding requirements to assure that trainees return, whether a job will be available or not. Quotas of all kinds are proposed.

More positive approaches, consisting of efforts to maintain good communication between foreign students in the United States and the institutions in which they are studying, and possible employers at home, are not discussed, at least at any policy level which would make a difference. Better collaborative methods of recruiting foreign students might be another approach. American faculty and/or student-brokerage organizations such as the Institute of International Education might undertake to set up counseling centers in various countries to work with governments, private business, and prospective students in identifying areas in which training is needed, and to recruit students in those areas. Such centers might, at the same time, help selected students maintain contact with prospective employers during their stay in the United States, thus helping to assure their being offered posts upon completion of their studies.

Another, perhaps complementary, approach might be for American educational institutions to form a consortium to deal with some of these problems as they affect institutional members. The consortium could help institutions in selecting students from abroad by providing field interviews on a regular basis, and could provide liaison services with foreign institutions and governments of use to both the institutions involved and the foreign students. Finally, such a consortium could help member institutions define the kinds of modifications appropriate in their degree programs to prepare foreign students for the kinds of responsibilities they will encounter upon return.

Admittedly, there are educational and cultural attachés and binational centers (often called "foundations") in many countries which try to assist with some activities. But these activities are usually underfunded and often ill-defined as to goal. They have traditionally been connected with propaganda organisms of the U. S. Government, and can hardly be said to have a focused and clearly defined professional counseling and follow-up function in the direction of building institutional capacity. The new (1978) International Communications Agency, of course, absorbs the Cultural and Educational Affairs Unit of the State Department, and this may lead to strengthening of the binational centers and related activities in the various countries.

The merging of educational and cultural responsibilities with the propaganda arm of the American overseas establishment may imply, however, that the primary reason we enter into collaborative educational and cultural activities is to encourage others to like us and imitate our way of life. Indeed, most studies of foreign students in the United States show that they go away respecting us for many of our traits, as individuals and as a nation, but that does not mean that they learn to like us, or that they accept our political and social behavior in the aggregate. It is neither realistic in terms of the evidence we have, nor reasonable, to expect exchange programs to operate as a one-way street.

Dramatic evidence of the need for a sensitive understanding of the complex nature of the foreign-student experience is presented in studies of black African students in the United States. Most African students generally admire the American economic and social system, but many are critical of our ambivalent policies toward white minority regimes in South Africa (e.g., Becker, 1973; Miller, 1967). These studies also show a clear mutual distrust between many African blacks and American blacks, which is accentuated by their encounters in the United States. Similarly, prejudices have been found among foreign, United States mainland-born, and Hawaiian-born Asians which affect interaction between these groups (Wakita, 1971).

The rather naive assumption that the prime purpose of educational and cultural exchange is to get everyone to like us as a people impedes the pursuance of a more broadly based policy of encouraging a more focused program geared to broadly-based goals of true intellectual collaboration with other countries. Intellectual collaboration leads to mutual respect, but not necessarily acceptance of everything American. And this is as it should be.

Such collaboration is important in developing a domestic understanding of the complex development and social-change problems in the world today. As a nation of immigrants, a pluralistic society which prides itself in being a leader among nations, the United States is provincial in its ways. Foreign languages are losing in popularity in colleges and universities, and international biases are hard to find in college and university curricula, except as afterthoughts for small numbers of students who wish to specialize in international affairs. Multinational corporations make nasty mistakes in dealing with their international clientele, in part because of the provincial training that its managers have received in American colleges and universities. These corporations might well invest a small percentage of their overseas earnings in appropriate cross-cultural counseling for its managers and their families. Such programs might help sensitize them to emotional and social stresses in the new environments and provide them with mental-health intervention principles and techniques. Such an investment would not only benefit the managers and their families, but would help improve overseas working relationships.

There should be a lively concern for internationalizing American college and university programs in virtually all areas, much as recommended in the President's Commission on Foreign Language and Area Studies (November 1979). One outstanding way of doing this, of course, is to encourage more collaborative faculty- and student-exchange programs. If anything, the reverse is happening. News reports from Ohio in the mid-1970s indicated that some in the state government believe that foreign students cost the state money, since out-of-state tuition does not cover the full cost of their instruction. One proposal was to establish quotas, and another to charge foreign students even more tuition than out-of-state students. Such an argument totally ignores the benefits to Ohio institutions and communities of having students from other countries and cultures on their campuses (v. Deutsch, 1970; the President's Commission, November 1979). It further ignores economic studies which show that foreign students generate large amounts of income for business and industry in a locality by virtue of the sums they bring to the locality in the form of living expenses and consumer purchases. For example, one study (Farmer & Renforth, 1971) showed that the state of Indiana received an income in 1971 of at least $17.6 million from foreign students, including $2 million which came from the federal government in the form of subsidies from the Agency for International Development (AID) and other organizations. The remainder was income from other countries in the form of expenses and purchases of nonsponsored students from their own resources. A similar study in New York (Kent, 1973) showed that the foreign student "industry" ranked eighteenth among the state's exports. The presence of foreign students was estimated to result in 6,571 jobs in New York State.

The Policy Challenges

The strength and weakness of educational and cultural-exchange activities both rest in the diversity of activities and organizations involved in them. Almost every government agency has some kind of program involving exchange of information and the training of foreign nationals. A myriad of private organizations encourage exchange activities in various ways, and many act as brokers in assisting students, faculty, and others in arranging exchange experiences.

Research activities are primarily based on university campuses, and are low-budget events conducted by students and faculty with little outside support. Few governmental agencies fund efforts to study the impact of educational- and cultural-exchange efforts, though there have been some modest efforts under the old Bureau of Educational and Cultural Exchange (now under ICA), the Bureau of External Research of the Department of State, and the Agency for International Development (AID). Funding organizations such as the Office of Education, the National Institute of Education, and the National Institute of Mental Health have placed little or no priority on such research.

One national professional association, the National Association for Foreign Student Affairs (NAFSA), provides a focal point for interests of foreign-student advisors, but this organization concentrates on problems of foreign-student selection, admission, and care while on university campuses. It rarely ventures into policy questions of the kind raised in this paper, not does it have the membership, visibility, or prestige to do so. The Institute of International Education, private but subsidized through a virtual monopoly on placement activities involving various funding organizations, concentrates solely on administering foreign-student scholarships.

There are, and have been, of course, various committees, commissions, and study projects under the sponsorship of the various government agencies, professional associations, and various associations of colleges and universities. (For a comprehensive annotated list of reports from such groups through 1972, see Spaulding & Flack, 1976).

The most influential over the years in exchange work has been the

U. S. Advisory Commission on Educational and Cultural Affairs, which operated under the Bureau of Educational and Cultural Affairs. The continuance of this Commission, now that the Bureau has merged into the International Communications Agency, is unclear. In the past, the Commission has been primarily limited to assessing government-sponsored exchange programs. Its 1963 report to Congress outlined far-reaching recommendations for improving the selection of trainees from abroad under government-financed programs, and the administration of such programs (Gardner, 1963). It also questioned the heavy reliance of such programs on "blocked" foreign currencies. If anything, the situation has deteriorated steadily since 1963, with many government-sponsored programs disappearing in countries with no more "blocked" currencies and with even the Fulbright–Hayes Program curtailed sharply.

It is unrealistic to suggest a new initiative by one of the existing vested interests in this area, with the idea that this alone would make a difference. What is needed is a constructive long-range perspective by the administration and appropriate legislation by Congress. A small portion of the money now allocated for political, economic, military, and intelligence activities abroad should be clearly set aside for educational- and cultural-exchange initiatives. These funds must be sufficient to encourage highly visible research and action, building on what has been done in the past, and focusing on how to improve our relationships with other nations through "people to people" and "professional to professional" dialog.

The International Education Act or something like it should be funded at significant levels to encourage American institutions of higher education to create sensitive American international leadership in all disciplines and professions. Technical-cooperation activities, both of multilateral organizations such as the United Nations and its specialized agencies and of the Agency for International Development, must receive significant new legislative support (many countries, including Sweden, France, and other European countries, contribute a much higher percentage of their gross national product to such efforts than we do). Support for programs to encourage follow-up dialog with returned students must be provided.

Finally, legislative authority for existing research and action programs of the various federal agencies should be reviewed with an eye to strengthening those portions which deal with these issues. For instance, the National Institute of Mental Health, the U. S. Office of Education, the National Institute of Education, and the Department of State all have units and programs which impinge on the problems outlined here. Legislation should strengthen the authorization and appropriateness of these units to proceed with major new efforts in the area of educational and cultural exchange and the study of the impact of these efforts. The President's Commission (November 1979) addresses this and similar issues, and its recommendations merit serious consideration.

In his Inauguration Address, President Carter spoke of the dialog with the world's nations on problems affecting us all. In addition, he spoke of our goal to further human and individual rights throughout the world. President Carter earlier elaborated on this theme in a 1976 address to the Foreign Policy Association. Ullman (1976) reports that President (then Governor) Carter "spoke of the need for the Western societies to learn from one another in such prosaic spheres as health care, urban planning, mass transportation, and measures to counteract unemployment, rootlessness, and alienation." Ullman comments that "in the short run, such sharing of ideas and techniques may be the most valuable because they are the most concrete achievements of tribalism. In the long run, it may be the effective protection of rights and liberties" (p. 19).

We can only hope that this vision of dialog can be translated into policy and programs that are concerned not only with "Western societies," but with all societies. Furthermore, such dialog will not achieve its goal if it is only closed dialog between government bureaucracies. It must be encouraged through broad and varied support of educational- and cultural-exchange activities involving people of all ages and in all sectors of society.

Such a program could be the underpinning of a long-term foreign policy based on principles of dialog and understanding. Its impact would surely be more positive than a foreign policy biased heavily toward weapons counts, economic blackmail, and political coercion.

References

American Council on Education. *Education for global interdependence*. Washington, D. C.: ACE International Education Project, October 1975.

Becker, T. Black Africans and black Americans on an American campus: The African view. *Sociology and Social Research*, 1973, *37*, 168–181.

Becker, T. Patterns of attitudinal changes among foreign students. *American Journal of Sociology*, 1968, *73*, 431–432.

Coelho, G. V. An investigation of the consequences of international educational exchanges. *Foreign Affairs Research Bulletin*, Department of State, October 1973.

Cohen, R. D. *The Functions of a Co-National Group of Foreign Students in New York City*. Unpublished doctoral dissertation, Columbia University, 1971.

Coombs, P. H. *The fourth dimension of foreign policy*. New York: Harper & Row, 1964.

Davis, J. F. The two-way mirror and the U-curve: America as seen by Turkish students returned home. *Sociology and Social Research*, 1971, *56*, 29–43.
Deutsch, S. A. *International education and exchange: A sociological analysis*. Cleveland: Case Western Reserve University Press, 1970.
Farmer, R., & Renforth, W. Foreign students in Indiana: Our intangible exports. *Indiana Business Review*, 1971, May/June 12–16.
Flack, M. J. *International educational and cultural relations and their treatment in international affairs textbooks*. Washington, D. C.: U. S. Advisory Commission on Educational and Cultural Affairs, Department of State, 1971.
Frankel, C. *The neglected aspect of foreign affairs*. Washington, D. C.: The Brookings Institution, 1966.
Gandhi, R. S. Conflict and cohesion in an Indian student community. *Human Organization*, 1970, *29*, 95–102.
Gardner, J. A. *A beacon of hope: The exchange of persons program*. Washington, D. C.: U. S. Government Printing Office, 1963.
Glaser, W. A. *Brain drain and study abroad*. New York: Columbia University. Bureau of Applied Social Research, 1974.
Heath, G. L. Foreign student attitudes at International House, Berkeley. *Exchange*, 1970, *5*, 66–70.
Hegazy, M. *Cross-Cultural Experience and Social Change*. Unpublished doctoral dissertation, University of Minnesota, 1968.
Ibrahim, S. E. M. *Political Attitudes of an Emerging Elite: A Case Study of the Arab Students in the United States*. Unpublished doctoral dissertation, University of Washington, 1968.
Jammaz, A. I. A. *Saudi Students in the United States: A Study of their Adjustment Problems*. Unpublished doctoral dissertation, Michigan State University, 1972.
Kang, T. S. A foreign student group as an ethnic community. *International Review of Modern Sociology*, 1972, *2*, 72–82.
Kelman, H. C. (Ed.). *International behavior*. New York: Holt, Rinehart & Winston, 1965.
Kent, J. Foreign students: An economic approach. *Post-Secondary Education in New York State*, 1973, *1*, 1.
Klein, M. H., Alexander, A. A., & Tseng, K.-W. The foreign student adaptation program: Social experiences of Asian students, *Exchange*, 1971, *6*, 77–90.
Klineberg, O. *International educational exchange: An assessment of its nature and its prospects*. Publication of the International Social Science Council. The Hague: Mouton, 1976.
Klineberg, O., & Hull, W. F. *At a foreign university: An international study of adaptation and coping*. New York: Praeger, 1979.
Lysgaard, S. Adjustment in a foreign society: Norwegian Fulbright grantees visiting the United States. *International Social Science Bulletin*, 1955, *7*, 45–71.
Miller, J. C. *African Students and the Racial Attitudes and Practices of Americans*. Unpublished doctoral dissertation, University of North Carolina, 1967.
Orr, J. D. *The Foreign Scholar Returned Home: A Review of Selected Research*. Unpublished doctoral dissertation, Columbia University, 1971.
Poole, I. Effects of cross-national contact on national and international images. In H. C. Kelman (Ed.), *International Behavior*. New York: Holt, Rinehart, & Winston, 1965.
Porter, J. W. *The Development of an Inventory to Determine the Problems of Foreign Students*. Unpublished doctoral dissertation. Michigan State University, 1962.
Pratt, C. The relation of culture-goals to the mental health of students abroad. *International Social Science Bulletin*, 1956, *8*, 597–604.
President's Commission on Foreign Language and International Studies. *Strength through wisdom: A critique of U. S. capability*. Washington, D. C.: U. S. Government Printing Office, 1979.

Rising, M. N. & Copp, B. M. *Adjustment experiences of non-immigrant foreign students at the University of Rochester, 1967–1968.* Rochester, N.Y.: University of Rochester, 1968.
Selby, H., & Woods, C. Foreign students at a high-pressure university. *Sociology of Education*, 1966, *39*, 138–154.
Selltitz, C., Hopson, A. L., & Cook, S. The effects of situational factors on personal interaction between foreign students and Americans. *Journal of Social Issues*, 1956, *12*, 33–55.
Spaulding, S., & Flack, M. *The world's students in the United States: A review and evaluation of research on foreign students.* New York: Praeger, 1976.
Tanner, S. *An Investigation of Friendship Patterns of Foreign Students.* Occasional Paper No. 1, University of Michigan, International Center, 1968.
Ullman, R. H. Trilateralism: 'Partnership' for what? *Foreign Affairs*, 1976, *55*, 1–19.
Wakita, K. *Asian Studies Survey, Spring 1970.* Los Angeles: Los Angeles City College, October 1971.
Win, U K. *A Study of the Difficulties Indian and Japanese Students Encountered in Six Problem Areas at the University of Southern California, 1969–1970.* Unpublished doctoral dissertation, University of Southern California, 1971.

V

Stressful Situations of New Settlers: Coping Strategies of Immigrant Women and New Ethnic Groups

Introduction

The adaptive tasks of young, unaccompanied students who plan a temporary stay abroad for study and training in the United States are different from the corresponding tasks of the prospective immigrant who intends to take up permanent residence in the host society. Their situation, resources, and motivations are different. This section examines specific spheres of stress that are faced by new settlers and their families. The problems of coping include changes of economic role, change in social status, cultural differences in perceiving and feeling close relationships, uncertainty of employment prospects, practical emergencies, and loss of emotional and social-support systems which have been provided by kin and friends. Although certain stressful situations are common in every immigrant's adaptation process, these studies of different cultural groups from South Asia and Latin America illustrate how the impact of these changes varies for different groups, depending on their preparation, their previous socialization experiences, and differential access to coping resources. The authors discuss the sociocultural, situational, and psychological variables which influence the use of coping strategies.

In Chapter 16, Cohen examines the coping strategies of Latin American aliens in Washington, D. C. In comparison with male immigrants, women appear to be able to cope much better with the strains of their uncertain legal status and precarious economic position. Cohen's study showed that men and women, both unemployed or underemployed, also exhibited high levels of emotional stress. Most of these Latin American immigrants (approximately two thirds are women) had moved to the United States in the expectation of gaining better employment in America and of raising the social and economic level of their family. Their goal was to reunite the family in the United States. They appear to tolerate the burden of unpaid, physically demanding labor as a way of coping with the "sacrifice" in order to achieve these goals.

Cohen suggests that cultural predispositions play an important role in shaping coping patterns. For example, the control of negative feelings *(controlarse)* involves resigning oneself to events and avoiding thinking too precisely of one's problems. This is an important coping mechanism. In evaluating her job, the immigrant values a respectful treatment by her employer even in an unskilled job. In other cases the source and status of their admission as immigrants are also related to the choice of strategies of adaptation.

The importance of one's cultural heritage is brought home in Chapter 17 by Saran, who has studied the Asian Indian immigrant population in New York City. He finds that, though structural adaptation to living in the United States is quite advanced, there is still a tendency to preserve the traditional observances of marriage, child-rearing, and family-role relationships. Thus, ties with family members in the United States and other compatriots are important in the daily round of living. Many of the respondents in the study rely upon their co-nationals for emotional, financial, and social support. The maintenance of Indian family values is considered to be extremely important, and those who have become superficially "Americanized" are criticized for becoming self-centered, selfish, and materialistic. As they overextend themselves in the new culture, they may be unable to fulfill traditional obligations, such as being present for the death of an important family member or at the marriage of a younger member of the family. The resulting conflict is another source of guilt and tension. Deviations from traditional Indian mores are criticized, though such deviations are gradually adopted by the younger generation.

In Chapter 18, Nair reports on his study of over 200 immigrants, mainly from Asia and Europe, in Toronto. He divides them into two categories: (a) *independent immigrants,* who applied for an immigrant visa on their own, and had to meet rigorous selection proecedures based on educational attainment and desirability of skills; (b) *dependent immigrants,* who have to produce a spouse's affidavit of support, indicating that a friend, relative, or employer is responsible for their care, accommodation, maintenance, or employment. Several differences are noted between these two categories. Independent immigrants make efforts to gain advance information about Canada before arrival, whereas dependent immigrants rely on information provided by relatives. Independent immigrants also obtain realistic information about available social services. Those from industrialized societies also have a clearer understanding of the role of social services, and they use these services to seek information about employment opportunities, accommodation possibilities, and language-training facilities. Although these institutions provide adequate assistance for immigrants, many immigrants tend to avoid contact with governmental agencies, and look to compatriots for help and advice. One of the important functions of an ethnic community-support system, and especially family members, is to provide a solidarity of mutual obligations and aid in the environment.

Another important support system is the close-friendship network. In Chapter 19, AmaraSingham analyzes two contrasting patterns of friendship among Indian women immigrants in Boston. Two distinctive styles appear to be related to the degree of commitment to living and settling down in the United States. According to the author, owning a house in the suburbs, and adopting the American emphasis upon individualistic rather than familial definitions of achievement and relationships, constitute important markers of the commitment to permanent residence in the United States.

One respondent initially had a small friendship circle. On moving to the suburbs, however, she began to make friends with women who did not share her cultural background. Friendships therefore became a facet of her individuality, rather than an extension of her family. The other respondent modeled her friendship circle on the close friendships she knew in India, and restricted herself to intimate relationships with coimmigrants who lived near her. Although this group is smaller and less homogeneous than her circle had been in India, it is nevertheless an extension of the socialization patterns which had been learned in India. In addition to retaining the concept of adaptation as a continuing process, the articles in this section focus on personality and situational variables which may determine strategies of coping in new cultural environments. These discussions emphasize that close friendship bonds and kinship ties are important in mitigating tensions arising from dissimilar or conflicting cultural-value orientations.

In summary, these studies of adaptation to new environments suggest that survival skills and competencies that are learned effectively in one society may not immediately or easily be transferred to another. Adaptation is facilitated if there are additional social supports and community networks that provide ready acceptance in a manner which respects the newcomers' positive contributions and presence.

16

Stress and Coping among Latin American Women Immigrants

Lucy M. Cohen

This chapter focuses on social and cultural factors and the adaptive patterns of Central and South American immigrants in Washington, D.C. It concentrates on two areas: (1) an understanding of linkages between levels of stress and their correlation to sociocultural characteristics, and (2) the identification of patterns of conflict resolution used by Latino newcomers to cope with the demands of a new urban environment. Since a sizable proportion of these immigrants are women, the findings presented draw heavily on the feminine experience.

In recent years, there has been a growing recognition that the period since World War II has brought new immigrants to the United States who do not fit the "ideal type" of peasant and immigrant depicted in major works about the settlement and adjustment of earlier newcomers. Passage of the landmark Immigration and Nationality Act of 1965 (PL 80–236, 79 Stat. 920) reminds us that Americans have now chosen a policy which gives priority to *what immigrants do,* rather than to *who they are.* For some regions of the world, such as Latin America, however, this policy obscures the realities of what it means to enter the United States.

Whereas the typical pioneers of immigration in the past have been

This article is adapted from a study sponsored in part by the Center for Minority Mental Health Programs, National Institute of Mental Health (Grant 21725).

Lucy M. Cohen • Department of Anthropology, Catholic University, Washington, D.C. 20064.

men, approximately two-thirds of the Central and South American newcomers are women. Moreover, they are not widows or young single persons who venture on long voyages to the promised land. Rather, they are mostly women who established households in their places of origin, and then left children behind, under the care of maternal grandmothers or other kin. For both the women and the men from these areas there is limited information available on their processes of migration to the United States, even though there has been a resurgence of research on the lifeways of the peoples of Hispanic heritage in this country.

The research upon which this chapter is based was drawn from a study of cultural influences on patterns of stress and illness among Central and South American immigrants. (See Cohen, 1975, 1977, 1980; Cohen & Fernandez, 1974.) The investigation focused on three areas: (1) the study of beliefs and perceptions about disease and the practices followed in the management of illness; (2) the identification of levels of stress by sociocultural characteristics; and (3) description of selected patterns of conflict resolution. This article presents findings about stress and conflict resolution to provide insight into the processes which contribute to vulnerability and symptom development among some Latinos, and resilience and mastery over stress among others. Research on the social-psychiatric impact of conditions of settlement and the identification of culture-specific coping patterns can broaden our theoretical perspectives on the problems of social and cultural change. Furthermore, as decision makers give increased attention to the world-wide implications of the phenomena of international migrations, their policies on prevention of family breakdown should benefit from systematized research on the subject. With sensitivity to the at-risk factors which contribute to the prevalence of stress among women and men, resettlement and counseling agencies can also become increasingly effective.

In the past three decades there has been a growing interest in epidemiologic analysis in the field of mental health to determine correlations between overall symptom scores and selected characteristics of population groups. These data have broadened our knowledge of the etiology and distribution of mental health problems. To my knowledge, however, there have been few efforts to study stress among people of Latin American heritage in the United States.

A specific aim of this research, therefore, was to measure levels of stress and to correlate stress with other sociocultural characteristics, such as occupational levels, marital status, and sex of respondents. In approaching this aspect of the study, the writer was influenced by the work of A. H. and D. C. Leighton and their colleagues in the Stirling County Study, which suggests that the development of symptoms of psychiatric disorder is a result of interference with a person's strivings for the satisfaction of certain basic needs, interference which may come from within the individual, or from outside environmental forces. (See Hughes, Tremblay, Rapoport, & Leighton, 1960; Leighton, 1959; Leighton, Harding, Macklin, Macmillan, & Leighton, 1963.) The development of symptoms during attempts to cope with distress is common. Psychoneurotic and psychophysiologic symptoms such as anxiety, depression, pounding heart, and "stomach troubles" have been found to be frequently exhibited by people who suffer mild emotional upsets. (See Dohrenwend & Dohrenwend, 1974; Hinkle, 1973; Levi, 1971; Meyer, 1951; Selye, 1956.) The identification of high-risk and low-risk groups among immigrants has theoretical implications, and it can also provide important knowledge for the extension of mental health services.

As findings regarding indicators of stress have begun to emerge, a critical complementary dimension that requires attention is the process through which conflicts resulting from interference with basic strivings are resolved. Conflict-solving mechanisms are guiding forces in the behavior patterns followed by individuals as they face the inconsistencies and contradictions of their lives. With the help of these mechanisms, individuals respond to the perception of a threatening condition, and they decide on potential avenues for its solution or mastery. (See Lazarus, Averill, & Opton, 1974; White, 1974.)

The culture of any human group offers its members guidelines about what to do in the face of the problems and difficulties they encounter in daily life. Defenses, mastery, and coping are mediating mechanisms which help individuals to deal with major and minor problems of adaptation. Following White's definition, a defense is an "adaptive response in which present danger and anxiety are of central importance." Mastery is an adaptive response to problems having a certain cognitive or manipulative dimension, but which at the same time are not heavily weighted with anxiety. Coping refers to adaptation under relatively difficult conditions. (White, 1974, pp. 48–49.)

Each culture provides a framework to guide individuals as to strategies that should be used when men and women face difficult problems. Differing demands and experiences in cultural systems may result in variation in the mediating mechanisms relied upon to resolve problems. As a result, for example, contrasts may be found in the ways in which different peoples express affect or emotionality, or in the extent to which direct or indirect action is used to resolve conflict.

The Study Population

The total of 97 respondents in the population selected for this study included 71 women and 26 men.[1] The population was drawn from two sources. The first was a community group of known seekers of health service from a multipurpose community center. The second was composed of the parents of children from the two schools in Washington with the highest proportion of Spanish-speaking children.[2]

Over half (53.1%) of the School parents had been in the United States six years and more, whereas only 16.7% of the Community sample had been in the United States for that length of time. Whereas most School parents were permanent residents or United States citizens, 41.7% of the Community respondents were undocumented workers.[3] For purposes of comparison the health status of the School group was unknown prior to research. The School parents were selected because they were assumed to be a more stable population than the Community respondents.[4]

Methodology

The major data-gathering instrument was a three-part interview schedule with sets of items on sociocultural components, biomedical

[1] Women who entered the United States in the year ending June 30, 1974 constituted well over half of the immigrants in the 20–39 age group from the Central American countries, Panama, and Colombia. In this same age group from Mexico, Peru, Ecuador, and the Dominican Republic, men formed slightly higher proportions than women (United States Department of Justice, Immigration and Naturalization Service, *1974 Annual Report*. Washington, D. C.: U. S. Government Printing Office, 1975, pp. 45–46). These proportions include those with established immigration status, and not other groups of aliens such as students, temporary visitors (e.g., tourists), or undocumented workers.

[2] Forty percent of the children in the first school were from households with parents of Latin American origin, and 53.4 percent of the children in the second school were of similar origin.

[3] The undocumented worker, also called illegal alien, is a category which includes: those who enter through border points without proper papers; visitors or students who overstay the terms of their nonimmigrant status; or seamen who desert ship (U. S. Department of Justice, p. 15). The usual Spanish word for this category is *indocumentado* (without documents).

[4] Over half of the respondents were from Central America (57.7% from El Salvador, Guatemala, and Nicaragua), and approximately 1/3 came largely from the Andean area of South America (34.0% from Colombia, Venezuela, Ecuador, Peru, and Chile). The rest (8.2%) were from Mexico, Puerto Rico, and the Dominican Republic. This choice was based on the proportions from Central and South America estimated for the Washington Metropolitan area.

information, and behavioral aspects. For the study of stress, it included the 20-item Health Opinion Survey (hereafter called the HOS) constructed for use in the Stirling County Study of psychiatric disorder and sociocultural environment.[5] The present investigation was used as a basis for assessing some of the issues involved in the adaptation of this instrument among groups of Latin Americans. For purposes of the present study, in consultation with other collaborators, the author translated the 20-item HOS. The translated versions were field-tested among persons with national backgrounds similar to those of the respondents in the study, and among a small group of health caregivers from these same countries.[6]

A group of forty persons from the total sample was studied over the period of one year, with the aim of developing a more detailed understanding of their way of life. Of particular interest were their perspectives about ongoing problems of disease and its management, and about the resolution of conflict. The author also conducted three months of field work in Colombia and El Salvador, in the places of origin of 14 of the respondents, and carried out semistructured interviews with immigrants who had gone back home,[7] and with relatives and friends of the Washington residents. Observation in small town and in urban health-care settings, and discussions with scientific and lay practitioners, offered information on the changing nature of health-care delivery in these areas.[8]

All interviews, with one exception, were conducted in Spanish by the author and two collaborating interviewers. Data from the schedules were coded and punched on data-analysis cards. There were seven cards per person. Frequency distributions and means calculated for the

[5] The 20-question HOS was derived from a set of 75 questions prepared by A. Macmillan for use in the Stirling County Study. The original test was built up from several sources, including questions from the Army's Neuropsychiatric Screening Adjunct, and others from post-World War II screening instruments which were reported to be useful neurotic discriminators. Questions concern chiefly psychoneurotic and psychophysiologic symptoms. (For more detailed history and assessment, see Macmillan, 1957.)

[6] For earlier uses of the HOS among peoples of Hispanic heritage, see Kellert, Williams, Whyte, and Alberti, 1967 and Rogler and Hollingshead, 1965. I am grateful to professor William F. Whyte, Cornell University, for his assistance in making available part of the HOS version used in Peru.

[7] These included two undocumented workers who had been deported and a woman waiting for her permanent resident visa.

[8] The author was born in Costa Rica, has kinship ties in El Salvador, and long-standing interest in Colombia, demonstrated through work experience in the country and research in the areas of culture change, medical care, and ethnohistory.

HOS were done on a PDP DECsystem-10 computer.[9] The field data and selected documentary material were content-analyzed to permit the identification of themes of central relevance for the major subject areas. Detailed case-by-case analysis offered a rich source for the study of the processes through which the immigrants adapted and faced their problems.

The scoring of the HOS was done according to procedures recommended by those who had worked in the Stirling County Study. The HOS has a simple scoring system with a total range of 20 to 60. There are standard questions as to symptoms, and standard answers—either Yes/No, or Often/Sometimes/Never. The Yes or Often ("sick") answer receives a score of 3, the No or Never ("well") answer a score of 1, and the intermediate answer (or no answer) a score of 2. The range of scores, 20–60, thus shows that the lower scores are usually associated with the absence of psychiatric involvement and the higher scores with evidence of "psychiatric disorder." (See Leighton & Cline, 1968.)

Three levels of HOS scores have been used in several studies to suggest possible clinical significance:

Normal range	20–29
Borderline (middle)	30–34
Increased stress (high)	35–60

Whether these levels should be modified for Latin American subjects is not known at present.

Patterns of Entry and Settlement

In this research, a number of women led the way in the movement of entry to the United States. Moreover, both the women and the men had already begun to establish their own households in Latin America prior to immigration, and thus they were separated from children, spouses, or other relatives for whom they had assumed some responsibility. At the time of this research, 48.5% of the total group was married. A total of 23.8% were divorced, widowed, or separated, and 16.5% were single parents. The only single male in the single-parent

[9] Interviews were conducted by Carmen L. Fernandez, M.S.W., Rita L. Ailinger, M.S.N., Ph.D., and the author. Antanas Suziedelis, Ph.D., and Mary Louisa Luna, M.A., assisted with computer analysis. Jane Nakayama Cole, M.A., offered research assistance throughout the study.

category was divorced, but he remarried toward the end of the study.

The history and sequence of migration followed by respondents and their "significant others" show that in 68.9% of the cases, a female had been the first of the family to come to the United States. Magdalena Torres,[10] a 33-year old immigrant, exemplifies the planning and initiative exercised by women as they become the organizers and counselors for other relatives who follow them to this country. She works as a beautician, though she entered as a domestic with an American family with whom she had worked in her home country. Six months after arrival she brought one of her sisters to Washington, and a second sister followed a year later. Three adult nephews—sons of the sisters—entered next. She then succeeded in convincing her mother to come to visit them. She and the two sisters and their families settled in apartments located in the same block. Magdalena and her husband were married in the United States. However, both were born in the same home country, and they had known each other there. Both had children by previous marriages. Her husband left his children by the first marriage in his home with their maternal relatives, and Magdalena brought her child to the United States after she had settled here. At the time of this research, she was involved in helping three other nephews to come to Washington.

The migration patterns of the undocumented workers were different in some ways from those of the established immigrants. Women who crossed the borders without documents did not usually bring young children with them, since this was considered too grave a risk. Subsequent to entry, such mothers brought their children to Washington only when they felt that their job and living situations were stable enough to permit it. The restrictions in immigration make it difficult for parents who feel unsettled in their designated visa categories to have their children join them.

Since Latinas frequently first enter the country after they have begun to establish their households, the issues of substitute child care and the anxieties experienced due to crises in the lives of absent children are realities to which these immigrants frequently addressed themselves. Findings about the place of residence of children showed that 43.0% of all parents had some children remaining in their countries of origin. Typically, the Community parents, who were in the early phase of settlement, had a greater proportion of children age 17 and under left in the places of origin than the established School parents.

[10] All names used in this article are pseudonyms.

The more settled the immigrants are in Washington, the greater the number of their children that are with them.

The women immigrants and couples with younger children left behind depended almost exclusively on the maternal grandmother for child care. Although there were not very many adolescents in the home countries, this was a stage when the caretaking role began to shift to a wider network of kin, notably a woman's married daughters or sister. Upon the mother's departure, a child moved to the maternal grandmother's household. This move often meant contact with a wider group of maternal kin than had been typical in the parental household. These findings suggest that the immigrant's separation from children in the early and late stages of childhood serves to solidify the child's kinship ties with the maternal line.

During field work in Latin America, the author visited some of the homes with children of immigrants in Colombia and El Salvador. Observation of child-care patterns suggested contrasts, as noted in the households of Esperanza Lopez and Prudencia Sanchez, parents of undocumented workers in Washington. These contrasts appeared to be influenced in part by the ages of the children left at home, by socioeconomic status, and by views about the meaning of migration.

Esperanza Lopez, a 55-year old woman, and her husband lived by themselves, but were caring for their daughter's three-year-old child. Their daughter and her husband had left for Washington a year after the birth of this first child. Esperanza had hired a young maid to care for and play with the youngster while she and her husband tended to their business enterprises in town. Esperanza and her husband said that they hoped their daughter would find some way to take this child to the States, or, if not, that the daughter and husband should return to their home town.

Prudencia Sanchez, who was 62 years old, lived with her husband, one unmarried son, and eight children, age 17 and under, who belonged to three of her children who had gone to work in Washington. During the author's visit, Prudencia expressed concern about her ability to cope with a recurrent intestinal problem and the signs of weakness *(debilidad)* found among some of the children. She also discussed her own long-standing problem with her nerves. Her hopes were that, as the children got older, they would join their parents in Washington. As a matter of fact, in the year following the author's visit to this home, the oldest child (an 18-year-old girl) and her new husband did come to Washington, but as undocumented workers.

These illustrations emphasize that the support of kin is crucial for immigrants with children left behind. Nevertheless, during periods of crisis, the substitute caretaker as well as the immigrant parents have to face the challenge of mastery over the problems at hand. Typical patterns of conflict resolution may be noted in reference to Esperanza and Prudencia.

When the author visited these two grandmothers in Colombia and El Salvador, she was instructed by their daughters in Washington to observe carefully the state of health of their absent children, and to identify troubles and worries with which their caretakers at home needed possible assistance. These visits raised questions for me about the impact of substitute parenting on the caretakers themselves. The grandmothers, as caretakers of smaller children and youth, carried multiple burdens which were not easy for them to assume. Esperanza, for example, was still grieving over the death of her mother, whom she had lost prior to my visit. Prudencia's responsibility for the eight children of her absent daughters and son in Washington, for an alcoholic husband, and for a son with the same problem, was recognized as the source of her continuous suffering from "nerves."

While grandmothers or sisters coped with responsibilities of substitute caretaking, immigrant mothers searched for ways of dealing with anxieties and distresses which they associated with separation. The use of medical practitioners and medicines for the relief of symptoms of anxiety was frequently cited. An immigrant woman, Cruz Nuñez, for example, described a two-week bout with insomnia, for which she hoped to receive medical relief from a general practitioner. Upon more detailed discussion, she noted that sometimes she "thought too much" (*a veces uno piensa mucho;* the connotation is to think too much of a problem). She had several sons in her home country under the care of a sister, and one of these boys was a school dropout. The sister at home had found a boarding school for this boy, but she needed money for his board and tuition. Cruz, however, was hard-pressed for money. Her problems had become accentuated by the recent unemployment of her 20-year-old son, who was with her in the United States. He had been laid off after a one-night absence from his job. Actually, he had missed work to attend his graduation from a high-school-equivalency program.

Thus, resettlement frequently calls for initiative and careful planning around the care of children. In the early stages of entry, parents tend to separate from younger children. The strains experienced by the maternal kin who assume substitute child-care responsibilities are associated with the burdens of multiple tasks. These may be resolved when the children of immigrants rejoin their parents in the new country. The newcomers, however, often try to cope with worries about absent children by "not thinking too much" about the problems of those left behind. This culturally influenced pattern of coping with anxiety by attempting to avoid disturbing thoughts may not always be successful. Some parents, for example, may develop physical complaints for which they hope physicians will offer symptomatic relief through the suggested use of prescription medicines.

Commitment to Work

Most of the immigrants enter with the goal of working in order to improve the family's socioeconomic status. Among these working-class Latinos there are marked contrasts in educational levels by sex. Most of the men have completed their education at the primary school level, in contrast to the women, who tend not to have finished primary school. These differences are reflected in the type of work and annual income of the male and female Latino immigrant, accentuated by the structure of the labor force in the United States.

Men and women tend to work in unskilled or semiskilled jobs, but there is greater occupational mobility for men than for women. Furthermore, the Latina who has been previously employed in her home country tends to be underemployed in the United States to a greater extent than men. Women tend to fall into the $3,999-or-less annual income category, whereas men are concentrated in the $4,000–$5,999 group.

Most men and women work full time; and of the full-time workers, a sizable proportion hold down "moonlighting" jobs as well. Those who work extra time are usually the immigrants with more limited knowledge of English. Women who work overtime regularly are mostly single parents with an extended household. The men have working wives, and are under heavy expenses to purchase homes.[11]

The respondents with heavy work and family responsibilities frequently pointed to their belief that *parents had to sacrifice themselves for their children or loved ones.* Difficult types of work or the burdens of long hours were tolerated by appeal to this belief, as noted in the

[11] At the time of this study, 73.1% of the population held full-time jobs, and 12.3% worked part-time or sporadically, leaving 14.6% who did not work at all. The latter included persons in training programs or actively seeking work. Only one person in the study was on public assistance.

following illustrations in the Estela Leon and the José Ramos families. Estela was a single mother living with her three daughters (22, 20, and 14 years old), a female cousin, and a grandson. Her two teenage sons remained under the care of an aunt in Central America, a woman who has cared for the children of all her siblings who have emigrated to Washington until the parents were ready to send for them.

Estela's annual earnings fell in the $3,000–$3,999 range. She worked on a Monday–Friday schedule, which started every day with the 6 A.M. to 3 P.M. shift in a cafeteria. She came home and left again for the 6 P.M. to 10 P.M. shift in a janitorial service, where she cleaned buildings. Her 22-year-old daughter worked as a waitress, supporting her year-old infant, and she helped with the expenses of the household. The 20-year-old daughter was in a special training program to improve her clerical skills. The three women were saving money to bring the remaining two boys to Washington. Estela indicated that she frequently felt tired by the rush of this schedule, and her responsibilities at home. She hoped, however, that when she succeeded in bringing her remaining two sons from her home country to Washington her sacrifices would be rewarded.

In the case of José Ramos, a man with a heart problem, the family had moved to a small home in the suburbs, and they had a number of financial obligations. Although he and his wife harbored some fears about the threats of a heart attack (José remembers that his father had died of congestive heart failure at the age of 35), they retained the belief that parents had to sacrifice themselves for the welfare of their children.

Self-sacrifice was a sustaining motive in tolerating the burdens of a heavy commitment to work. This motive of self-sacrifice also helped Latinos to cope with the difficult-to-tolerate aspects of a job. The most important aspect of working conditions for the immigrants was their relations with employers and/or colleagues, in terms of *buen trato* (proper and good treatment) or *mal trato* (ill treatment, or lack of consideration).

Buen trato reflected the display of appropriate *respeto* (respect), but above all the according of *dignidad* (dignity) to an employee. This is a core value of traditional Latin American society. *Dignidad* gives worth and respect to persons, regardless of their status in the social hierarchy. This value was particularly meaningful in relation to the cultural background of immigrants and the types of jobs in which a good proportion of them worked in Washington.

When the bonds based on *buen trato* had been established and intensified, employers or supervisors frequently became the trusted advisors of the immigrants, and were sought out for solutions to various problems, depending on the particular phase of settlement. Among household employees, both live-in workers and day workers often described the mode of treatment by the women for whom they worked. Employers who had accorded *buen trato* to immigrants during the initial years in Washington not only had helped with translations and the preparation of documents, but also had suggested resources to meet various problems, including personnel for needed health care. In addition, such household employers often gave valued information about other homes in which a family wanted hired help, thus assisting Latinas in their efforts to find jobs for relatives and friends.

When immigrants complained of what they considered *mal trato,* they often pointed to problems associated with social status, sex-role relations, or contrasting cultural backgrounds as the source of negative experiences. Some Latinas felt forced to tolerate *mal trato* because they could find no alternative work options. *Mal trato* under these circumstances became a source of suffering *(sufrimiento),* which some respondents associated with the plight of the poor. It should be noted, however, that some immigrants had limited tolerance for the lack of *buen trato,* and left their jobs in search of more favorable working environments *(buen ambiente).*

Eugenia Suarez, an elementary school teacher in her country of origin, had dreamed of earning enough money to buy a house there for her family. She was unmarried, and enjoyed helping her parents and siblings. On arrival in Washington, she found that she could not be a teacher because of her deficiencies in English. A friend suggested that she work as a waitress in a club. This first work experience taught her that she would have to approach work in the United States ready to withstand the "suffering" associated with continuous orders from disrespectful bosses and clients. She decided not to lose status in this host society in which the value of honor and the concept of *respeto* in sex-role relations appeared to be absent. To protect herself from men who made promises to secure a resident visa for her in exchange for sexual favors, she decided to adopt a special fictive role. She chose that of a "poor" mother who had been forced to leave her children in her place of origin and had come to earn a living for them. She feels that this role has helped her to maintain *dignidad* and to protect her honor, not an easy thing to do in Washington, a city where she believes that it is difficult for a Latina to maintain *pudor* (modesty, shame). Nevertheless, she has attributed a rather sudden onset of *reumatiz* (rheumatism) to the strains imposed on her by the lack of respect toward her shown by the men at her job.

Mercedes Lopez was less tolerant of work than Eugenia Suarez.

One day she mentioned to me that she did not feel well because of the circumstances under which she had quit her job as a restaurant table girl. A new male supervisor had been assigned to the group of table girls who worked together. They did not like him because he sat "drinking coffee" instead of helping them during the rush hour serving periods. The girls complained to the manager, who yelled at them, saying that if they did not like the situation, they were free to go, and pointed in the direction of the door. Mercedes left the job the same day. She expressed regrets about this action, because up to that point the manager had shown *buen trato* toward employees. Retrospectively, she thought that the issue was, perhaps, that Latinas worked too hard, and that, in contrast to blacks in the city, who received praise for their efforts, the work of Latin Americans went unrecognized.

These data indicate that successful settlement in an individualistically oriented society requires a heavy commitment to work. Latinas frequently refer to the altruistic principle of self-sacrifice as a supportive force for the tolerance of difficult conditions at work. In light of the reality that they are less educated and earn lower wages than men, self-sacrifice appears to be a useful coping mechanism. The cases of Eugenia Suarez and Mercedes Lopez suggest, nevertheless, that when employers threaten such core values as *respeto* and *dignidad,* women experience strongly felt negative reactions, and the impact of self-sacrifice is reduced. Eugenia believes that her rheumatism was precipitated by *mal trato,* while Mercedes prefers the strains of temporary unemployment to the toleration of a disrespectful employer.

Sociocultural Factors and the Prevalence of Stress

The problems of the family and work which immigrants encounter during settlement can thus contribute to increased sentiments of impairment. The patterns of adaptation of Latinos are shaped by their cultural tradition, the specific role demands of their positions in the life cycle, and their individual life experiences. To understand specific relations between demographic and sociocultural characteristics and adaptation, this section focuses on sociocultural factors and stress. The chapter concludes with a discussion of *controlarse,* a major component which Latinos draw upon to govern the management of stress.

A number of mental health surveys show a high prevalence of psychiatric symptoms correlated with certain demographic and sociocultural conditions. A. H. and D. C. Leighton indicate that the majority of these disorders are minor, involving persons who are impaired to no more than a mild degree, rather than the severely incapacitated or psychotic. They emphasize the importance, nevertheless, of giving attention to data which identify levels of impairment, since even minor disorders may interfere to a significant extent with the expectations and activities of daily living. (See Leighton, 1959, 1976; Leighton *et al.*, 1963.) The author's specific interests, however, were related to the identification of high-risk and low-risk groups in stress situations, to complement qualitative findings about culture, symptoms of malfunctioning, and their management.

Comparisons between the symptom patterns of immigrants in the Community group and those in the School parent group were of special interest. On the whole, respondents in the Community were believed to be at greater risk because they had lived in the United States for a shorter period than the School parents. (As indicated earlier, over half, or 53.7%, of the School parents had been in the United States six years and more, whereas only 16.7% of the Community group had been in the country that long.) Four out of 10 persons in the Community group (41.7%) were in the unsettled undocumented worker status, and this factor could only help create insecurity or anxiety. Most School parents, on the other hand, were no longer faced by the demands of initial settlement in the city. Although the health and mental health status of the School parents prior to this research was unknown, it was believed that, as established immigrants, they would have lower levels of stress than the Community group.

Table I shows that more than four times as many respondents in the Community sample of recent immigrants as in the School parent group were rated in the high-stress levels. Twice as many Community respondents as School parents were in the intermediate-stress level. It seems important to emphasize, nevertheless, that the impairing impact of recent entry needs to be understood along with *specific* sociocultural variables which contribute to the emergence of psychiatric disorder, as discussed below.

As expected, respondents with a reported health problem had a higher level of stress than those who were not experiencing such problems (Table II). A finding of special interest among the higher-

Table I. Distribution of Stress Levels by Groups (Percentages)

	Type of Group	
Stress level	Community (n = 48)	School (n = 49)
High stress	18.8	4.1
Intermediate	25.0	10.2
Normal range	56.3	85.7

Table II. Distribution of Mean HOS Scores by Total Group, Sex, and Presence or Absence of Health Problems

Attribute	Mean score	Highest score	Lowest score
Mean score for total group (n = 97)	27.8	54.0	20.0
Mean score by presence or absence of health problems			
"Yes" health problems	30.0		
"No" health problems	23.7		

Mean score by sex and groups	Male mean score	Female mean score
Overall group	27.7	27.8
Community group	31.1	28.0
School group	23.8	26.5

stress groups was that they were active users of medical resources for the resolution of somatic problems, as well as for complaints identified as "nerves." More detailed examination of the life-styles and concerns about health of the higher-scoring respondents shows that Latinos typically do not seek professional mental health services for their crises. Nevertheless, they recognize symptoms of psychological distress, and these are expressed to their significant others or to representatives of the professional medical system, particularly to physicians in private practice. The tendency for Latinos to express symptoms of psychological distress in "general-health" rather than in specialized "mental health" terms is associated with their cultural tradition, and reinforced by the presence of the general physician as the most available resource for alleviation of discomfort.

With regard to the immigrant's sex and response to stress, men and women in general do not show major differences in mental health risk (Table II). However, there are factors about length of residence and household organization which influence the responses of specific categories of immigrants. Men who are long-established, living in nuclear families, tend to be the best adapted, in contrast to the recently-arrived male migrants, separated from their families, who are the highest stressed of all groups (Table II). It appears that separation from families and the unsettled status of the recently-arrived male worker contribute to this contrast between recent arrivals and the settled men. However, women do not show as marked contrasts in stress by length of residence as do men. For many women, the early period of entry faces them with separation from children and spouses and the tasks of initiating the reestablishment of households.

Women as mothers, however, follow paths within the household which differ from the family careers of men. Latinas may be part of a conjugal unit, or they may be heads of single-parent households due to the circumstances of widowhood, separation, divorce, or unmarried parenthood. Men in Latino society are seldom single-parent heads of households. Mothers are expected to exercise greater emotional self-reliance than fathers. These cultural expectations, which are discussed at greater length in the following section, may have enabled women in this study to cope somewhat more successfully than men with the absence of spouse and children.

The childless never-married, who by and large were the more highly educated men and women of the study, had higher stress levels than the married, the previously-married, or never-married *parents* (Table III). Most were underemployed. In addition, they did not show the strong sense of purpose exhibited by those with spouse or children, who tolerated difficult conditions in order to attain desired improvements for their families.

An area of special interest is the comparison of occupational and stress levels, in view of the high commitment of men and women to work. With the exception of the one person in a professional occupation, the 54.6% who worked in unskilled jobs scored higher than those

Table III. All Respondents: Distribution of Mean HOS Scores by Marital Status and Education

Demographic characteristic	Mean score
Marital status	
Childless never married	29.7
Single parents	28.6
Married	27.8
Widowed	26.6
Separated	27.0
Divorced	26.0
Education	
None	31.3
Some primary	29.7
Complete primary	26.6
Some high school or technical	26.4
Complete high school or technical	26.9
Some college	29.1
College graduate	31.0
Unknown	22.0

Table IV. All Respondents: Distribution of Mean HOS Scores by Occupation and Work Characteristics

Occupational characteristics	Mean score
Occupation	
Major professional	31.0
Small business	25.8
Clerical	24.3
Skilled	25.4
Manual, semiskilled	26.5
Unskilled	28.9
Not working	27.2
Interest in change of occupation	
No information	25.5
Yes, want to change	28.6
No change desired	26.5
Not working	27.2

in other categories (Table IV). Those who wanted to change their occupations (two-thirds of the total working population) had higher stress levels than those who felt satisfied with their jobs (Table IV). These findings should be emphasized, inasmuch as they indicate important relations between work and mental health, both for recent immigrants and for settled Latinos. Plans for change within a job, or between jobs, are central topics of concern in the lives of a sizeable proportion of respondents.

Data regarding educational levels and stress, shown in Table III, are of related importance in considerations of occupational satisfaction. The fact that persons with the least education (representing largely Community parents) as well as those with the most advanced schooling (representing largely School parents), experienced greater stress than the midlevel group underscores the need to direct attention to the study of work and the adaptive patterns of disadvantaged groups. If Latinos with high educational achievement experience stress that is associated with blocked mobility, mental health experts should recognize the importance of the problems which face such individuals in their efforts to enter the occupational mainstream of American society. Attention should turn also to the impairing symptoms experienced by most Latino women and men in unskilled occupations, who are acutely aware of the limitations of their jobs as compared to their own abilities, and experience marked difficulties in efforts to improve their employment levels.

Literature on Latino mental health points to the supportive functions of the family for the containment and management of psychological disorder. (See Fabrega, 1970; Jaco, 1959; Madsen, 1969.) This study supports these findings. The data emphasize the critical importance, however, of understanding similarities and differences in the support systems of different types of households and the specific ways in which Latino men and women cope with stress. This subject is further discussed in the following section.

Controlarse and the Management of Stress

Controlarse (control of the self) is the dynamic theme which guides Latinos as they attempt to master the adaptive challenges of their environment. This theme is based on certain guiding principles about the containment of feeling. Through the practice of *controlarse* (control of the self) and *sobreponerse* (overcoming oneself), Latinos cope with stress-inducing situations. To understand how women and men cope with the problems of life, it is useful to discuss the theme of *controlarse* and the mechanisms used by Latinos to deal with symptoms of anger, anxiety, and depression.

Controlarse is a central mechanism for the regulation of behavior. It enables a Latino to exercise discipline over unpleasant feelings, thoughts, and moods. Through control of the self, Latinos keep in check negative feelings associated with unpleasant events *(disgustos)* or troubles and upsetting situations *(contrariedades)*. *Controlarse* helps to hold back outbursts of feeling such as anger *(corajes, enojos,* or *rabias)*, or the reactions of fear which result from unexpected experiences.

Animo decaido (low spirits) is one of the frequent first indicators of depression. The persistence of depressed feelings leads to states of sorrow *(pena)*, suffering *(sufrimiento)*, and feelings of being disgraced *(desgracia)*. Descriptions of the suffering woman *(mujer sufrida)*, or the disgraced man *(hombre desgraciado)*, refer to those who have met with sorrow-laden events. Although a Latina may receive the sympathy of family and friends for the unfortunate events which she has met, she is expected to exercise control over her feelings, and to raise her spirits.

Control of one's emotions and moods leads to various states, such as *resignarse* (to resign oneself), *no pensar* (not to think; in this context, to avoid thinking of a problem), or *sobreponerse* (to overcome oneself). Resignation reflects acceptance of a sorrowful event and consent to fate, while *no pensar* refers to the avoidance of confrontation and the desire to suppress disturbing thoughts and feelings. *Sobreponerse* is the effort to overcome reactions to stress-conducive situations; it represents a Latino's willingness to confront a problem, and a desire to alter his reaction to disturbance.

In the process of socializing their sons and daughters, Latino parents place priority on teaching children proper conduct through emphasis on the containment of feelings. Girls, for instance, who have to learn how to elicit respect and to maintain proper distance in interpersonal relations with boys, should govern their general demeanor by their ability to suppress their feelings. The belief that boys tend to express aggression overtly, leads likewise to emphasis on the exercise of moderation in the display of aggression.

Men and women in conjugal relations emphasize the avoidance of a direct expression of conflict. This ideal is attained through a mutually shared belief that, when interpersonal conflicts occur, they should avoid the overt expression of negative feelings. A Latino who loses control of his ability to govern disturbing thoughts, feelings, and moods frequently reports changes in personality which are described as modifications in *caracter* (character). For example, women who feel that they are unable to restrain their anxiety speak of changes in *caracter.* Men who experience an increasing difficulty in controlling their feelings of anger *(enojo)* over unpleasant situations note that, as a result, their character has changed *(tengo el caracter alterado).*

The dynamic aspects of these concepts can be understood by giving careful attention to the common, as well as to the contrasting, expectations of the feminine and the masculine ideals of *controlarse,* as illustrated in two areas: (1) the behavioral problems of schoolchildren, and (2) conflict between men and women in conjugal relationships.

Descriptions of the behavioral problems of children who live with their parents in Washington offer insights into the ideal roles for which they are being socialized. Problems of concern to parents reflect the cultural expectations of behavior for adult women and men which are linked with the concepts of containment and control of negative sentiments. There are contrasts between expectations about proper conduct and the regulation of behavior of boys and girls, as noted in the cases of the children of Olga Jimenez and Hilda Molina.

In rearing a daughter of elementary school age, Olga expressed a central concern with providing an environment that nurtures an appreciation for the value of *respeto* (respect). In recognition of this ideal, Blanca, her 10-year-old daughter, was required to maintain proper distance and control of self in relation to boys. *Respeto* was to become a major behavior dynamic upon attainment of full adolescence and adulthood, as noted in the following events.

Blanca's mother, Olga Jimenez, was called by her daughter's teacher to discuss reports that Blanca was "not studying." Schoolteachers felt that the girl's increasing loss of interest in her studies should be treated in a mental health center. Her problem had been brought to the attention of a school counselor who had, in turn, referred Blanca and her parents to the center. At the time of this research, however, Olga had not taken Blanca there. She was more preoccupied with the effects of the family's living conditions on Blanca than she was with her failing school record. She felt that the apartment where they lived was "too closed in" for a youngster, especially since the manager did not allow children to play in the hallways. Moreover, Olga was deeply worried because she had heard that at Blanca's school there were a number of male students who had not been brought up to "respect" girls. Consequently, she was seriously considering the possibility of sending Blanca to a boarding school, where she believed Blanca would not only be protected but would also have more companionship. She and her husband would have to "work and sacrifice" to send Blanca to a good school, which she defined as being one with teachers who are concerned over the proper behavior of boys and girls to each other. Olga had become so worried over Blanca that whenever she spoke of her, she experienced the onset of headaches and increased nervousness.

Mothers discussed a different set of problems for their sons, in contrast to their daughters. Undesirable behavior for boys included rebelliousness *(conducta rebelde)*, lack of discipline *(indisciplinado)*, a tendency to fight *(peleon)*, and nervousness or excitability *(nervios)*. The etiology of these problems was sometimes ascribed to physical dysfunction such as weak blood and head injuries, or to heredity. At other times, it was linked to the influence of an estranged parent or a relative. The types of problems described in the case of Hilda Molina and her son, Roberto, offer perspectives on the parental views regarding the nature and management of boys' behavior difficulties.

Four years prior to the study, Hilda Molina, a single mother, had brought her only son, Roberto, to the United States. They lived with her sister and her sister's husband. At the time of this research, she was worried because at age nine he was repeating the second grade. He could read neither English nor Spanish, and she had been called to talk with school personnel who wanted to help Roberto. She wondered whether he suffered from some form of congenital retardation, or whether his behavior had resulted from a sharp blow on the head which he had received from playmates in the first year after their arrival in Washington.

Hilda had concerns about Roberto's nervous mannerisms and his *rebeldia* (rebelliousness) towards her. She had taken him for examinations and tests in several well-known children's health centers in the city, and he had been treated mainly for allergies. The school counselors had referred him to a local psychiatric center, but at the time of the study he was not in active treatment at this facility.

Throughout their contact with health centers and mental health resources, Hilda and her relatives had hoped that someone would prescribe the correct tonics and foods to fortify her son. The family believed that, with good physical health, defined mainly as a strong "constitution" and the prevention of weak blood, he would control his rebelliousness and improve his learning.

Descriptions of the behavioral problems of these elementary school age students point to several aspects of the nature and management of conflicts. Parents hope that their children will develop the ability to exercise control and containment of certain negative feelings. Marked differences are evident, however, in the behavioral expectations for boys and for girls. Feminine ideals about the protection of sexual sanctity (women's source of honor) call for the early insistence on conduct to prepare girls to elicit respect and deference. The cultivation of these qualities requires training in self-containment, particularly in the presence of males. The discipline of boys, in contrast, is centered to a much greater extent on the containment of the overt expression of aggression.

Differences between the views of parents and those of school authorities about the nature and management of problems were noted in the cases of Blanca Jimenez and Roberto Molina. Their parents, like other immigrants in this study, emphasized the supposed links between physical symptoms of weakness or hereditary defects and the behavior problems of children. They hoped that nervousness and deficiency in school performance would be outgrown as their children attained optimal levels of physical health, measured in particular by "strong blood." School personnel, who had greater concern with educational performance and achievement, frequently referred such children to community counselors and agencies.

The differences in behavioral expectations of boys and girls continue into adult life. A leading source of stress between spouses is the contrast between the feminine and the masculine concepts of the nature and exercise of containment of feelings and control of the self. Adults who have marriage problems are expected to try as much as possible to guard against expressing negative sentiments, and to keep in check feelings of hostility toward a mate. On the one hand, women are expected to act as moderators in tense situations, and to contain emotions such as hostility to a greater extent than men. Following cultural tradition, women's behavior is supposed to bring stability to a conjugal relationship. On the other hand, the practice of control of the self among men calls for the governing of strong feelings such as those associated with the expression of anger. Men are expected, however, to depend not only on their own control, but also on the moderating influence of women. An aspect of strength in women's character is thus based on independent self-mastery, and energy left over to help men; while masculine control of the self is to some extent dependent on the influence of women.

This study shows, however, that among women these concepts of sex-role relations are in a state of change. For instance, Juana Quesada indicated that drinking in itself was not the only source of her husband Melchor's diminished control of himself, and of his shifting moods. She felt that he had other personal problems which required attention.

Juana and her husband expressed contrasting views about his emotional outbursts. Juana was much troubled by his frequent fits of temper, but he indicated that his loss of control took place only when he drank too much beer. Juana labeled her husband as neurotic, and felt that something was the matter with his nervous system, but Melchor insisted that alcohol was the only explanation for his frequent bouts of anger. Their contrasting views were becoming a source of stress in their marriage, although they both contained their feelings to a degree, and tensions surfaced mostly during his drinking episodes. At the time of this study, Juana was increasingly concerned about Melchor's view of his problem, particularly because his marked shifts in mood and irascibility were not limited to the periods when he drank beer.

To cite another instance, Lucía Díaz was a woman whose husband expected her to show control over her disturbed emotions. She feared increasingly, nevertheless, that she would not be able to cope with her anxiety and depression. Lucía did not share her husband's view that she had to assume the major responsibility for improvement of their marriage through the exercise of control over her troubled feelings. She and her husband, Tomás, had a number of fights concerning management of money, their relationships with relatives, and sexual incompatibility. One day, after a strong disagreement, he left her, and she felt as if the world had come to an end for her. When he came back after a few weeks, Lucía was happy, even though she did not like his advice upon his return. He told her, for example, that she ought to "conquer herself," she ought to avoid "thinking" of their troubles *(El me dijo que me debo sobreponer, que no debo pensar)*. But Lucía found it difficult to pursue this course. She sometimes wondered whether someday she would become as distraught as her mother, who had died in an "insane asylum." She consulted various physicians who prescribed medicines to calm her, but these medicines did not relieve the sense of sorrow *(pena)* and emotional strain *(sufrimiento moral)* experienced when she realized that her marriage might terminate in separation. She was feeling overwhelmed by the burdens of too much suffering and too much affliction *(mucho sufrimiento y tanta aflicción)*.

During the course of this research, Lucía developed a number of organic and psychological symptoms for which she consulted several general practitioners. Some treated her physical symptoms only, and others suggested that all her problems were psychological. These contrasting ways of dealing with her symptoms led her to doubt the power of professional medicine. She increased her participation in religious services, and hoped that her rediscovered faith would serve as an anchor for the resolution of her problems. Moreover, a central preoccupation throughout this period was the role of fate and heredity in her illness. She was losing hope in her own ability to understand and to face her husband, and she was developing fears about the inevitability of following in her mother's footsteps.

Lucía was most distressed because she could not meet the cultural expectation that voluntary control over her feelings would resolve her conjugal difficulties. She could not heed her husband's advice that she avoid thinking of the problems. Shortly before the completion of this research, she took a heavy overdose of aspirin. After this suicidal gesture, she continued to search actively for advice among lay and scientific practitioners of medicine, and among various religious ministers and counselors.

The exercise of control over unpleasant or negative feelings in order to face the difficulties of the surrounding world is a neglected dynamic aspect of behavior among Latinos. The present research shows that, as Latinos and Latinas have engaged in efforts to alter their life situations, they have overcome difficulties through this mechanism. In the traditional manner, boys and girls are expected to learn to face the temptations and problems simply through control over their disturbing feelings and thoughts. It should be noted, however, that the attainment of this behavioral ideal is seen as dependent, in part, on the maintenance of good health. A strong and healthy body is believed to be the foundation for the proper regulation of behavior.

As to conjugal relationships, tension and contradictions often are resolved through mutually shared expectations about masculine and feminine forms of containment. Serious strains occur, however, when husband and wife have different ideas about the reasons for their inability to govern their disturbed sentiments. Changing concepts of role relations between the sexes accentuate these problems.

Summary and Implications

As increasing numbers of Latino women assume leadership in immigration movements, attention needs to be given to the conditions which contribute to stress in their lives, and the dynamic ways through which they cope with the challenges involved in resettlement.

Women Lead in Resettlement

The present study has shown that Latinas are careful planners, and that they take initiative as pioneers throughout their resettlement. Women are key organizers and counselors for other relatives who follow them. Furthermore, since they migrate typically after they have established households in their places of origin, they have to coordinate their plans for resettlement with relatives at home who assume the roles of substitute caretakers for children.

Immigrant mothers frequently cope with the anxieties associated with separation from children. They deal with distresses by attempts "not to think" about disturbing problems. This is a culturally influenced mode of dealing with upsetting feelings. Some parents consult physicians for physical complaints with the hope that prescription medicines will offer relief for physical as well as psychological problems.

Researchers and policy makers should give increased attention to the questions raised by the realities of present-day separation of parents and children which takes place during early phases of international migration. The study of patterns of caretaking and its meaning for adults and children should become an area of focal interest in Latino studies, since a number of Latinas, in particular, enter the United States while their children are young. Indeed, the woman's reliance on her relatives for substitute child care suggests that families with strong bonds in the maternal line would be most likely to have women who lead in migration and resettlement. Those concerned with the development of policy and counseling services for newcomers might consider the increased sponsorship of programs of intercountry collaboration, so that both the immigrant and those left behind can benefit from coordinated family-based program development.

Latinas Work in American Society

Almost all Latinas in this study work full time. Their high investment in work should be given attention, in view of the popular belief that Latino mothers tend to remain in the home. Furthermore, Latinas come with more limited formal education than men, and so they tend to work largely in unskilled and semiskilled capacities. They cope with the difficult-to-tolerate aspects of their jobs motivated by the belief that parents have to *sacrifice* themselves for the future of their children and their loved ones.

An altruistic motive of sacrifice for the benefit of the group would appear to conflict with the typical individualistic orientation which

Americans tend to hold toward jobs. For the Latino, self-sacrifice is tempered by concepts about the ideal type of relationship between employers and employees on jobs. Immigrants retain the belief that *dignidad* and *respeto* are important qualities of social relationships. Work is not simply governed by impersonal rules and material rewards. Supervisors who display *buen trato* toward a Latino frequently become the trusted advisors of the immigrant, and they are sought out for the solution of problems. Thus, employers and supervisors frequently become key intermediaries through whom the immigrants learn about the host society. They act as helpful counselors outside of the immigrant's network of family and friends.

Researchers and policy makers concerned with the processes of adaptation through which immigrants master tasks in their host society should give attention to the central place of work in their lives. Those interested in theories of adult socialization, in particular, should increasingly examine the dynamics of learning which shape the lives of immigrants as they draw heavily on their experiences in the workplace to orient themselves to the new society.

Sociocultural Factors and the Prevalence of Stress

Analysis of HOS scores by sociocultural characteristics indicates that selected aspects of the experiences of Latino immigrants merit attention, namely, length of settlement, health status, respondent's sex, marital status, educational levels, and occupational satisfaction. Several specific areas are of importance.

1. Respondents drawn from the Community group who tend to be the more recent newcomers show higher stress scores than the longer-established. However, certain sociocultural variables, along with recent entry, contribute to the emergence of psychiatric disorder. For example, the more recently arrived man who is separated from his family is higher-stressed than women in similar conditions. Such findings should lead to a more careful assessment of similarities and differences in ways by which Latino men and women cope with stress and use mutual support as parents, spouses, or adults with children. This area has received relatively limited attention in the literature on Latino mental health.

2. Material regarding the impact of educational and occupational levels on groups of Latinos bears attention among researchers and those concerned with programs of prevention. Data which show that those subjects who wish to change occupation (two-thirds of the total working group) have higher HOS scores than those who are satisfied with their jobs are significant, particularly when examined along with

information about the poorly and the well-educated. Persons with the least education (representing largely Community parents), as well as those with most advanced schooling (representing largely School parents) experience greater stress than the midlevel group. These findings underscore the need to direct attention to the impact of education on the mental health of the members of disadvantaged groups. This receives support from data gathered in the Stirling County Study, and in North Carolina.

In Stirling County, stress levels fall as education increases up and through high school. The lowest risk of such disorder for both men and women occurs among persons with 11 or 12 years of schooling. But the risk rises again with additional education (Leighton *et al.*, 1963). The same trend is observed for blacks in a North Carolina study which used the HOS among patients of public health nurses. The authors suggest that these blacks may have been unable to apply their education, and this may have led to frustration (Leighton & Cline, 1968).

Among the Latinos in this research, there are trends somewhat similar to the studies cited above. If Latinos and blacks with high educational achievement experience stress associated with blocked mobility, mental health experts should recognize the importance of the problems which face such individuals in their efforts to enter the occupational mainstream of American society. Attention should turn also to the impairing symptoms experienced by most Latino men and women in unskilled occupations, who are acutely aware of the limitations of their jobs as compared with their own abilities, and experience marked difficulties in efforts to improve their employment levels.

3. The Health Opinion Survey is useful to help identify characteristics associated with high or low levels of stress among Latinos in various types of households. A more extensive use of the HOS should offer a basis for definitive analysis of characteristic patterns of stress symptoms for this cultural group for comparison with other populations among whom the HOS has been used. Interpretations of scores appear to be particularly meaningful when accompanied by complementary ethnographic study of the sociocultural environment in which respondents live and work.

Controlarse and Coping with Stress

Control of the self *(controlarse)* is identified as a central mechanism for the regulation of behavior, which Latinos use to cope with symptoms of anger, anxiety, and depression. A number of researchers have characterized Latin Americans as persons who are passive endurers of stress (see Díaz-Guerrero, 1967; Holtzman, Díaz-Guerrero, & Schwartz, 1973; Reichel-Dolmatoff & Reichel-Dolmatoff, 1961; Samora, 1961.) and who tend to avoid direct interpersonal conflict. Latinos are said to bear disease and troubles through denial, courage, and acceptance. Studies conducted by Díaz-Guerrero (1967) among people of Mexican heritage, for example, show that the passive endurance of illness and stress is considered a virtue sustained by values such as harmony, protection, dependence, formality, and cooperation. Self-sacrifice is expected in all members of the family, together with submission, dependence, politeness, courtesy, and *aguante* (the ability to hold up well even in the face of abuse).

Studies of conflict resolution in Latin American cultures often emphasize the dynamics of resignation and conformity, rather than control of the self and mastery over difficult circumstances. Resignation is, however, only one of the behaviors which can result from an ideal that leads to containment and suppression of feelings. *Controlarse* has two complementary dimensions. Latinos can contain their feelings and *either* resign themselves to their unkind fate *or* strive to overcome stress-inducive situations. Among the immigrants in this study, there is emphasis on the practice of *sobreponerse,* the ability to conquer and overcome one's disturbing feelings.

Mental health practitioners who work with Latino groups should examine the contrasts between the feminine and masculine concepts of the exercise of containment of feelings and control of the self. The Quesada and Díaz cases illustrate the tendency for women increasingly to call into question traditional beliefs about the nature of aggressive feelings among men and the self-containment of troubles among women. These modified beliefs should have impact on the Latino woman's role as wife and mother. In conjugal conflicts, wives and husbands may express contrasting views about the etiology of problems, and about how to cope with them. In relations to her children, a Latino mother may introduce new concepts about how growing girls and boys should cope with the challenges of the new setting. Unfortunately, there is limited information on these changing patterns of role relations among Latinos in the United States. In light of the increased interest in understanding patterns of resilience and vulnerability among adults and children, this area should receive high priority.

Concluding Comments

In a recent article, Coelho and Stein (1977) have drawn our attention to the challenges and costs to human adaptation posed by the present-day uprooting of populations throughout the developing world. As an anthropologist concerned with the relations between cultural factors and mental health, the author of this study has identified dimensions of the family and of work which influence the expression and management of stress among Latinos. Investigators and action-oriented professionals should be increasingly involved in understanding how stress and coping are manifested among men and women in various cultures. We need to respond creatively to the reality that uprooting has become a central characteristic of our rapidly evolving world.

Acknowledgment

The author acknowledges the helpful consultation offered by Dorothea C. Leighton, M.D.

References

Coelho, G. V., & Stein, J. J. Coping with the stresses of an urban planet: Impacts of uprooting and overcrowding. *Habitat*, 1977, *2*, 379–390.

Cohen, L. M. Health status of Central and South Americans in Washington, D. C.: Sociocultural factors and health-care policy. In *Actas del XLI Congreso Internacional de Americanistas*, Mexico, 1974, Vol. III. Mexico: Instituto Nacional de Antropología e Historia, 1975.

Cohen, L. M. The female factor in resettlement. *Society*, 1977, *14*, 27–30.

Cohen, L. M. *Culture, disease and stress among Latino immigrants*. Washington, D. C.: The Smithsonian Institution Press, Research Institute on Immigration and Ethnic Studies, 1979.

Cohen, L. M., & Fernandez, C. L. Ethnic identity and psychocultural adaptation of Spanish-speaking families. *Child Welfare*, 1974, *LIII*, 413–422.

Díaz-Guerrero, R. *Psychology of the Mexican*. Austin: University of Texas Press, 1967.

Dohrenwend, B. S. & Dohrenwend, B. F. (Eds.). *Stressful life events*. New York: John Wiley, 1974.

Fabrega, H. Mexican Americans of Texas: Some social psychiatric features. In E. B. Brody (Ed.), *Behavior in new environments: Adaptation of migrant populations*. Beverly Hills, Calif.: Sage Publications, 1970.

Hinkle, L. E. The concept of stress in the biological and social sciences. *Science, Medicine and Man*, 1973, *1*, 31–48.

Holtzman, W. H., Díaz-Guerrero, R., & Swartz, J. D. *Personality development in two cultures*. Austin: University of Texas Press, 1973.

Hughes, C. C., Tremblay, M., Rapoport, R. N., & Leighton, A. H. *People of cove and woodlot*. New York: Basic Books, 1960.

Jaco, E. Mental health of the Spanish-American in Texas. In M. K. Opler (Ed.), *Culture and mental health*. New York: Macmillan, 1959.

Kellert, S., Williams, L. K., Whyte, W. F., & Alberti, G. Culture change and stress in rural Peru. *Milbank Memorial Fund Quarterly*, 1967, *XLV*, 391–415.

Lazarus, R. S., Averill, J. R., & Opton, E. M. The psychology of coping: Issues of research and assessment. In G. V. Coelhó, D. A. Hamburg, & J. E. Adams (Eds.) *Coping and adaptation*. New York: Basic Books, 1974.

Leighton, A. H. *My name is legion*. New York: Basic Books, 1959.

Leighton, A. H. Conceptual perspectives. In B. H. Kaplan, R. N. Wilson, & A. H. Leighton (Eds.), *Further explorations in social psychiatry*. New York: Basic Books, 1976.

Leighton, D. C., & Cline, N. F. The public health nurse as a mental health resource. In T. Weaver (Ed.), *Essays on medical anthropology* (Southern Anthropological Society Proceedings No. 1). Athens: University of Georgia Press, 1968.

Leighton, D. C., Harding, J. S., Macklin, D. B., Macmillan, A. M., & Leighton, A. H. *The character of danger*. New York: Basic Books, 1963.

Levi, L. (Ed.). *Society, stress and disease*. New York: Oxford University Press, 1971.

Macmillan, A. M. The health opinion survey: Technique for estimating prevalence of psychoneurotic and related types of disorders in communities. *Psychological Reports*, 3 (Monograph Supplement 7), Grand Forks, N. D.: Southern Universities Press, 1957.

Madsen, W. Mexican-Americans and Anglo-Americans: A comparative study of mental health in Texas. In S. C. Plog & R. B. Edgerton (Eds.), *Changing perspectives in mental illness*. New York: Holt, Rinehart & Winston, 1969.

Meyer, A. Pathology of mental diseases. In E. E. Winters (Ed.), *The Collected Papers of Adolf Meyer*, vol. II: *Psychiatry*. Baltimore: The Johns Hopkins Press, 1951, pp. 289–310.

Meyer, A. The life chart and the obligation of specifying positive data in psychopathological diagnosis. In E. E. Winters (Ed.), *The Collected Papers of Adolf Meyer*, vol. III: *Medical Teaching*. Baltimore: The Johns Hopkins Press, 1951, pp. 52–56.

Reichel-Dolmatoff, G., & Reichel-Dolmatoff, A. *The people of aritama*. Chicago: University of Chicago Press, 1961.

Rogler, L. H., & Hollingshead, A. B. *Trapped: Families and schizophrenia*. New York: Wiley, 1965.

Samora, J. Concepts of health and disease among Spanish-Americans. *American Catholic Sociological Review*, 1961, *XXII*, 314–323.

Selye, H. *The stress of life*. New York: McGraw-Hill, 1956.

White, R. W. Strategies of adaptation: An attempt at systematic description. In G. V. Coelho, D. A. Hamburg, & J. E. Adams (Eds.), *Coping and adaptation*. New York: Basic Books, 1974.

17

Patterns of Adaptation of Indian Immigrants: Challenges and Strategies

Parmatma Saran

Social change is much talked about today, not only in the field of the social sciences, but in popular magazines and the mass media. Toffler (1971) dramatized this issue in his famous book, *Future Shock*, warning Americans about what might come in the future as a shock.

In contemporary societies, rapid industrialization, urbanization, migration, and scientific-technological advancement have accelerated the pace of social change and personal mobility. Coelho and Stein (1977) have identified the emotionally stressful aspects of the impact of uprooting on the quality of life in different population groups and communities around the world.

This study was initiated in early 1975. It can be viewed as an attempt to understand the patterns of adaptation of Indians living in the New York metropolitan area. This effort was supported by a summer grant by the Center for Urban Ethnography, University of Pennsylvania, which resulted in a paper presented at the Eastern Sociological Society's meetings in New York City on April 20, 1975. Later, a larger paper entitled "New Ethnics: The Case of the East Indians in New York City" was presented at the Smithsonian Institution's Research Institute on Immigration and Ethnic Studies' national conference on "The New Immigration: Implications for American Society and the International Community," held in November, 1976 in Washington, D.C. Subsequently, the paper was revised and published in *Society* (1977, *14*, 65–69), with the title "Cosmopolitans from India."

Parmatma Saran • Department of Sociology, Baruch College, The City University of New York, New York, New York 10010.

The act of migration alone is a very important aspect of the larger phenomenon of social change. It is one of the most significant changes which can take place during an individual's life experience. The decision to leave one's home environment and start a new life in a completely different sociocultural setting is determined by many dynamic factors, and the very act of migration has a variety of psychological consequences for those who actually move.

This paper attempts to describe, analyze, and understand the pattern of adaptation of Indian immigrants living in the New York metropolitan area. More specifically, our objective is:

1. To examine the consequences of migration for the Indian immigrants residing in the New York metropolitan area.
2. To understand the pattern of adaptation of Indian immigrants.
3. To identify whether or not change in sociocultural environment has any effect on their emotional health.
4. To identify the kinds of strategies Indian immigrants have developed for coping with the new challenges they face.

Relevant Studies

There is a large number of studies dealing with the experiences of earlier immigrants who came to the United States from Europe. Handlin's *The Uprooted* (1973) and Thomas and Znaniecki's *The Polish Peasant in Europe and America* (1958) are classics in this area.

How do these immigrants assimilate to the new culture? What exactly do we mean by "assimilation"? Gordon (1964) and Glazer and Moynihan (1963) have sufficiently dealt with these issues.

Although there is an established body of theoretical and substantive writing about "older" immigrant groups, there is a lack of ethnographic and survey data available about recent immigrants. There is no major study dealing with Indian immigrants in the United States. Some work has been done on Indian immigrants in Great Britain by Desai (1963), Elkan (1960), Eames and Robboy, (1980), and Aurora (1967), in East Africa by Morris (1956), and in Trinidad by Klass (1961), showing patterns of adaptation of Indian immigrants and their experiences after migration.

In the past twenty years empirical studies dealing with Indian students in the United States have been done by Coelho (1958), Lambert and Bressler (1958), and Gandhi (1967).

After a decade or so, there is renewed scholarly interest in the Indian immigrant experience in America. A major contributing factor probably is that recent changes in the United States immigration laws have enabled Indians in various trades and professions to bring their families to settle in the United States, as noted below.

Saran's work (1977) is probably the most recent publication on Indian immigrants in the New York area. Fisher's (1978) work on the role of Indian organizations in maintaining ethnic identity is another important contribution. Similarly, there are some working papers dealing with the Indian family (Nandan, 1978) and with political orientation and discrimination (Mohaptra, 1977), and an historical account of East Indian immigrants to the United States (Jensen, 1980). A major survey designed to obtain demographic data on Indian immigrants, conducted by Leonhard-Spark and Saran (1980) will be most helpful for determining the size and profile of the Indian community in the New York metropolitan area.

Background of this Study

This is a longitudinal study focusing on patterns of social and psychobiological adaptation among Indian immigrants residing in the New York metropolitan area. A survey is also under-way to obtain a background and behavioral profile of Indian immigrants. Some data from this survey have been used for this paper. (See fn., p. 375.)

Data Used

This paper is based primarily on five in-depth interviews selected by the author from approximately fifty that he has conducted for the larger study. An interview guide was used for gathering information on the immediate reactions of these immigrants after coming to this country, their patterns of adaptation, especially in the areas of employment, education, and family, and strategies employed by them to cope with any strains they have experienced. Each interview took about three hours, and was conducted in private, either at the interviewer's or the interviewee's residence or office. No tape recorder was used in any of these interviews, but notes were transcribed as soon as the interviewee had left.

The interviewees were told that the interviewer would summarize their interviews and report them in his study without any reference to names or any disclosure of identity. All interviewees clearly understood this, and gave their permission to the author to do so.

The people interviewed were not completely unknown to the author. Because of his active participation in the affairs of the Indian community, he has developed contacts with a large number of Indians. He deliberately chose to interview those about whom he already had some information, so that the reliability of the information received could be checked.

The five cases selected for the purpose of this paper cannot be claimed as representative of the entire population. However, they do typify a segment of the population, and may be considered appropriate for portraying a general account of the consequences of migration faced by Indian immigrants.

For an exploratory study the methodology used is adequate to shed some light on the phenomenon of uprooting and coping, which is the main theme of the volume for which this chapter has been written. Data gathered on the basis of participant observation and survey study were also used.

It may be useful here to identify the main areas of difficulty and distress reported by the five cases at different stages of their stay in the United States.

In the first phase of their stay, difficulties were experienced because of the differences in cultural background, language (not significantly), food, and loss of family and friends in India.

In the later phase of their stay, difficulties have been experienced in the areas of employment, lack of acceptance by the local community, and a general sense of alienation because of racial and cultural background.

Indian Community: Profile and Attitude

Starting in 1969–1970, because of the change in the immigration laws (the passage of Public Law 89–236, 1965), scores of Indians (approximately 250,000 are in the United States at present) came with their families to settle in the United States. A large number of these immigrants stayed in and around New York City. Approximately 40,000 are living in the tristate area at present.

Progressively, Indian enterprises began flourishing in the city. The rapidly growing number of Indian restaurants, grocery stores, appliance stores, sari stores, temples, and cultural and social organizations, among many other activities, clearly indicates the emergence of a permanent Indian community developing in the city.

It must be noted that, because of the current immigration laws, those Indians who come to the United States are professional, or at least skilled people. Leonhard-Spark and Saran's study reveals that 50% of the Indian immigrants living in the New York area are between the ages of 31 and 40, 29% are below 30 years of age, and the remainder are above 40. Their educational level is very high: 75% have graduate degrees, 18% have completed four years of college, and the remainder have had some formal education. Thirty-four percent earn more than $25,000 per year, 35% between $15,000 and $25,000, 16% between $10,000 and $14,999, and only 11% less than $10,000. Eighty-five percent work full-time, 9% work part-time, and 11% do not work at all. The nonworking population is overwhelmingly composed of housewives. The majority of those who work are professional (91%). Of this 91%, 24% are engineeers, 15% physicians, 18% managers, administrators, and business owners, 6% accountants, 5% biologists or chemists, 3% are in teaching, and the rest are in other professions. Almost 30% own their own homes, and the others live primarily in rented apartments, and some in rented houses.

More than 80% are married, and the average family size is four. Some 25% are young couples without any children.

Almost 70% come from large cities, 20% from small towns, and the rest from villages or rural areas. The largest number of them came to the United States between 1965 and 1971 (over 50%), about 45% have come since 1972, and only 5% came here before 1965. Twenty-one percent are already naturalized citizens.

Data based on participant observation and some interviews (Saran, 1975) reveal that the attitudinal profile of Indian immigrants remains traditional, and there has been little change in their attitudes since coming to this country. In the area of marriage and family, the traditional values are still dominant. The majority of unmarried men prefer to go back to India for a traditional marriage. Parents who have growing children hope that the children either will go back to India for marriage, or will find suitable Indians for marriage in the United States. In those relatively few cases in which an Indian has married an American or other non-Indian, the degree of acceptance is minimal, and sometimes total rejection is also evidenced. The parents' attitude toward their children also generally remains traditional and authoritarian. Because of peer-group influence and exposure to school, children demand that they should have more freedom and greater equality. This has created some conflict and tension within Indian families. The religious values and attitudes of Indian immigrants are also not affected because of migration. Interestingly enough, families engage here in more religious activities than they did in India. In the area of economics, also, traditional Indian ways are retained, for example, saving is greatly emphasized, and they tend neither to borrow nor to use credit as much as their American counterparts do.

We also found that there is a very strong stigma attached to mental illness. Very seldom would an Indian see a psychiatrist, even if advised to do so by a physician. Those few who have seen a psychiatrist, either for themselves or for a relative, requested that it not be discussed in this research, for fear of being somehow identified. In health matters, the Indian immigrants have changed their attitude. They have become more health-conscious since coming here. Most Indians see doctors for medical checkups, and follow their advice. There is little or no evidence of the use of ayurvedic or homeopathic medication. Although basic food habits have not changed, there is greater concern about the nutritional value of the food eaten, and among adults, especially women, a desire to avoid items which are fattening. The average size of an Indian family is small, slightly smaller than comparable families in India. It seems, therefore, that Indian immigrants are more family-planning oriented than their counterparts in India.

It is interesting to note that the structural assimilation of the Indian immigrants is relatively easy as compared with that of many other immigrant groups, whereas the cultural assimilation of the Indian immigrants is minimal. A look at the Indian community clearly suggests that there is a strong desire to maintain their cultural heritage. The notion of ethnicity remains strong, and is also perceived as desirable. The growing interest in Indian culture, music, food, religious values, yoga, and meditation among Americans, especially some well-known Americans, is inspiring, and reinforces a sense of ethnic pride and cultural identity, particularly for youngsters in the immigrant population.

Our investigation (Saran, 1976) suggests that Indians may be classified into three categories with respect to their cultural assimilation into American society. It must be pointed out that there is an overlapping of some of the traits in these categories. The categories are:

1. Those (a very small proportion) who have become completely Americanized, or at least claim to have done so. In their attitudes and behavior we have observed the following traits: the Indian community is not their reference group; they generally have changed their nationality and show full commitment to their adopted home; often they are critical of Indian values and have a sense of ridicule for India; their primary group relationship is outside the Indian group; most of their friendships and leisure time activities are outside Indian groups; their food habits have also changed considerably; they, however, still participate in Indian organizations, only using the English language.

2. Those (a reasonably large proportion) who carefully maintain their Indian heritage, and at the same time accept new values and consider themselves as part of the mainstream of American life. The

Indian community remains their reference group; most of them have not changed their nationality, but even those who do become American citizens show a strong sense of commitment to India; they tend to be more objective, and, though basically subscribing to Indian values, have made some change, or at least recognize certain positive aspects of American society; basically their primary group relationship is within the Indian community; they have close ties with local communities and have close American friends; their leisure-time activities are divided within and outside the Indian community; mostly they eat Indian food, but they also cook American food at home, and sometimes eat out; occasionally they go to temples and Indian movies, but they actively participate in Indian organizations, using both English and their native language.

3. Those (the largest proportion) who are highly conscious of their Indian heritage and want to keep it intact, they generally resist new values, and live as marginals in this society. This category includes the majority of the most recent immigrants: the Indian community is their only reference group; even if they think that they would live here permanently (and most of them do not), they do not think of changing nationality; they see only bad things in American society, and remain very critical of American values; their friendship patterns and leisure-time activities are exclusively Indian; they go only to Indian movies and concerts; they are quite unaware of local politics and happenings; they actively participate in Indian organizations, but they belong mostly to regional organizations; they eat only Indian food, and seldom eat out; they frequently go to temples and participate in religious activities.

Before we present the five cases for further analysis, it would be useful for us to see which of the above categories are represented by these cases, and what specific patterns they illustrate. Case I falls under category 2, case II falls basically under category 2 but it has some features of category 1, Case III is also under category 2 but has some features of category 1, case IV falls under category 3, case V also falls under category 2, but it has some resemblances to both categories 1 and 3. In terms of attitudes and values we find these five cases somewhat different from each other. However, in so far as their coping patterns are concerned, there is not much difference. For example, none seek professional help in times of stress, and all rely heavily on friends and family in times of need. There is also a lack of awareness of some serious problems they may have, and a tendency to underestimate them or treat them philosophically. There are certain differences in their coping patterns which will be elaborated later.

Case I

Rahman is 27 years old and came to New York City in 1974. He goes to school and also works part-time as an accountant. He is single and shares an apartment with a male friend. Upon his arrival he lived with his sister for several weeks. He comes from a large city in India. He has not visited India since his arrival in the United States; however, his parents have been here for a visit. He also has two married brothers who live with their families in this country.

Rahman's immediate reaction after arriving in this country was favorable. Impressed by the scientific and technological advancement and the high standard of living, he felt that his decision to come here was right and that he would be able to make an easy adjustment to this society. The first few weeks he stayed with his sister and her husband. Although living with them was no problem, after a time he decided to find his own place, and also started looking for a job. Soon after finding employment, Rahman moved to a small apartment.

Within three or four months after his arrival here, Rahman was beginning to realize that things were not going as he had expected. He felt that because of the language problem (he speaks English well, but still thinks that he has problems in communicating with Americans), differences in color and physical characteristics, and cultural differences, he was at a disadvantage. His reactions and attitudes toward this country also changed somewhat. He started to worry, and became skeptical about his future in the United States.

Within the next three or four months, Rahman developed a skin allergy, and was very ill for a week. He had swellings all over his body, and became very stiff. He could not even move. This was a most shocking and devastating experience for Rahman. He feels that American doctors are very selfish and unconcerned for their patients. Things got so bad that he had to move back to his sister's home again. Rahman was particularly pleased with the care and affection he received from his sister's mother-in-law, whom he characterizes by saying, "Unlike most Indians, she cares, she is not selfish; nice lady." Rahman also thinks that Indians change too fast after coming here, and become selfish. Although he remains generally satisfied with his sister's and brothers' relationships with him, he says, "You know they don't realize, but they have changed a lot. Our relationship is not the same now as it was in India." In any case, he remains grateful to them for whatever help they have rendered to him.

During this period of crisis, friends were also helpful, but "They couldn't do anything except consoling and telling me that everything would be all right." Rahman then himself said, "Things are so different here, everyone has his own problem. This is the nature of this society. You cannot blame people."

Rahman's feelings about job and school are mixed. He recognizes that the educational system is better here, and makes you work hard. Efficiency and hard work are demanded at the job, but often these demands are not reasonable. Once he told one of his supervisors, "Leave me alone. I know what I have to do and I will do it, but don't push me all the time. It makes me nervous."

Rahman's social life is minimal. The young man with whom he shares an apartment is a nice fellow, but they don't have much in common. The reason they live together is that they have developed a good understanding and have learned to adjust. Rahman is basically an introvert and feels very lonely. Sometimes he goes through severe depression. He smokes and drinks a lot when he is depressed. He talks to no one about it, including his family and friends. Once he casually mentioned his depression to a fellow student, who advised him to see a counselor, but Rahman thinks that this will go away as time passes.

Sometimes Rahman gets so worried that he wonders whether coming to this country was the right decision. His major concern is to avoid getting involved with an American girl. His parents back home and his family here have repeatedly told him that marrying an American girl would be a disaster for him as well as for the family. "They have brainwashed me," says Rahman. He is not totally negative toward American girls himself, but he has tremendous respect for his parents (although he does not feel very close to them), and he will never do anything that will hurt them. In addition, he feels that there is so much difference in culture, values, and attitudes that it would not be easy for him to adjust to an American girl. His contact with the Indian community is minimal, and he has, thus far, had no opportunity to meet any Indian or Pakistani girls.

Loneliness seems to be a serious problem for Rahman. "I want company. If I am lonely, I get mad." However, everyone, including Rahman, remains ambivalent about it. Rahman volunteered the information that he is very short-tempered. He is trying to change, though, because he knows that he cannot survive unless he gets along with people. He also feels that his problems have to do with the very authoritarian parents he has, and his complete sense of dependence on family. "Back home I never had any responsibility, now I make my own decisions. I think and try to see what is best for me. I realized one thing after coming here—that you are on your own. As a matter of fact, it is good. I am now more confident and independent."

Rahman's stay in this country has not been very easy. He has a great sense of loneliness and is not confident. He attributes his problems to his authoritarian parents, and, even though he shows a sense of respect for them, he lacks admiration and closeness.

Unlike Case I, Case II reveals that her adjustment in this country was rather smooth. According to her, this has largely been owing to the presence of her family and the very cordial and friendly relations that she has always had with them.

Case II

Sushma is 29 years old and is married to a college professor. They have a five-year-old child and live in an apartment which they own. She comes from a big city in India, and came to this country some ten years ago to join her parents. She has some professional background, but at the present time is basically a housewife, and works for a small business that she and her husband have started recently. She has been to India for visits several times since she came to New York.

Sushma's immediate reaction after coming to this country was favorable. She had a fairly accurate idea about the United States, and she found things as expected. However, she was somewhat disappointed to see dirt, filth, and crowds in the city. She found people friendly, and was able to make friends easily. The first two or three months, when she was not doing anything, things were difficult and she often became bored. As soon as she started school she felt much better. She found the atmosphere in school very conducive to learning, with more openness in the relationship between students and teachers. One thing she did not like at school was the use of drugs. She was invited by some friends to try drugs, but she completely refused to take any part in it. Sushma feels that her strong family ties and Indian background were instrumental in coping with peer pressure. Living with her parents at the time was also helpful.

Sushma's attitude has not changed much since she cam here. It remains positive and favorable. However, she is aware of the fact that there are certain built-in differences between Indians and Americans. As she explained, these differences are basically cultural, and in terms of attitude toward life in general. For example, "Americans are materialistic; they are not close to the family." She does not find much difficulty in relating to Americans or non-Indians. "As a matter of fact, I have some close American and non-Indian friends. Of course I have Indian friends." Despite this acculturation, her Indian identity is very strong. "I identify myself as an Indian," was her reply. She is very particular and conscious about rearing her child as an Indian. She speaks her native language with the child, and thus tries to cultivate Indian values in her.

Sushma comes from a liberal family background. Her relations with her parents have always been very friendly and intimate. Her parents never objected to her dating, even with Americans or non-Indians. She was married here by her own choice, but says that her decision was not based on romantic love alone. "It was not heart alone, but mind also played a part in my decision. I was aware of some criticism, but I knew what I was getting into and was very confident." Her parents remained neutral and accepted her decision. This situation, however, did not cause any serious conflict or strain. "I don't get upset about things easily." Things were quite normal at the time of her daughter's birth. However, some anxiety was felt when her husband was operated on for suspicion of some serious disease. She feels that the doctors were not very cautious and made a hasty decision. But, despite some confusion caused by this incident, she maintained her calm.

There has been no occasion for her to seek professional help in time of stress, but even if there were such an occasion, she would not seek help from professionals. "Psychiatrists go for unnecessary analysis. They make things more complicated," Sushma observes. When asked, "What do you do when you are confronted with a serious problem?" her answer was, "I talk with my family and close friends." In the area of marital relations compromise is the best thing, "both parties must learn to accommodate as much as possible. There are no set rules. Each couple has to work out things for themselves as they see fit. For example, my husband has a very hectic schedule so I don't demand that he share in household work. These things depend entirely on circumstances."

Sushma thinks that her coming here at a young age was in her favor, and adjustment to this society was much easier for her than for many Indian girls she knows who came here after marriage, with young children. However, the fact that she was not overwhelmed by America has to do with the very close ties and relations she has maintained with her parents, and the Indian values she has acquired from them. Even after marriage, she maintains close relations with her parents, and believes that this has helped her to maintain her Indian identity, and to cope with life in an easy manner.

As we examine case III we shall find many similarities with Case II. Both feel that their adjustment was easy, and they have assimilated quite well, but still are particular about maintaining their Indian identity. Both preferred to marry Indians even though they dated

Americans. They are self-confident, do not get upset easily, and believe that compromise is the best way to deal with marital problems. The major difference between the two is that Case II is very close to her family and Case III is not; but both have good friends, and they rely on them.

Case III

Kapadia is 39 years old, and first came to this country as a student almost sixteen years ago. After completing his education he returned to India, and came back to the United States as an immigrant in 1967. He has been married for about ten years, and has two children. He lives in his own house in the suburbs. He works as a junior executive for a large corporation. He comes from a middle-sized city in India, and has been there several times since he immigrated.

Kapadia was sufficiently exposed to America before he first came here that his immediate reaction was not one of surprise at all. He found things as expected. He found Americans very hospitable and friendly. The thing he liked best about them was their openness. Generally his reaction was favorable. Sometimes he missed his family and did not like American food, but he also pointed out that this was quite normal.

Although his attitude toward American society remains positive, he observes, "Let us face it. We are different because of our looks and culture. I look different. People think I am different. Sometimes it may be to your advantage, sometimes not." He thinks that, unlike other Indians, he has no problem with language. Consciously or unconsciously, he has picked up "American English," so he has no problem in communicating with Americans.

The thing which bothers Kapadia most is the lack of recognition of his potential at work. "Money is okay, but the job is not responsible. This is a constant source of strain and frustration to me." Kapadia cannot say whether it is discrimination, but he thinks that it has something to do with his being different. "I resent it. I want to prove myself. I want to show that I have ideas, I can do things. That is why I have joined Indian organizations. My activities in these groups give me reassurance and confidence."

He dated girls, but mostly non-Americans. Even though he likes American girls, he avoided dating them. They are too independent, in his view. "We (Indians) are too emotional," he points out. "I was falling in love like crazy."

Kapadia is happily married to an Indian woman. Theirs is an

arranged marriage, but they had an opportunity to meet beforehand. Kapadia made a trip to India for the marriage. He has had to learn to adjust after marriage, he says. "Compromise is the best technique. Two people cannot live together without making compromises." He has not sought professional help in resolving marital difficulties, and does not talk to friends or relatives about it. "We talk between ourselves. The important thing is love, and, more importantly, convictions. I can't stand divorce or even separation. I have changed my views about women. Now I believe that women are also important and are capable of doing things. This I never believed until I got married."

Kapadia has not experienced any problem thus far in rearing children in this country. However, one day he found out that his children didn't even known what their religion was, so now he makes it a point to take his children to religious functions whenever possible, even though he is not religious *per se*. His wife has also influenced his thinking about rearing children, and he now thinks that the father has to play an active role, so he tries to spend time with the children, and talk with them as much as possible.

A few years ago Kapadia's younger brother came and stayed with him. This was before Kapadia's marriage. His brother's expectations were too high, and he wanted Kapadia to do everything for him. Even after getting a job, the brother was unwilling to contribute to their expenses, which led to a serious misunderstanding between them, so that his brother moved out without informing Kapadia. This was a source of anxiety and unhappiness. Some time back Kapadia's father died, but this had been expected, as he was not well and was also old.

Kapadia does not seem to maintain close ties with his family in India, even though he visits them often. Naturally, he does not share much with them in times of stress. He does not get upset easily unless it is a very serious matter.

Kapadia has a small circle of friends, but basically he is an introvert and a family-oriented man. He seems to be well-adjusted. However, lack of recognition of his potential at the job remains a source of frustration. He believes that "Maintaining one's heritage is important; at the same time, we must also mingle with Americans. Only then suspicions can be removed and we can learn how to adjust and live in this society."

Unlike Cases II and III, we find clear differences between Cases III and IV. Case IV found adjustment more difficult in the United States. She remains an outsider, and does not associate herself with the new society. She maintains close ties with her family and puts greater emphasis on religion. Both cases lack faith in seeking professional help, and rely on informal sources in dealing with conflict and crises in their lives.

Case IV

Rajni is 35 years old and lives in a rented apartment. She is single, and came to this country in 1972. She works as an analyst in a medium-sized corporation. She has not been to India since she came here. She lived in a large city in India.

Initially, Rajni had never wanted to come to the United States. She had a good job with the central government and was quite satisfied. Rajni's younger sister, who was already here with her family (she has since returned to India), insisted that Rajni come here for higher studies and professional training. "My immediate reaction was that of shock. I didn't like the way people acted because of my background. Basically, it was a state of confusion for me," says Rajni. Gradually she learned from her sister and brother-in-law about this culture, its customs and traditions. "I understood but never accepted it," she remarks. Her sister and brother-in-law returned to India in 1974 and she also wanted to go back, but again her sister insisted that she must complete her studies and training. "I could never imagine that I would survive without them, but gradually things worked out okay," Rajni observed. At this time she pointed out that she comes from a traditional conservative family, and has always lived a very sheltered life. Even after coming to the United States she was entirely dependent on her sister.

Now, she feels much better. Sometimes she even thinks that in some respects life is better here. There is more openness and understanding. However, she feels that people who come from other parts of the world, especially Western parts, find it much easier to adjust and they are also more readily accepted. "For us, it is not easy. Our background is very different." She does not think that language creates any problem. "There is no problem in communicating with educated people. Only some local people (shopkeepers, etc.) don't understand and think we are wrong." On the whole, Rajni's attitude toward America is more favorable now than it was when she first came. She understands things better, even those she does not appreciate.

Rajni intends to marry. She will marry only an Indian, either in India or the United States. "I am a family type." However, she does not go out, and thinks that establishing any relationship before marriage is not good. "When you get involved with somebody you lose your objectivity. Meeting somebody a couple of times with your family is okay. It gives you some idea about that person. If you think he is okay you can get married." This is her idea of selecting a marriage partner. She further explains her point by saying, "Even if you have lived with the person you are going to have problems. So if you think that this person is good and you have faith in the institution of marriage, things will work out all right " Once Rajni's sister introduced her to a young man, and they met twice in her sister's presence. But then this man wanted to go out with Rajni alone, which she refused. She does not understand why these Indians forget their customs. She also feels that one has to be more careful with Indian men than with Americans. "Our own people misunderstand, try to take advantage of your being here independently." She prefers arranged marriage, and strongly approves of it.

She thinks that the educational system is better in this country. Students have direct contact with their teachers, and it is more competitive. She enjoys her school work.

In the area of employment, she thinks that she deserves a better job. She does not think that there is any discrimination at the job. She thinks that she herself has not explored all the job opportunities to find a better position. One thing she has learned: "If you are polite, people take advantage of you. I have become more aggressive now, and fight back. You cannot survive without this," she feels.

Rajni's social life is very limited. She has few Indian friends. She prefers to make friends with families, rather than with single people. She knows some Americans, but the relationship is very formal.

How does she go about meeting new challenges in a new society? Rajni's answers are simple. She talks to friends and seeks their advice. Sometimes she writes to her sister, but prefers not to, because it may cause unnecessary anxiety for the sister. She is very close to her mother, but never writes to her about any problems, for the same reason. There has been no need for her to seek professional help because of any problem, but she does not believe in it anyway. Sometimes she prays and goes to a temple. "I have always believed in God. It gives me strength. It helps me."

Although Rajni is aware of the new realities of life, she operates pretty much in her traditional conservative framework.

We find in examining Case V that, unlike Case IV, he has attempted to participate in the mainstream of American life, and is professionally successful. But, despite this success, the sense of frustration in Case V is not any different from that in Case IV. Their reliance on family and friends is very strong, but case V is the only who believes in the value of professional help in times of crisis, even though he has never sought it. This change in his attitude as compared to the rest of the cases may be related to his profession of medicine.

Case V

Sudarshan is 42 years old, and lives with his wife, three children, and parents in his own house. He came to this country some 10 years ago. He lived in a small town in India, and has been there several times since he came to the United States. He is an M. D.

His immediate reaction after arrival was favorable. He was quite impressed by the material and scientific achievements. Although local people were friendly, he himself was shy, and tried to keep his relations confined to a few Indian friends. His main problem at that time was food, as he was then a strict vegetarian. Now he eats meat when necessary, but still prefers a vegetarian diet. Eating outside the home was not easy. "I would pick up my food and sit at a corner in the cafeteria because I did not want to create much curiosity among my colleagues about my food. At that time rarely were Americans vegetarians." Communicating with them individually was no problem, "but whenever we sat in a group I felt out of place."

Sudarshan's attitude since then has changed somewhat. He is not particularly impressed by the material achievements of this society. Although he lives comfortably and enjoys a high standard of living, he does not think much of it. He is quite successful in his profession, but thinks that his being Indian has some disadvantages. "You have to be better than your American counterpart to get the same position," is his feeling. He mingles socially with Americans, but thinks that he and his wife have certain disadvantages because of their cultural background and lack of knowledge about local affairs. "We are not really accepted by local people, whereas in India we would be at the top of things. Sometimes we feel like fish out of water," observes Sudarshan painfully.

For a few years they lived in an apartment. This experience was not pleasant. One neighbor always bothered them and complained that they had too many visitors. When his parents came from India and Sudarshan was not home, none of the neighbors showed any courtesy. His parents had to wait in the hallway. Sometime ago, he bought a house in a nice community. Although they are comfortable, they feel that they are not quite welcomed by the community.

Sudarshan is also experiencing some difficulties that his 12-year-old daughter is facing at school. Girls of her age have boyfriends, so his daughter feels isolated in her social life at school. "When they are young there is no problem. They mix all right, but as they grow up they figure this girl is different," was his observation. So, his daughter has mainly Indian friends, and she enjoys meeting them on weekends.

His family life is quite happy. He thinks that both he and his wife have changed their attitudes to some extent, and have learned to enjoy life more than they did in India. "Emotionally, we feel younger." Sometimes there may be the usual arguments between husband and wife, but nothing more than that. He does not think that their coming to this country had any adverse effect on their relationship. He has brothers and other relatives in this country. Because of some sense of competition and jealousy, especially among the wives, some strain is caused in their relationship.

"Conflict is the natural consequence of migration. Like any other group of immigrants, we also face them. We always think of our native land and want to retain its value and culture, but again we must be realistic," maintains Sudarshan. For example, he says, "I have raised my daughter like an Indian and we would be happy if she marries an Indian, even if the boy does not come from our region or background. We would not be happy if she marries an American, and I don't think she would, but anything can happen." Sometimes Sudarshan also feels that it was not realistic for him to bring up his daughter completely like an Indian. It would have been easier for her to adjust if he had raised her somewhat differently, i.e., a little bit more in the American way.

Most of the stress and strain experienced by Sudarshan is because of his professional ambitions. He is very successful, but he wants to do still better. But then he also realizes that he has certain handicaps because of his cultural background. "You know I am considered to be a good second man but not number one. I am not aggressive enough, so I try to limit my goals to avoid frustration, otherwise it can be very stressful." He does talk to his friends about these problems, but not to relatives. "I have no inhibitions about seeking psychiatric help if it is needed, but so far there has been no occasion."

Sudarshan is very successful in his profession, and has established a good reputation. But he is not quite at ease in the new environment in which he is living. His attitude about life is somewhat philosophical, and he uses it to deal with any stressful situation he confronts.

A careful examination of all five cases suggests that there is some variety in so far as their coping patterns are concerned. They all rely heavily on family and friendship groups, and are somewhat philosophical in their attitudes in dealing with the new realities of life. With one exception, they all show distrust in the profession of psychiatry, and feel that it is unnecessary.

Discussion

Consequences of Migration

The results of the present investigation and the case examples noted above suggest that the consequences of migration for the first generation in the New York City area are not very significant in any fundamental way. They report having achieved in some areas, experienced loss in other areas. They certainly have done well in terms of improving their standard of living and their economic and professional status. However, the quality of their lives has not necessarily improved. There is a sense of alienation, frustration, anger, and, worst of all, loneliness. The institutional and community support which was available to them in India is not there any more. In the absence of these institutional supports, with their unwillingness to avail themselves of professional help available here, they are in a dilemma, and unsure how to meet the challenges of the new life.

Although strong ties with the family are still maintained, and the extended family remains very strong, it is not the same as it was in India. Even though there have been some changes in the attitudes of husbands and wives toward each other since coming here, the traditional values remain dominant. Strong conviction in the institution of marriage, a willingness to compromise, and a strong sense of commitment to marriage are all very important in tolerating stress and managing conflicts.

The recognition of the fact that they are on their own has worked in two ways. In some cases it has given a sense of independence and confidence, and in others it has created a sense of fear and insecurity.

Immigrants generally think there has been no significant change in their attitudes since coming here. Yet their relatives and friends who come here to visit, or even new arrivals, have consistently pointed out that after coming here people become more selfish and self-centered. It seems that this behavioral change is in response to the new conditions in which they live, in which being very friendly, polite, and nice is not considered very desirable.

We also note that, though the overwhelming majority of Indian immigrants are doing well in their respective professions, they remain dissatisfied with their jobs and feel that they are not getting what they deserve. Although it is possible that there is some discrimination involved, their sensitivity may be heightened because of the rising level of expectation characteristic of an immigrant population like the Indian one.

Our hypothesis is that the Indians after coming here tend to become more conscious of their Indian identity, and try to preserve this insofar as it is possible.

Patterns of Adaptation

The profile of Indian immigrants points out an interesting phenomenon in the context of their structural adaptation. Since they already have a high educational and professional level, their structural assimilation, unlike that of many other immigrant groups, is relatively easy. They do well in school, have steady jobs, own property, and are active in civic and organizational activities. However, in the social and cultural sphere of life they tend not to adapt to the American mainstream. For example, going back to India for an arranged marriage is a very common phenomenon. Even those who marry here consider family, caste, religion, and economic factors important, rather than love alone. For those who come here with families, the attitudes and values have not changed much. Although children have somewhat more freedom than they would have had in India, the basic authoritarian structure of the family is maintained. A girl who came here at a young age and considers herself to be sufficiently assimilated thought that love alone was not the basis of her choice in marriage. She also feels that, if the husband has a very heavy schedule, he should not be bothered with household work. However, there is an increasing sense of equalitarianism between husbands and wives. Husbands do confess that they have changed their attitudes toward their wives, that they respect them, and also share the responsibilities of rearing children.

In their social life, the immigrants mainly confine their relations and friendships to Indians. Their pattern of close friendship is often confined to people who come from the same region in India. They have mainly secondary group relations with Americans, and only a few have personal intimate relations with American friends. Many Indians see American friends as a bad influence on their children, and feel that children must be protected from these exposures. Their concept of time, also, has not changed. Although they are punctual on the job and in meeting their professional obligations, they fail to understand why one should follow time so strictly.

On the whole, the Indian immigrants have adapted to new conditions wherever essential, but the basic personality structure and value system is generally not affected. Although some Indians feel that there should be more interaction with local communities, and that they must learn to adapt to a changing way of life, this remains a minority view. The majority of Indian immigrants are of the view that there is no need to change and adapt. "We are better off as we are," is the widespread feeling of the community. The notion of cultural pluralism is a more attractive option, and some argue that, physically and culturally, Indians are so distinct that they can never be fully assimilated in this society.

Effect on Emotional Health

There are many studies which shed some light on the phenomenon of migration and mental health. Yap (1951) has done a comparative study of the problem of acculturation and mental illness. Kiev (1963) draws attention to the association between the patient's delusions and the beliefs of West Indian immigrants, especially with regard to religion and magical concepts. Odegaard's (1932) classical work on Norwegians emigrating to the United States indicates the comparatively high rates of mental-hospital admission among the immigrants as compared with the nonmigrating population in Norway, and suggests that the migrating process makes manifest the basic instability. Paranoid reactions have been described as common not only among refugees, but among most immigrant groups (Pederson, 1949). Gordon (1965) suggests that the most obvious environmental stresses to which the West Indian immigrant group is subject are socioeconomic in nature. Glass (1961) states that working-class West Indians have greater difficulties in becoming integrated than those who are middle-class.

A number of studies have shown that there is some relationship between migration and mental disorder. Murphy (1965) suggests that, with all the variables controlled, the phenomenon of migration itself is not a primary causative determinant.

Saran and Sarma (1970) report that only a small percentage of the Indian immigrants in New York City reported psychosomatic illness after coming to this country. Pitchumoni and Saran (1980), found that nearly 90% of their respondents report that they feel more mental tension in the United States than in India. However, none indicated that they had ever consulted a psychiatrist.

Although tension and strain are experienced by a large number among the Indian population living in New York City, the degree of tension is not serious or significant enough to cause them to seek professional help. It is also clear that Indians attach a strong stigma to mental illness, and avoid seeking professional help for it. Some informants said that, though some members of their families had sought some professional help, it was not known to anybody outside the family. It is also felt that problems can be resolved without going to psychiatrists. It is clear why the incidence of psychological problems reported is so low.

In private conversation with some Indian psychiatrists, two important points emerged. Some of the psychiatrists have seen Indian patients. The first point is that the level of tolerance among the Indian immigrants is rather high. "Don't get upset easily" is a frequent reply when they are asked about strains or stresses they have experienced. The second point is that Indians who have migrated to the United States are qualified people; therefore, unlike other immigrant groups, their adjustment is relatively easy.

Strategies for Coping with Challenges

People differ in the ways and means they adopt to cope with their strains, and the Indian population is no exception. However, reliance on family and friends for emotional and social support seems to be a very common strategy. Even though there is a lack of formal organizations or service agencies established to provide assistance and help for Indian immigrants, the existence of a large number of social, cultural, regional, and religious organizations provides an informal network which is readily available to give assistance in times of need.

It has also been observed that Indians tend to compartmentalize their activities; that is, activities in one sphere of life are separated from those in another, or at least there is little influence of one over the other. As a result, strain in marital relationships does not necessarily affect performance on the job, or relationships with friends. Again, the notion of "don't get upset easily" prevails.

A look at the five cases reported in this study gives an idea of the variety of strategies employed.

Case I does not do anything about his problems. He just thinks that with the passage of time things will get better. He talks neither to his family nor to friends. However, it seems that his family is aware of some of his problems, and tries to give him moral support in times of crisis.

Case II has not experienced any serious strain. She is very confident, and attributes this to her closeness with her family. She is opposed to psychiatrists, but talks to friends about her problems. She recognizes that compromise is essential for adjusting in married life.

Case III has also not experienced any serious strain since coming here. His major source of strain is the lack of recognition of his potential in his job. To compensate, he is involved in community affairs, and has taken positions of responsibility in them. His success in this area provides him with reassurance and self-confidence. He also believes that making compromises is the best technique for resolving differences.

Case IV is very dependent on her sister and family. She does talk to friends about her problems. She has also changed her personality traits to deal with new challenges. She is more assertive now, and exerts herself. She goes to temple and prays. This always helps her.

Case V experiences a major source of strain in his job. Although he is very successful, he thinks he deserves more. He is both realistic and philosophical about life. He limits his goals to avoid frustration. He knows that he is different; therefore he is always going to have some handicap. He accepts this as a fact of his new life and surroundings, and has made peace with himself.

Conclusion

It can be concluded from this study that Indian immigrants living in the New York City area have not faced any serious consequences of migration, especially in the social-psychological spheres of life. It must be kept in mind that their stay in this country has not been of great duration, and that the children in most families are still very young, so that the conflict with children which is likely to occur because of peer influence has not yet been experienced in a major way.

It is also clear that Indian immigrants have felt the strain of a highly competitive and individualistic way of life, coupled with a lack of the institutional and community support more readily available in the Indian context. There are some cases of divorce, mental illness, suicide, and similar pathologies, even though they are not characteristic of the larger community.

An examination of the five cases presented also suggests that Indian immigrants have developed different strategies for coping with whatever strains they experience. There are no formal institutional arrangements available for support, except for families and friendship groups. In addition, an informal network has come into being as a result of the establishment of a large number of social, cultural, regional, and religious organizations.

A careful examination of the life-styles and patterns of adaptation of these immigrants also suggests that there is a potential for strain and conflict, which in some cases is evidenced. In the light of the data which suggest that this population is reluctant to seek professional help in times of distress and crisis, it is even more important for those who are interested in the problems of immigration, mental health, and quality of life to take steps for providing some assistance to this group of immigrants, and making their process of adjustment easier.

Some Suggestions

On the basis of this study, a number of practical and research suggestions may be made.

It is clear that, though there is need, the Indian community is at a disadvantage, since there is a total absence of any institutional and organizational support available to them for coping with conflict and strain.

Professional and social agencies especially concerned with the quality of life and the mental health of new immigrants must make efforts to establish service agencies, counseling groups, and information pools for assisting new arrivals in particular in a variety of ways.

This study suggests that, unlike older immigrant groups, structural assimilation of the Indian population is relatively easy. We also find that there is growing support to maintain ethnic identity. We also find that the socioeconomic and cultural background of this group is very different from that of earlier immigrant groups. Therefore, it is necessary to develop new models and theoretical approaches for studying such groups as the Indians. A comparative study of other new immigrants, especially Asian groups, may be very useful for both practical and theoretical considerations.

It is hoped that a longitudinal study of the kind this author has initiated, and from which data were used for this paper, is particularly relevant for the testing of many hypotheses in the area of ethnicity, cultural pluralism, migration, and mental health.

Acknowledgment

The author is thankful to Ms. Avrama Gingold for typing and editorial assistance and to Ms. Rebecca Greenberg for assisting in library work.

References

Aurora, G. S. *The New Frontiersmen: Indians in Great Britain*. Bombay: Popular Prakashan Press, 1967.

Coelho, G. V. *Changing Images of America: A Study of Indian Students' Perceptions*. Glencoe, Ill.: Free Press, 1958.

Coelho, G. V., & Stein, J. J. Coping with the stresses of an urban planet. *Habitat: An International Journal*, 1977, *2*, 379–390.

Desai, R. *Indian Immigrants in Britain.* New York: Oxford University Press, 1963.
Eames, E., & Robboy, H. *The British Midlands* and its Punjabi community. In P. Saran & E. Eames (Eds.), *New ethics: Asian Indians in the United States.* New York: Praeger, 1980.
Elkan, W. *Migrants and Proletarians.* New York: Oxford University Press, 1960.
Fisher, M. P. *Ethnic Identity: Asian Indians in the New York City Area.* Unpublished doctoral dissertation, Graduate School of the City University of New York, 1978.
Gandhi, R. *Little India: Localism and Cosmopolitanism in an Indian Student Colony.* Unpublished doctoral dissertation, University of Minnesota, 1967.
Glass, R. *Newcomers: The West Indians in London.* New Haven: Harvard University Press, 1961.
Glazer, N., & Moynihan, D. P. *Beyond the Melting Pot.* Cambridge, Mass.: The M.I.T. Press, 1963.
Gordon, E. B. Mentally ill West Indian immigrants. *British Journal of Psychiatry,* 1965, *111,* 877–887.
Gordon, M. *Assimilation in American life.* New York: Oxford University Press, 1964.
Handlin, O. *The uprooted* (2nd ed.). Boston: Little, Brown, 1973.
Jensen, J. M. East Indians. In *Harvard Encyclopedia of ethnic groups,* New Haven: Harvard University Press, 1980.
Kiev, A. Beliefs and delusions of West Indian immigrants in London. *British Journal of Psychiatry,* 1963, *109,* 356–363.
Klass, M. *East Indians in Trinidad: A Study of Cultural Persistence.* New York: Columbia University Press, 1961.
Lambert, R., & Bressler, M. *Indian students on an American campus.* Minneapolis: University of Minnesota Press, 1958.
Leonhard-Spark, P., & Saran, P. Indian immigrants in the United States: A portrait in number. In P. Saran & E. Eames (Eds.), *New ethics: Asian Indians in the United States.* New Haven: Harvard University Press, 1980.
Mohaptra, M. K. *Orientations of Overseas Indians toward Discrimination in American Society.* Paper presented at the 6th Annual Conference on South Asian Studies, University of Wisconsin, Madison, November 1977.
Morris, S. Indians in East Africa: A Study in plural society. *British Journal of Sociology,* 1956, *3* 194–211.
Murphy, H. B. M. Migration and major mental disorders: A reprisal. In M. B. Cantor (Ed.), *Mobility and mental Health.* Springfield, Ill. Charles C. Thomas, 1965.
Nandan, Y. *The East Indian Family in American City and Suburb.* Paper presented at the 6th Annual Conference on Ethnic and Minority Studies, University of Wisconsin, LaCrosse, April 1978.
Odegaard, O. Emigration and insanity. *Acta Psychiatrica et Neurologica* (Supplement 4), 1932.
Pedersen, S. Psychopathological reactions to extreme social displacements. *Psychoanalytic Review,* 1949, *34,* 344.
Pitchumoni, C. S., & Saran, P. Health care in a new social and cultural setting. (In preparation.)
Saran, P. *Indian Immigrants In and Around New York City.* Paper presented at the Eastern Sociological Society meetings, New York City, April 1975.
Saran, P. *New Ethnics: The Case of the East Indians in New York City.* Paper presented at the Smithsonian Institution's Research Institute on Immigration and Ethnic Studies' National Conference on "The New Immigration: Implications for American Society and the International Community", Washington, D. C., November 1976.
Saran, P. Cosmopolitans from India. *Society,* 1977, *6,* 65–69.

Saran, P., & Sarma, A. V. N. Ecological and sociological influences on patterns of illness and treatment. *Indian Journal of Social Research*, 1970, *11*, 31–41.

Thomas, W. I., & Znaniecki, F. *The Polish Peasant in Europe and America* (2nd ed.). New York: Dover, 1958.

Toffler, A. *Future shock*. New York: Bantam Books, 1971.

Yap, P. M. Mental diseases peculiar to certain cultures: A survey of comparative psychiatry. *Journal of Mental Science*, 1951, *97*, 313–327.

18

New Immigrants and Social-Support Systems: Information-Seeking Patterns in a Metropolis

Murali Nair

The modern immigrant no longer has the freedom of the early settler, who usually simply transferred his own ways of doing things. Upon arrival in a new country, today's immigrant faces a task of integration into a new culture; there are psychological stresses and strains, which have become magnified with the complexity of modern social organization. Present-day immigration has a desocializing effect on individual life, involving complex disorganization of the individual's role system, and some disturbance of social identity and self-image (Cf. Eisenstadt, 1955). This chapter argues that the possession of adequate information about the social-support systems of the new country lessens the desocializing effect of immigration. The conceptual theme of this study is that psychosocial adaptation is a function of information seeking which facilitates, and is facilitated by, use of social-support networks. Some information is sought before the arrival of immigrants in a new country, and some after arrival.

Study Methodology

This article presents part of the results of a research study conducted in metropolitan Toronto regarding the new immigrant's use of

Murali Nair • Graduate School of Social Work, Marywood College, Scranton, Pennsylvania 18509.

social service agencies (Nair, 1978). The purpose of that study was to find out how far new immigrants depend on social-service agencies for assistance in coping with their "settling-in problems"—their psychosocial adaptation in the new country. The study assumes that, in many cases, the new immigrants do not have close friends, or family members, to depend on in the new environment. Relatively little seems to be known about the immigrant's use of the social agencies that are available to assist him. Are there barriers of language and culture that prevent the immigrant from using the help offered? Is a social agency such a strange organization to the immigrant that it may not, in its present structure, be accessible to him? When he lacks an orientation toward social agencies and services, is it simply because he is not aware of them? Where else do immigrants go in the new country for assistance in coping with their problems, when they have no knowledge of the network of support systems in the new environment? The aim of this study is to find answers to these questions that arise when one begins to consider the adaptation of the immigrant.

The study population were immigrants who had come to Toronto less than two years before. It is especially during this early period that the immigrant has to cope with adjustment problems.

Immigrants were divided into (a) those who sought assistance from social-service agencies and (b) those who did not. A quota-sampling method was used to select immigrants. Fifty immigrants were selected from each of two public social-service agencies and each of two private agencies—agencies that had been set up primarily to provide services to new immigrants to help them cope with initial settlement problems in the new country. To these 200 cases were added 75 immigrants who never had been to an agency for assistance; they were selected from the membership lists of ten immigrant associations.

An interview schedule was used, with the wording and ideas as simple and concrete as possible. Where this was not possible and a concept had to be expressed, the wording was often taken from statements that new immigrants had made during the pilot interviews. Items used in the interview schedule were based in part on responses derived from these pilot interviews, but mainly on the author's personal experience in working with immigrants, plus general suggestions from the literature. The researcher administered the open-ended questionnaire personally in 1977, in all the 275 interviews with the immigrants. In the interview situation the questions were followed by probes, which provided a much better indication of whether the respondent had any information about the issue, whether he had a clearly formulated opinion about it, and how strongly he felt about it. In cases where immigrants could not communicate in even a little

English, the researcher used interpreters, except where he himself spoke the immigrant's language. Altogether, 200 immigrants were interviewed at four different social-service agencies, and 75 nonusers of agencies were interviewed at their homes. In most of the home-visit cases, the interviewer took a volunteer belonging to the same nationality group as the immigrant, to instill a sense of trust in him and to interpret.

The schedule began with questions related to information-seeking patterns during the preimmigration phase, that is, from the time a person started thinking about immigrating to a country until the actual departure from the country of origin. It focused on the immigrants' experience in seeking assistance in their home countries, the use of social-support systems in gathering information, the type of assistance they sought from these agencies, their awareness of the existence of social agencies in the new country, and their reasons if they did not use agencies after their arrival. Information was also collected on the immigrants' countries of origin, their cultural and linguistic background, how they handled their personal problems in their home countries, their education and occupation, and whether they came to Canada as "independent" or "dependent" immigrants.

Tabulation of the data shows an equal number of immigrants in the sample who came from the developing countries (53%) and from the industrialized countries (47%). Half the sample had no relatives or friends in Toronto, while the other half had at least one close relative or friend who was in Toronto before the new immigrant's arrival; 56% had English, the others had no English, or so little as to need the help of an interpreter in seeking assistance from the social agencies; 62% had college-level education, and 38% less; 63% were employed before they emigrated, and 37% were unemployed (including housewives, aged parents, and children).

Social-Support Systems

According to Caplan (1974), a "support system" is an enduring pattern of continuous or intermittent ties that play a significant part in maintaining the psychological and physical integrity of the individual over time. This support comes from important figures in the environment, including relatives, friends, and neighbors. Such a network often meets the needs of human beings for relatedness; provides recognition, affirmation, and protection from social isolation; and offers the means for identification, and for socialization to the norms, values, knowledge, and, belief systems of the particular culture. It serves as a mutual-aid system, essential for adaptation and for coping with stress (Germain & Gitterman, 1976).

Studies done on the network of support systems that people use in order to cope with different psychosocial problems address themselves to support within the person's existing environment, (Caplan, 1976; Collins & Pancoast, 1976; Craven & Wellman, 1973), or in an environment which is not much different from the old one (Coelho, Hamburg, & Murphey, 1963). But, in the case of immigrants, most of the time these networks of support are not readily available in the new country.

Individuals help others, in part, in order to insure that they, in turn, will receive help when they need it. Since this behavior is functional for society, it is supported by social norms (Gouldner, 1960). Studies of the migration of people from rural to urban environments within the same country offer evidence that ties of kinship and friendship are important (Craven & Wellman, 1973). People do not always enter the city as individuals. Often friends and relatives in the city have induced them to migrate. When people announce their intention to move, friends and relatives frequently supply them with names and addresses of persons to contact in the new place (Collins & Pancoast, 1976).

Why People Immigrate

What would make them choose a particular area to migrate to? Lee (1969) observes that prospective immigrants tend to assess the area in which they live, against the areas that they might go to, in terms of negative as well as positive factors. He finds that people's knowledge of an area of possible destination is seldom exact. He also finds that people's assessment of their own native land is very subjective. People may underestimate their ability to improve a situation at home, and overestimate the benefits of an area of destination. He suggests that the final decision, including the assessment of ability to overcome obstacles, is never a completely rational decision.

The Toronto study shows that younger people who have just joined the labor force tend to emigrate more frequently than elderly people.

The characteristics of people who emigrate—their education, skills, age and so on—tend to be intermediate between those of the area of origin and those of the area of destination. People who are slightly better skilled than the average population in an underdeveloped country tend to emigrate, and these people also tend to be slightly less skilled than the average population in their country of destination.

Sometimes there is a chain-immigration process, in which the migrant follows relatives or friends. Often there are few jobs at home, and one hears that there are plenty of jobs abroad. In addition, there may be other frustrations or dissatisfactions at home, and hope of resolving them through emigration. There are many other possible factors, including a sudden whim or impulse. Chance plays a role. There are usually several forces at work; rarely can an individual's motivation be reduced to a single factor.

Phases of Information Seeking

As to the important information about services available to new immigrants in the country of destination, there are articles by Cachero (1974), Bernard (1977), and Nair (1978) about how the different government and private agencies disseminate it, and how the dissemination could be improved. Here the focus will be on how the new immigrants themselves go about the process of collecting pertinent information about different services available to them. The process of collecting information about the country of destination is quite different (a) in the preimmigration phase, which usually starts as soon as a person thinks about applying for immigration, and (b) in the early postimmigration phase, during the initial adjustment in the new country.

Preimmigration Phase

The prospective immigrant gets general information about the new country from sources such as local libraries, newspapers, documentary films, newsletters, promotion materials put out by the foreign embassies, friends and relatives who had been there, travel agents, etc. This general information usually includes something about its history, its economic and political structure, its culture, its life-style, the different types of social services available to newcomers, etc. The Toronto study shows that people who are aware, already before their immigration, of the existence of different types of services in the new country find it less difficult to find jobs, good housing, etc. after they arrive.

Some prospective immigrants are in touch with relatives or friends in the new country, and get information from them. They are here referred to as "dependent immigrants," the term used by the Canadian immigration authorities for those who have been assisted in their application for immigration by a close relative in the country of destination, who has agreed to take full responsibility for their care, accommodation, and maintenance, and to give assistance in finding employment, if required, for a period of five years. In contrast, "independent immigrants" are persons who have applied for immigrant status entirely on their own. The Toronto study shows that knowledge of the new country is greater among independent than among dependent immigrants.

Independent Immigrants

People who have no close relatives in Canada must, nowadays, go through a rigorous selection process. Education and a skill or a profession count in their favor. In addition, there is a personal interview with the visa officer in the home country. The officer expects the prospective immigrant to have general information about her or his new country. Usually the applicant has collected information on its geography, important cities, climate, type of government, political parties, population, industries, education system, and other basic aspects. Usually it has been collected through reading books or pamphlets, watching documentary movies, talking to friends who have been to the country, or listening to lectures.

By virtue of this selection process, only the educated, skilled, or professional types of people in the Toronto study had come as independent immigrants; usually such people know more about the new country than dependent immigrants do. The study found that 75% of the independent immigrants had acquired information about service availability in the new country before they arrived; they knew much more about the nature of services available to them than did the dependent immigrants. A good proportion of them (60%) get it from reading books and pamphlets. Moreover, once the visa office confirmed their acceptance, some of these immigrants checked with the visa officer on the accuracy of the information they had collected about services from the secondary sources.

Another interesting finding was that people who emigrate from industrialized countries (England, France, West Germany, Italy, the Scandinavian countries, Greece, Portugal, Holland, etc.) have a clearer idea of what types of services they can expect from the agencies in the new country regarding employment, housing, education, health, etc., than people coming from developing nations (East African countries, India, Pakistan, the Philippines, the West Indies, South American countries, etc.).

Independent immigrants from developing nations who lived in big cities knew more than those who were from the rural areas of industrialized countries. Of those who came as independent, 78% of the city residents from industrialized countries and 30% of the city residents from developing nations had information, on arrival, about services relative to employment, health, education, housing, and the like. Independent immigrants from industrialized countries generally sought and obtained most of their information from the visa officers (85%), in contrast to the independent immigrants from developing nations (18%), who relied mainly on travel agents, books and pamphlets, and friends who had been there.

Accordingly, independent immigrants in general are more knowledgeable about the new country, and find it less difficult to integrate themselves into the new society, than dependent immigrants. This finding is supported by the data in an official report, *Three Years in Canada* (1974).

Dependent Immigrants

Prospective immigrants who have close relatives in the country of destination—dependent immigrants—are admitted by criteria for selection that are very liberal compared with those applied to the independent-immigrant category. They do not need much education or occupational skill to be admitted.

People in this group depend heavily on their relatives for information, and this source is often not trustworthy. Often the relatives do not tell their families back home about the difficulties or bad experiences they have had in the new country. They are likely to give a distorted picture of the various services available. Sometimes the families back home do not even have a clear idea about the kind of work they are doing. When they write home, they are likely to comment on the new country's wonders, and to urge the others to immigrate. According to them, the new country is paradise itself. One new immigrant who came from a developing country told the author:

> Relatives who live in foreign countries who come home for a visit, exaggerate a great deal. They would say, "In the new country there are more cars than people; it's so easy to buy a car, even servants can have one," and similar dazzling stories.

Contrary to the general belief, this study shows that, in most cases (65%), potential dependent immigrants have very little factual information about the new country (especially about the services available in the areas of health, education, employment, etc., before they emigrate). Although 10% of the dependent immigrants studied had already visited the new country at least once before, compared to 2% of the independent immigrants, the dependent immigrants, in general, never bothered to collect accurate information about social services available in the new country. They expected their relatives to take care of them in all respects.

Language and level of education are two important factors correlated with the seeking of information back in the home country, and only 26% of the dependent immigrants communicated in English, compared to 85% of the independent immigrants; and only 32% of them had any kind of college education, as compared to 91% of the independent immigrants.

Dependent immigrants receive general information about the new country from letters written by their relatives as well as the relatives' personal visits to the old country.

The study shows that, in general, independent immigrants seek more accurate information than do dependent immigrants. During the preimmigration phase, the visa officers give out less service-related information to the dependent immigrants, since they assume that it is not needed. Dependent immigrants do not bother to improve their English before immigration. Once they arrive in the new country, the dependent immigrants' first thought is family reunion, whereas independent immigrants stress looking for a job.

Postimmigration Phase

After the immigrant arrives, one of the pressing problems he must deal with is adjusting his illusions to what the realities are. He must cope with the immediate practical problems of housing, education, employment, and, in many cases, the task of learning the language. Along with these practical problems, the immigrant must try to be psychologically comfortable with the culture of the new country, and to adjust to it. He is in danger of culture shock; there are conflicts with the culture of his homeland. Weinberg (1961) has observed a remarkable similarity between the needs of the new immigrant and those of the newborn human being: the need for belonging, the need to be loved, understood, and supported, but not to be dominated, pampered, or spoiled. These needs are similar to those enabling the child to develop into a sound, mature person, satisfactorily integrated with his family, community, and society.

In the postimmigration phase the immigration officer at the port will be the first contact for the new immigrants in a strange land. He processes the final part of their visas and accepts them as landed immigrants—as legal residents. The immigrant can—if he/she wishes—ask virtually any sort of question of these officers regarding different services available to him/her; and the officers, in most cases, are equipped to give all kinds of information and to make referrals to social agencies. But, as the Toronto study shows, many immigrants avoid contacts with public officers who could serve them; those from Iron Curtain countries have often had bad experiences with bureaucrats, and people from non-Western countries often expect officials to be corrupt. If an immigrant has relatives or close friends, they are likely to be waiting to receive him at the port of entry, and this type of new immigrant does not have to worry about accommodations or transportation for the time being. In contrast, the independent immigrant usually has to depend on someone else for information regarding transportation, accommodations, shopping, banking facilities, etc.

New immigrants who come from countries where social-service delivery systems are highly developed seek assistance from port-of-entry agencies such as the Travelers Aid Society, church-related organizations, or governmental agencies. During the first few months, and in some cases the first few years, the new immigrants' reference groups seem to be the "people back home" and/or immigrants from their old country who are already in the new country. To lessen the disorientation, insecurity, and anxiety of being in an alien culture, some groups of recent immigrants initially look to their own subculture to assist them in securing employment and low-rent accommodations, and for all kinds of other services. This study shows that, during the first three months of their stay in the new country, many seek information and referral assistance from their own people, or from agencies or associations run by people of the same origin: immigrants who cannot communicate fluently in English (90% of the sample), people who are identifiable as visible minorities, such as Chinese, Japanese, Koreans, blacks and Indians (65%); and some religious groups, such as Jews and Ismailis (70%).

This finding supports the Triseliotis (1972) argument that, even under ideal conditions, newcomers in any area take some time before they feel emotionally ready to seek relationships outside their own group and participate in the life of the local community. No doubt this process can be accelerated if appropriate programs are set up by outsiders in an atmosphere of acceptance and toleration. Original cultural traditions and values become very important to new immigrants, who feel secure with people of this origin.

This study shows that services requested by new immigrants from the immigrant-service-oriented agencies are primarily related to employment, accommodation, and language problems. Immigrants have difficulty in finding jobs, in communicating with employers, and in finding their way around the city and a place to live. But they often fail to seek help; their inability to communicate in English results in their not knowing about community services, and explains much of their failure to use them.

Recently-arrived immigrants approach their ethnic travel agencies and real-estate offices to seek basic information about the availability of social services. The cost of this aid to the immigrants may be a few dollars for translating a letter into English, or it may be a large sum to get a job or other counseling, whereas the same aid is available free from governmental agencies or from volunteer organizations in the community. These new immigrants are unaware that they have a right to the services, and so fall back on private entrepreneurs. Ferguson (1964) found, working with the Italian and Portuguese immigrants of Toronto, that their most pressing social need was for someone or some social agency to help in completing application forms, to interpret government documents, and to put them in touch with general information sources.

Some immigrants never bother to go to traditional social-service agencies, and, instead, rely on radio, television, local newspapers, and word of mouth. The concept of seeking assistance presents different pictures to different immigrants. For those from developing countries, social services that exist in the home countries are of a rudimentary kind (Triseliotis, 1972). Social work, with few exceptions, is an unknown activity, and the notion of a trusting professional relationship is alien to their cultures. Moreover, this study shows that only 15% of the newcomers in the sample who were from developing nations replied that they possibly would ask for or receive help with personal or emotional problems from a social agency, compared to 75% of the immigrants from westernized countries. A majority of the former prefer to discuss their personal, financial, or emotional problems with someone within the family, or a very close friend.

In general, new immigrants prefer to seek information from social agencies which are closer to where they are living (80%). They feel more comfortable if the staff in their agencies speak their language and understand their culture (65%); and they seek out agencies which deal with immigrants on a group basis, rather than the ones dealing with them on a case-by-case basis (65%). One of the new immigrants told the author that, though he speaks English a little, he finds it hard to think in English, and difficult to express all his needs to the social worker in the agency; but, if a social worker understands his culture and language, the immigrant can express all his needs to the worker in his own language.

Comment

Dependent immigrants, who have friends and relatives in the new country, seek less information about social services available to them; instead, they depend heavily on the network of support systems. This study shows that the support systems in most of the cases were not able to provide accurate information to the new immigrants about the new country's service availability. Independent immigrants, on the other hand, have no close relatives or friends, and out of desperate necessity, look for as much information as possible about the new country; and the study showed that, in most cases, the information gathered was correct.

Another interesting finding was that people from developing countries depend less on social-service agencies for coping with settlement problems than do immigrants from industrialized countries. Of the immigrants from developing countries, those from urban areas find it much more comfortable to approach an agency for service than do those from rural areas.

Immigrants who have college-level education and English-language skill use agencies more than non-English-speaking, less educated people. Immigrants who plan to enter the labor market seek outside help more than others.

Irrespective of command of English and occupational status, among immigrants who are identifiable as visible groups, such as Chinese, Japanese, Koreans, East Indians, blacks, and Filipinos, a higher proportion seek information from their own ethnic groups during the first few months of their stay in the new country. Some religious groups, like Jews and Ismailis, also prefer to seek assistance from fellow countrymen or their ethnic and religious associates. Regarding these groups, one immigrant-agency social worker in Toronto made the following comment, which seems accurate:

> As a person proceeds from the position of new arrival to recent immigrant to older immigrant to Canadian of foreign origin to simply a Canadian with perhaps a bit of an accent, his attitudes, needs, desires, and interests undergo a very large reformation. While he is a newcomer seeking his first Canadian job he sees himself as virtually the brother of anyone else from the same country of origin; after he has found employment to his liking and thus achieved some sense of security in the new country, he tends to be more particular in his choice of associates.

In theory, immigrants have the same right to services as native-born people. In practice, however, they do not receive the same services, for reasons such as these:

1. Many services, whether governmental or voluntary in nature, are unknown to them. Often the same service does not exist in their native country, or it is delivered in a different way.

2. When immigrants are aware of available services, they are unable to use them because there is a language barrier and the service-delivery systems are geared to serve native-born Canadians whose needs and cultural patterns differ from those of most immigrants.

For some social workers, according to one immigrant, "difference is threatening." They should learn that not all East Indians speak the same language. Another immigrant, from a European country, commented: "Nurses in Canadian hospitals should know that quiet hospitals, free of visitors, are depressing to patients from European countries accustomed to the lively bustle of a hospital thronging with visiting papas, mammas, and aunts carrying nursing babies."

Conclusion

Seeking correct information is always the difficult part of sound adaptation. Moreover, when an individual decides to emigrate or has made up his mind to emigrate, whatever the negative things he or she hears about the country of destination will not be heard properly—he is much like a traveler who sees a mirage in the desert. He needs to experience, at first hand, what the country is like. By that time it is too late to return to the country of origin. He will have sold his house, resigned his job, said goodbye to his close friends and his family, and left the home country for good. Now he finds out the harsh reality that surrounds him in the new environment—a different culture, language, and climate, unemployment, unfamiliar faces, and no one to rely on for helping with day-to-day problems. Moreover, new immigrants arrive still oriented to the notions that prevail in the old country as to the respective responsibilities of different social-support systems and social-service agencies.

When a newcomer seeks out various types of information to help him settle effectively in a new environment, he is trying to resocialize himself in the new setting. Bar-Yosef (1968) argues that, in striving for adaptation in a new country, the new immigrant knowingly or unknowingly is making an effort to reestablish his role set, to rebuild the connections between self-image and role-image, and achieve a real social status and acceptance in his new country.

Information seeking is one aspect of adapting to a new environment. We all do it in our day-to-day life among familiar faces in a familiar environment. The process gets complicated when we move from one part of the world to another, into an entirely different system of doing things. Ordinarily we continue to use our regular way of doing things, and we may get the pertinent information through trial and error. The situation gets complicated when the new immigrants cannot understand the language of the new country, and have no friends or relatives there.

Recommendations

Even under ideal conditions, newcomers in any area take some time before they feel emotionally ready to seek relationships outside their own group and to participate in the life of the local community. No doubt this process can be accelerated if appropriate programs are set up in an atmosphere of acceptance and toleration at local agencies, staffed by workers who understand and appreciate the new immigrants' language and culture. We should keep in mind that original cultural traditions and values become most important to those people, like immigrants, who do not feel secure, and are frequently rejected by those in the wider community. One of the social workers familiar with immigrants and their problems of communication in English told the researcher:

> To a new immigrant, unhappiness is when the social worker talks to you and you don't understand him. More unhappiness is when you talk to the social worker and he or she doesn't understand you. Because the language differences are immediately recognizable, they generate an immediate impact on the new immigrant, which affects his attitude toward the agencies, as a recipient of services, and socially as a member of the local community. As a result, his perspective on his own future becomes that of a "disadvantaged person."

To make the operational aspects of information more effective, the visa offices in foreign countries should organize predeparture courses in which prospective immigrants receive suitable information on the country of destination and its social-service and other social characteristics.

The inability of non-English-speaking immigrants to communicate effectively is undoubtedly one of the major problems encountered in obtaining employment, housing, social and other essential services. A ices. A relatively easy solution would be for all agencies to hire workers who not only speak the various languages, but also are familiar with the culture and background of their clients.

It is essential that social agencies make available information regarding their services. Service agencies must not assume that their presence is known to all potential users of the service. Consideration should be given to providing information both in written form and through outreach programs. Information regarding services should be translated into major immigrant languages and presented in a form which will be understandable to the new immigrant. Agencies that have made efforts to translate material frequently have made the mistake of assuming too much prior knowledge on the part of the new immigrants. Recognition should be given to the fact that many immigrant groups are not accustomed to receiving information in written form. Service agencies should consider the establishment of outreach programs, utilizing volunteers where possible, to inform potential clients of the existence of their services. Information to new immigrants should be objective, impartial, uniform, and realistic, complete and direct.

The social-work profession, because of its humanitarian concern and familiarity with a wide range of problems and needs, can play an important role in the planning and provision of services designed to meet both material needs and needs in the area of social functioning. If we are concerned about the provision and equitable distribution of these services, then research into the immigrants' use of social-service agencies becomes relevant, with policy implications for how and to whom such services are provided. There is a growing concern to improve the training of professionals who work with immigrants. This is becoming a specialization in social work. Especially the issues of emotional stress, adaptation, and cultural factors with their various implications are being given more systematic consideration in social work with immigrants (cf. Cheetham, 1972; Kent, 1972). The need for training is not restricted to social workers in the immigrant-serving agencies, but is necessary as well for workers in agencies open to the host population, to which immigrants may be referred.

References

Bar-Yosef, B. W. Desocialization and resocialization: The adjustment process of immigrants. *International Migration Review*, 1968, *II*, 27–45.

Bernard, W. S. Services for foreign born. in J. B. Turner *et al.* (Eds.), *Encyclopedia of social work*. Washington, D. C.: National Association of Social Workers, 1977.

Cachero, L. A. Informing the migrant. *International Migration*, 1974, *VII*, 169–179.

Caplan, G. *Support systems and community mental health*. New York: Behavioral Publications, 1974.

Caplan, G. *Support systems and mutual help*. New York: Grune & Stratton, 1976.

Cheetham, A. *Social work with immigrants*. London: Routledge & Kegan Paul, 1972.

Coelho, G. V., Hamburg, D. A., & Murphey, E. Coping strategies in a new learning environment. *Archives of General Psychiatry*, 1963, *9*, 433–443.

Collins, A., & Pancoast, D. *Natural helping networks*. Washington, D. C.: National Association of Social Workers, 1976.

Craven, P., & Wellman, B. The network city. *Sociological Inquiry*, 1973, *43*, 67–73.

Eisenstadt, S. N. *The absorption of immigrants*. Glencoe, Ill.: Free Press, 1955.

Ferguson, E. *Newcomers in transition*. Toronto: International Institute, 1964.

Germain, C., & Gitterman, A. Social work practice: A life model. *Social Service Review*, 1976, *50*(4), 601–609.

Gouldner, A. The norm of reciprocity: A preliminary statement. *American Sociological Review,* 1960, *25*, 232–240.

Kent, B. The social workers' cultural pattern as it affects case work with immigrants. In J. P. Triseliotis (Ed.), *Social work with colored immigrants and their families.* London: Oxford University Press, 1972.

Lee, E. A theory of migration. In J. A. Jackson (Ed.), *Migration.* Cambridge: Cambridge University Press, 1969.

Nair, M. D. *Immigrants' use of social service agencies in a Canadian metropolis.* Unpublished doctoral dissertation, Columbia University School of Social Work, 1978.

Nair, M. D. Social services to new immigrants: A passport to successful adjustment. *Migration Today,* 1978, *VI*(1), 6–11.

Three years in Canada: Report of the longitudinal survey on the economic and social adaptation of immigrants. Ottawa: Information Canada, 1974.

Triseliotis, J. P. The implications of cultural factors in social work with immigrants. In J. P. Triseliotis (Ed.), *Social work with colored immigrants and their families.* London: Oxford University Press, 1972.

Weinberg, A. A. *Migration and belonging: A study of mental health and personal adjustment in Israel.* The Hague: Martins Nijhoff, 1961.

19

Making Friends in a New Culture: South Asian Women in Boston, Massachusetts

Lorna Rhodes AmaraSingham

Introduction

Since World War II there has been a large movement of students and professional people across international boundaries. The economic consequences of this movement have been much discussed, particularly as they relate to the immigration of professionals from Third World countries to the Western nations. Less attention has been paid to the effects of immigration on the personal lives of individuals. As Marris points out in Chapter 5 of this volume, the uprooting of an individual—no matter how voluntarily—from the social matrix in which his or her life has taken on its meaning is a seriously disruptive event. Uprooting "breaks the thread of continuity" provided by a commonly held social life, and inevitably loosens the ties which have

The study on which this paper is based was carried out while the author was a research fellow in the Laboratory in Social Psychiatry, Harvard Medical School, and was supported by a training grant from NIMH. An earlier version of the paper was presented to the conference on "The New Immigration: Implications for American Society and the International Community," sponsored by the Research Institute on Immigration and Ethnic Studies, at the Smithsonian Institution, in November, 1976.

Lorna Rhodes AmaraSingham • Bethesda, Maryland 20016.

bound a person to the shared values of his community. The reconstruction of meaning in a new context becomes one of the first tasks of the immigrant.

In recent years there has been a substantial increase in the number of Asians participating in this international movement of professionals. The number of Asians coming into the United States has increased dramatically; whereas in 1961 21,529 Asians entered this country, the figure had risen to 139,469 in 1975 (out of a total immigration for that year of 386,194). Between 1969 and 1975, 55.9% of all professional immigrants were from Asia. In 1975, of 44,915 Asians naturalized in this country, 11,220 were professionals.[1] Although still a small "minority" in the country as a whole, Asian professionals are becoming increasingly significant in certain fields, particularly medicine and public health.

At least superficially, many of these new Asian immigrants are "westernized." Most come from the urban, Western-educated classes in their home countries. They do not leave as refugees, but as deliberate seekers after greater economic or social security who are already familiar with many aspects of Western society.[2] Since they do not enter American society at the bottom of the economic ladder, they are not subject to the stresses of poverty and low status suffered by earlier generations of immigrants.[3] On the other hand, they come from societies whose social realities differ markedly from those of the United States, and they enter a country which (unlike Great Britain) has few enclaves of Asians already established. For most Asian immigrants, the dense and supportive social networks which characterized their life at home must be exchanged for a mobility and individualism which invest social relationships with unfamiliar meanings. The greatest discontinuity faced by the Asian immigrant is probably in this area of social support, where the contrast between the connectedness left behind and the isolation of the new country is most acutely felt.

Friendship is a somewhat neglected area in the study of immigration.[4] Yet it is the loss of friends as well as family that gives "uprooting" its meaning of a severing of ties. Friendship is important as one of the initial losses which makes immigration stressful; it is equally important as one of the major ways in which a new arrival can compensate for

[1] Compare with 4,878 professionals out of 50,268 Europeans naturalized. All figures are from the *Immigration and Naturalization Service Annual Report* (U.S. Department of Justice).

[2] Many Asians, however, come from former British colonies, and are much more familiar with British than with American culture and society.

[3] See Saran (1977) for a description of Indian immigrants in New York City.

[4] See DuBois (1974) and Eisenstadt (1974) for discussion of the general dimensions of friendship across cultures.

this loss. The creation of new, sometimes deep and lasting relationships can be seen as a positive outcome of the stress and loss inherent in uprooting; the search for new supports can, if fulfilled, become one of the satisfactions of the new way of life.

Yet this search can also be one of the more problematic of the tasks facing the new immigrant. As Paine, (1960) points out: "The rules of relevancy [of friendship] may be largely hidden from all [people] outside the relationship" (p. 510). The symbolic meanings and rules of behavior appropriate to friendship are not the most visible or easily learned aspects of culture, and newcomers to a society are at a particular disadvantage in discovering them, because they come with their own assumptions about the nature and meaning of friendship. To overcome this disadvantage and make new friends among strangers is one of the more difficult of the immigrant's tasks, and the way that this task is entered into is an interesting dimension of the experience of uprooting.

The central concern of this paper is the role of friendship in the lives of women from South Asia who have emigrated to the United States. Two case studies of young South Asian women who have recently come to this country will be presented; the cases illustrate, first, the specific context in which new friendships are entered into, and second, the different and sometimes contrasting ways in which friendship formation takes place. These case studies are the result of a study of South Asian immigrant women which was carried out in Boston and New York from 1974 to 1976. During this period intensive interviews were conducted with twenty women from India and Sri Lanka.[5] All of the women were from the urban middle class; most were 20 to 30 years old[6] and married to professionals who came to this country for professional and social reasons.[7] All spoke English as their primary language, and had some college education.

In South Asia, women in particular grow up in a close or enclosed

[5] Interviews were loosely structured at first, but later, as the issue of friendship emerged as primary, they became focused on this question. Interviews were taped, and all quotations in the text are taken directly from the tapes. The two women presented here were visited extensively in their homes, and interviews sometimes included their families or friends. Every effort has been made to preserve the anonymity of these women, and names, places, and distinguishing features have been changed.

[6] DuBois (1974) discusses the importance of adolescent friendship. The women presented in this paper were just emerging from their teens when they came to the United States, and some of their attitudes toward friendship have probably been influenced by the period of intense friendship which preceded immigration.

[7] Most of these professional immigrants hold immigrant visas allowing them a prolonged stay in the United States and the option of becoming citizens after five years.

network of kin and friends.[8] These relationships provide support for their role as women in the community, and contribute to their sense of identity and self-worth in personal relationships. When a South Asian woman emigrates, she loses direct access to these emotionally significant ties; even though she probably has a nuclear family from the beginning of her stay in the new country, she is usually unaccustomed to relying solely on her husband for emotional support. Unlike her husband, who may have educational or career goals and a group of colleagues to sustain him, she finds herself relatively isolated; friendship with other women becomes one of the major ways available to her for coping with loneliness and the loss of a community of shared concern.

The two women selected for presentation in this paper—here given the names "Kamala" and "Sushila"—are similar in many ways. They are about the same age, and come from middle-class homes in India; both came here shortly after marriage in order to be with their husbands, and both had been in the United States for five years at the completion of this study. Their initial attitudes toward friendship and kinship are similar, and are typical of most of the women in my study; they represent a common ground of values and expectations which are, I think, shared by many new arrivals from Asia.

At the same time, Kamala and Sushila have been chosen because they have taken very different paths in reconstructing a social life in the United States; their coping styles clearly illustrate the major strategies employed by the women in my study, and are particularly clear examples of typical contrasts among the women. "Kamala" has been chosen to represent what will be termed a "boundary-emphasizing" style which is characterized by an attempt to create strong and exclusive relationships with co-nationals. "Sushila" illustrates an "outward-reaching" style characterized by efforts to create relationships with Americans.

The explication of these two patterns of friendship involves three general areas of emphasis, each of which represents a dimension of the process of coping with a new social environment.

1. Kamala and Sushila left India with similar *values* and attitudes regarding friendship. These initial values are elucidated by Coelho (1955), and emerge also in the description of each woman's early adjustment to this country. The problem of either asserting or changing

[8] Most writers on friendship focus on relationships between men, and note the relative immobility and social constriction of women in many societies (DuBois, 1974, p. 27; Leyton, 1974, p. 97). Perhaps in order to discover the quality of friendship among women it is necessary to look at everyday social exchanges centered in the home.

certain values around friendship is central for both women, especially as each comes to perceive conflicts between these values and what are felt to be essential features of American social life.

2. This paper is concerned in large part with the everyday *acts* of friendship through which each woman initiates and nurtures the social relationships which are important to her. An important point is that the evolution of these women's coping styles proceeds, not through the larger cultural resources which might be available to women immersed in their own society, or through the economic and intellectual spheres more likely to be open to men, but rather through social transactions around the events of daily life in the home. Women who are restricted to the context of home and family develop their relationships within this context, and this means that relatively "minor" and everyday events—childcare, visits, shopping—take on importance as contexts for, and expressions of, friendship.

3. Each of the two styles indicated here has its own emotional *rewards* and *costs*. The very fact that social support is no longer a given of the social world, but must actively be sought, is a source of stress; further disappointments and potential losses are built into each woman's adaptive style. The process of bringing together expectation and reality in a new social setting is inevitably problematic; Kamala and Sushila display both the cost of immigration in terms of abbreviated or lost relationships, and the rewards inherent in seeking a new social life.

The social-science literature on friendship frequently refers to the relationship between kinship and friendship. Eisenstadt (1974) points out that this is an ambivalent relationship in all societies, because "while friendship and kinship appear to be both symbolically and organizationally distinct—even opposed—many of their characteristics are ideally similar" (p. 139). There seems to be a wide range of variation in the approaches which cultures provide for this ambivalence; in some, friendship is almost completely assimilated to kinship, with similar bonds and obligations: in others, friendship is regarded as an escape from the obligations of kinship, and its private, particularistic character is stressed.

DuBois (1974) notes that "friendship must be considered in the context of socially stressed values, particularly the images of self provided by the society to . . . its members" (p. 16). There have been several attempts to relate differences in types of friendship to differences in types of society. Cohen (1961) systematically compares friendship types in a number of societies, and reports a correlation between "inalienable" friendship and "solidary" societies; his categories of "casual" and "expedient" friendship are correlated with nonnucleated and individuated social structures. In his view, there is a continuum from bonded, permanent, and socially recognized friendships in more stable and traditional societies to casual and less formalized relationships in more atomistic and competitive societies. DuBois (1974) has also categorized friendships along a continuum from exclusive to casual, and suggests that these types of friendship are correlated with age and with the structure of sexual relationships in any given society. She points out that "friendship" in English usage refers to a wide range of types of relationship, whereas the equivalent word in other cultures may have a more narrow range of meaning.

Studies of middle-class friendship in the United States point to rather contradictory conclusions about American friendship patterns. Paine (1969) suggests that the very fact that United States society is so mobile and individualistic makes possible particularly close and private relationships, largely inaccessible to outside influences, and often perceived as permanent (p. 513). Kurth (1970), on the other hand, discusses what she calls "friendly relations" in American society—that is, casual, expedient, and transient relationships—and suggests that, as the society becomes more fragmented, these relationships predominate. Jacobson (1975), examining changing friendship patterns among unemployed Americans, emphasizes the importance of labeling in the acceptance or rejection of "friends" in times of stress.

The literature suggests that there are tangible differences in friendship patterns among societies, and that these are tied up with attitudes toward kinship, the degree of stability of social groups, and the values which are placed on relationships in general. The division between public and private space and the degree of social distance among individuals are also important determinants of friendship patterns, which differ markedly among societies (Lewin, 1948).

From the point of view of a South Asian immigrant, the question of whether these differences in social patterns are deeply ingrained in the culture or superficial impressions based on limited knowledge is largely irrelevant. Most immigrants do not have initial opportunities for close contact with Americans, and what is visible to them are the more fragmentary aspects of American life. They must deal with a felt discontinuity between the values associated with friendship in their home countries and what they perceive as the transient, unstable, and expedient relationships favored by Americans.

According to Coelho (1955), friendship appears as a central social value of great emotional significance for immigrants from South Asia. Basing his conclusions on a study of male Indian graduate students in the Boston area, Coelho identifies three major values which together constitute the "ideal" of friendship in India. This friendship ideal is shared by the two women in this study, despite their different coping strategies and friendship styles.

First, Coelho says, "friendship represents a close interdependence and interlocking of interests. The question of undue interference in each other's affairs or insensitivity to each other's privacy does not seriously arise" (p. 3). This kind of merging does not depend on "common interests" in the Western sense (see Mehta, 1970, p. 122), but on a common substratum of basic shared experience. The kind of "trial balloon" which is described by Suttles (1970, p. 129)—the use of household decoration, dress, and personal style to "alert people about one's real self"—is not felt to be a necessary prelude to friendship, because these aspects of selfhood are already felt to be similar.

Second, according to Coelho, "there is always time to reciprocate favors and benefits" (p. 3). Friends do not have to negotiate for the return of favors. Reciprocity is so much a part of the relationship, and so much is assumed to be in common between friends, that it is taken for granted that time will "even up" any exchange of favors. Time is never felt to be "running out." This lack of a sense of urgency is reinforced by the fact that even in urban areas Indian children grow up with little expectation of geographic or social mobility which might remove them from their original social circle.

Finally, "friendship cannot be broken" (p. 4). This is the assumption on which the women I talked with feel that their friendships at home were grounded. They do not imagine that they or their friends could change over time in ways that might disrupt the friendship. When friendships do rupture, as Coelho says, this occurs as a "gradual loosening, never a snapping of ties" (p. 4). The term friendship is, then, loaded with a sense of connectedness or bondedness that is almost symbiotic.

The prominence of the themes of permanence and interdependence in Coelho's analysis suggests that the Indians' cognitive map of friendship includes an assimilation of friendship into the kinship system. Relations with friends take on the same emotional sense of permanence and merging of interests as those with kin, as is indicated in the following statements of my informants:

> Friends are like an extension of your family.
> Relatives do not break up, why should friends?
> Friends are more like family than family.

Although dissension, quarrels, or estrangements among relatives are recognized, these possibilities are not important to this comparison, nor is there evidence of ambivalence over issues of privacy or independence. Rather, the values described by Coelho—permanence, lack of bargaining, and mutuality of interest—are considered fundamental to kinship, and are also extended quite literally to the relationship between friends. This equivalence not only elevates the friendship relation to that of "almost kin," but also provides a guideline for expectations and obligations within the relationship.

The case studies which follow have been divided into three phases—the first year, the second and third years, and the fourth and fifth years. The phases are intended to suggest something of the process of friendship formation and re-formation over time in the lives of the women described. It will be noticed that the "phases" are far more meaningful in the description of Sushila, who has greatly changed in her attitudes, than in that of Kamala, who has stayed more or less the same. Since this contrast between the two women is one of the main points to be made in comparing them, the relatively static quality of Kamala's life is shown by the lack of change from one phase to another in her case, whereas for Sushila the phases serve to indicate the importance which the passage of time has had for her.

Kamala: Boundary-Emphasizing Pattern of Friendship

Kamala is a 25-year-old housewife who lives, with her Indian husband and two young children, in an apartment in a large northeastern city. She is middle-class, English-speaking, and from a middle-sized city in India. She was married shortly before leaving for the United States, where her husband is completing his training as an engineer. She has lived here for five years, and has not gone home for a visit, though her parents have visited her here. Her children were born in this country. Her friendships are entirely with her co-nationals, many of whom live nearby. She says that she and her husband plan to go back to India "in five years."

The First Year

Kamala's way of life before she came to the United States was very "family-centered." If she, or anyone in her family, "felt lonely," they would visit one another. She describes her close friends as "family friends." They were "no different from family," and in fact her closest friend was the daughter of her father's best friend. She could talk with "family friends" just as with her mother or sister, and expected of them the mutuality and reciprocity that she expected of her family. Unlike Sushila (described in second case study) who is quite articulate about the characteristics of this kind of friendship, Kamala takes for granted the conditions set out by Coelho, and assumes that the closeness of "family friends" needs no elaboration.

When Kamala and her husband arrived in the United States, they stayed for the first two months with an Indian family they had known at home. This family was part of a community of immigrants from the same general area in South India who formed a small enclave of about fifty families. Most of these families lived in the same part of the city, and were in close contact with one another. All of them were aware of the arrival of a new family, and Kamala and her husband were soon invited to parties and get-togethers. Some of the members of this group of immigrants were already known to them, and others were met for the first time in this country.

The family with whom Kamala and her husband first stayed, and the families whom they met subsequently, provided an initial orientation, as well as emotional support and practical guidance. They helped them to find an apartment nearby and to buy a car, showed them how to go shopping, and provided them with information about how things get done in America. Kamala had to learn to cook, and was helped by other women who had also learned upon coming to America. Staying with friends and, later, learning to cook meant that she did not learn to eat American food.

The circle of co-nationals whom Kamala met in her early months here soon resolved itself into a group of six or seven couples whom she calls "good friends," and three whom she calls "very good friends." Of her very good friends, one was known to her before coming to the United States. Kamala's first year of developing a social life with these friends helped mitigate the loss of her friends and, even more importantly, of her family left at home. These friends provide each other with information and gossip about home and about each other, cook and baby-sit for each other, and create for each other an atmosphere which resembles that of "home." They practice a casual, comforting kind of reciprocity which provides a sense of security in a strange environment. Another woman who had a similar community of friends during her early years in the United States said that "we formed a close community; anyone could just walk into the other person's house. If it hadn't been for that, I think I really would have wanted to go back." Kamala, too, says that her initial homesickness was alleviated by her closeness to her co-nationals.

Kamala distinguishes between good friends and very good friends according to the degree of mutuality and intensity of reciprocity accorded them. Good friends are those who can be "visited whenever we feel lonely," but who must be telephoned slightly ahead; they are also the people with whom parties are often arranged. Kamala exchanges recipes with these women, gossips about home, and maintains, through them, wider networks of people who provide information about home. Her very close friends are those with whom a more complete reciprocity is maintained. One important kind of exchange is baby-sitting; these friends can leave their children at one another's homes in order to go shopping or to a movie. They can borrow household goods and money from each other. Transportation is often shared. These reciprocal gestures do not require negotiation over "paying back." Close friends can talk about their problems with husband or children, usually without fear of gossip. They do not need to telephone ahead, but can just visit whenever they feel like it.[9] When I asked Kamala whether she ever worried about "dropping in" on someone who was tired or busy, she simply said, "No, at any moment, it's OK."

By the end of her first year here, Kamala was a full member of an intimate circle who duplicated for each other many of the conditions of home, of which perhaps the most important was the sense that they could "just walk in on each other at any time." This security was grounded in daily acts which constantly demonstrated reciprocal relationships, such as baby-sitting exchanges and shopping trips. At frequently-held dinner parties a fairly rigorous separation of men and women was maintained (as in the home country), and this, too, cemented the sense of solidarity among the women. Kamala had learned, by the end of the first year, how to use the new environment to fulfill her need to maintain a home-like atmosphere (for instance, she knew just where to find rice imported from India), but she had not explored it on her own; she had never been into the downtown section of the city, or used a bus or subway.

The Second and Third Years

Kamala's situation and her network of friends have remained essentially the same in the five years that she has lived in the United States. From the beginning, the values of permanence, intimacy, and interdependence of interests were held in common by her and her friends. Their bonds were created by a shared life, not by "common interests" developed independently of their life within the home.

[9] Saran (1977) says that the immigrants he studied in New York have changed from spontaneous visiting to telephoning ahead, and that this is a general change affecting the whole community. Kamala, however, seems to move in a circle which has not made this change, perhaps because the use of the telephone serves to define the "inner space" of the home differently than in India, where phones are rare.

Kamala believes that "talking behind someone's back" is the one thing which is likely to lose her a friend. Gossip is far more important than breaches of reciprocity (such as borrowing money and not paying it back) as a threat to friendship, and Kamala has several times "broken off" briefly with friends who gossiped about her. Although she perceives gossip as a serious threat to the intimacy of the relationship of "close friends," the fact that it is so important is a manifestation of the kind of closure that exists in a community where everyone is a "known person." This closure allows for fairly tight social control, and ensures that deviations from the norm will be "talked about." Informants who have moved away from such tight communities complain about the restrictive effect of gossip, but Kamala, who accepts the premise of closure, simply says that the gossip here is "just like home."

During her second two years in the United States, Kamala developed, through talk with her friends, a set of opinions about the dangers confronting her and her family in this country. Of primary importance are the bad influences on her children's behavior which she feels surround her here. In a discussion with two of her friends at which I was present, all three women said that these influences are increasingly difficult to avoid as their children get older. Disrespect for elders seemed most important to them; noting that American children do not address adults as "aunt" and "uncle," they said that these children have "no respect." They mentioned the American children they see who are out late at night, and they expressed apprehension about drug-taking and drinking. Kamala fears that her children will "turn against her" or "forget her" as they get older. She and her friends are reluctant to leave their children with hired baby-sitters; as one woman said, "If you leave them, how will they know who their parents are?"[10]

Another worry for Kamala is the danger of violence which she feels surrounds her at all times. She does not go out alone, use public transportation, or walk alone even from her building to a nearby car. She does not distinguish clearly between the news and the other programs on television, and accepts the level of violence she sees as directly representative of the streets outside her home. Her husband emphasizes the dangers, and is insistent that she not make herself vulnerable in any way.

Although Kamala does not particularly dwell on it, other South Asian women mention the frequency of divorce in America, and the

[10] In the home country children are frequently left with servants; American baby-sitters are frightening in part because they are not "known," a fear which is exacerbated by the general notion that the surrounding environment is dangerous and intrusive.

fear that their husbands will be tempted to leave their families. They feel that Americans "have no family life," and fear that their own families will somehow be influenced by this perceived lack of concern for home and family. Kamala believes that nothing approximating her own closeness to her family and friends exists among neighboring Americans.

These pressures and dangers which are believed to exist outside the network of family and friends clearly help to reinforce the importance of that network. In fact, though Kamala feels that she is simply recreating as much as possible of her life in India, this resemblance rests partly on the exaggeration of certain aspects of her life there. For one thing, in order to create a new network of friends, she cannot be as selective as she might have been in her choice of friends at home. Some people chosen for her inner circle represent a compromise—people with whom, for reasons of community, class, or education, she would not have become friendly in her own country. She says that if they all went back tomorrow, they would probably remain friends; this assertion suggests the emotional importance which the idea of permanence has, even when the context has changed.

There are other ways in which the maintenance of this network involves a kind of exaggeration. Another South Asian woman says "we become *more* friendly here." Loneliness encourages more visiting and more interdependence among a smaller and less homogeneous group of people than in India. There is considerably more exchange of meals, and gossip has an intensity not present in the larger circles of home. The exigencies of child care (including either exchanges of baby-sitting or a determination to take the children everywhere) involve a more explicit merging of family and friends than occurs at home, where children can be left with servants. The emphasis on danger and intrusion from outside increases the importance of "known" people, and frequent contact and reciprocity intensify their dependence on one another.

The Fourth and Fifth Years

Kamala does not have any American friends. Although she describes her neighbors as "friendly," she says, "we don't go to their houses; we only talk to them in passing." Her husband's colleagues are also "friendly," but she does not know them, and never socializes with them. In this context, "friendly" to her means simply "courteous," and she does not interpret it as an overture to further contact. Although she says that "there are some Americans who are very nice," she also feels that "you just can't talk to them." The dangers that surround her in America seem to her to justify her aloofness from Americans.

Although Kamala wants her local network to remain the same, some changes have occurred in spite of her; especially in the last two years some of her friends have moved away. Kamala makes a strenuous effort to keep in touch with these friends, telephoning them often; sometimes she and her husband travel long distances to spend a weekend with friends who have moved away. Kamala is reluctant to discuss the possibility that these friends might change or become less close; she holds on to the idea that these relationships are unchanging and permanent. It is clear, however, that the mobility of American society poses a threat to her network. For the time being, at least, she has responded to this threat by acting to reaffirm her closeness to the friends she is already dependent on.

Sushila: "Outward-Reaching" Pattern of Friendship

Sushila is a 26-year-old housewife who lives, with her Indian husband and two preschool children, in a suburb of a large northeastern city. She is middle-class, English-speaking, and from a medium-sized city in India. She married shortly before coming to the United States. Her husband works as a doctor in a hospital in a nearby town. She and her husband have been in this country for five years; two years ago they moved out of an apartment and bought their own house. They have gone back to India once, and Sushila's parents and her brother-in-law have visited them here. Sushila has friends in the United States who are also from India, and who live in different parts of the city; she also has American friends, and has made extensive contact with her suburban neighbors. She says that she and her husband plan to go back to India "in five years."

The First Year

Like Kamala, Sushila was initially a guest in the home of someone she and her husband had known in India. Through this friend they were introduced to other Indians, and went to parties where they met people from their locale in India. Unlike Kamala, Sushila felt from the beginning that the Indians she met here were "different" from those at home. She felt, somewhat uneasily, that they seemed to have "changed after coming here."

Within a few months, Sushila and her husband moved into an

apartment in a suburb near her husband's place of work. They were not in daily contact with other Indians, and she felt isolated and lonely. The Americans she encountered during this early period were incomprehensible to her, and she felt that she had "nothing to say" to them; as an example of the gulf separating her from her neighbors she describes her dismay on discovering that a neighbor's visiting daughter had stayed at a motel rather than at home.

Sushila says that she was very homesick during her first year in the United States; much of this homesickness derived from the sharp contrast between her isolation here and her memory of her life in India. Before leaving India (during her late teens and early twenties) she belonged to a group of girls her own age who formed a close and supportive circle. The group consisted of about ten close friends, with subgroups of greater intimacy within it. These friends were in constant contact, and met every day at a favorite outdoor spot or in each other's homes. They kept each other's secrets, and provided a haven from parental disapproval. Mutual dependence was intense, and they "used to help each other in every way"; for instance, all the friends contributed to one who was in financial difficulty.

Although this group of friends sometimes formed a refuge from the restrictive demands of family obligation, relationships within the group were nevertheless modeled on the expectations pertaining among family members. Within the family, visiting relatives were always greeted with hospitality. "Every other day someone would come and just ring the bell, and sometimes you might be in the middle of a family argument, or somebody might be sick or something, but most women put up a happy face." In the same way, friends could come and go freely in each other's homes; Sushila calls this "informality," and says of these friends that "it was nice to know that you could just go and knock on their door and be comfortable that they won't mistake (misunderstand) you." In many ways friends were incorporated into each other's families, visiting, staying over, helping each other's siblings. Sushila remembers an incident from her visit to India (after emigrating) as particularly revealing of this kind of relationship. On her first night after her return, her family and friends all gathered at one house, and there were so many of them that they all slept together on the floor in one room, talking and laughing all night.

Sushila spent her first year here in a kind of limbo, mourning the relationships she had left behind and unable to imagine anything comparable emerging from her new social environment. She describes herself during this period as "shy" and "traditional," and she felt somewhat cut off from other Indians who were, she felt, more "modern" and "Americanized." She concentrated during this time on her relationship with her husband and on the more impersonal aspects of life in America, such as the opportunity to save money.

The Second and Third Years

One night late in her first year in the United States Sushila met an American woman who was to become her "close friend." Her relationship with this friend, who acted for a time as a kind of "local guide," was extremely important to her; other "outward-reaching" women among my informants experienced similar initial relationships with Americans.

Sushila met her "local guide" at a night class to which she had gone with her husband; this woman, the wife of one of her husband's classmates, persisted in talking to her in spite of her shyness, and offered to take her shopping. She began to help her by introducing her to people, taking her on outings, and seeing her almost every day. Gradually Sushila felt that they became "very close"; "we got to know each other so well, that one time I cried on her shoulder." Through this friend she says that she "changed *so* much"; once her initial vulnerability was out in the open, she felt free to follow her friend's advice and "Americanize" her appearance, giving up the sari, cutting her long hair, wearing slacks. She became confident about shopping and driving by herself, and began to try American foods.

In addition to these changes in outward behavior, she found that this new relationship demanded an important change in attitude; she had to give up the idea that friendship depended on similarity. Her "local guide" did things that she did not approve of, such as smoking and drinking. Sushila says that at first "I hadn't come to the stage to accept those things. But then I realized, all those things, it's all personal. If a person wants to be friendly, it doesn't matter." Separating out the "personal" in this way enlarged her concept of "friend," so that more types of people could be potential friends. In turn, her new friend demonstrated this kind of acceptance of her by learning to cook Indian dishes.

Although this friend opened up the possibility of new relationships with Americans, Sushila was very troubled at this time by the apparent lack of permanence in American friendships and marriages. Even after this period of outward "Americanization" she feared that "Americans feel you can always find another friend. You don't have to make an effort."

During Sushila's second year in the United States, some of her old friends from India also moved here. The opportunity to visit them occasionally made her particularly aware of the contrast between the kind of relationship she had with them and what she feared was the transience and noncommitment of Americans. Her description of her relationships with these old friends is revealing of her expectations of friendship. When she visits one of these friends, she feels that "I can be so free, I don't have to worry about anything. I can talk without having a second thought." During these visits the women sleep together in one room and the husbands in another, so that they can talk all night. Sushila feels that she "wouldn't mind doing anything" for these friends, including lending them money, or offering extended hospitality. She feels that these relationships will always be the same. Like all of my informants, she has difficulty thinking of anything that might cause her to "lose" one of these friends; as another woman said: "Friendship and love do not change."

The freedom which Sushila feels with these friends involves a similarity of style which precludes misunderstandings over such daily events as childcare or cooking; this similarity also means that she can do any of the childcare or cooking or housework for one of these friends. She talks of being able to "just be" with her close friends. Other South Asian women, too, speak of feeling with close friends that "you know what to expect," "you don't have to say anything to them," "you can just sit," indicating an acceptance which is not based on any deliberate attempt to create an interesting social persona. There seems to be no boundary around the "private" part of their lives which has to be protected from intrusion by a friend; in this context the question of whether "dropping in" on someone is an intrusion becomes meaningless.

During her second two years here, Sushila also saw more of the other Indians whom she had met here. Two or three Indian women became "very close" to her, a closeness indicated by her freedom in feeling that she can just call them to come to visit any time she is "bored." But though she feels free to "just sit" with these friends, she follows rules for interacting with them which indicate a difference between them and the old friends with whom she feels no constraints. She is much more formal about reciprocity, and does not involve herself in exchanges of favors or baby-sitting with them. She has noticed that, in addition to greater "formality" and a new emphasis on privacy, Indian couples who have been in the United States for a while also begin to act differently toward each other. For instance, some of the other Indian husbands had begun to help their wives in the kitchen.

Sushila was much affected by this observation, and began to want her husband to help too. When she hinted to him about it, he did help. But she found that she still was not happy; she says, "He used to do it, but to see him washing dishes bothered me." The contrast with her traditional expectations was more painful than she had expected. She also noticed that other husbands gave compliments to their wives. This is an aspect of Americanization mentioned by all the women I talked with;[11] Americans seem to them far readier to comment on each other's appearance, household decoration, and cooking than Indians. To some women this seems artificial and meaningless; but Sushila began to want such attention, and finally told her husband that he never appreciated her. He said "I married you, didn't I?" but added that he could never make such "artificial remarks." She says that she realized that her wish was a childish one. Her marriage had begun on the premise that the distribution of labor and the relative merits of the partners were already settled simply through the existence of the marriage itself. "Actually, he hadn't changed in any way, it's just that I was looking around *my* world, the world I am in now, wanting him to be like that." But she is still ambivalent: "It really bothers me; this thing I haven't solved yet."

During this period, then, Sushila sorted out three kinds of relationships possible in America. That with her "local guide" was most important in introducing her to the possibilities of life in this country; by the time she entered her third year here, she felt quite comfortable in the local American environment, and had become curious to learn more about life here. She did not continue to see her "local guide" with the same frequency once this initial change had begun; however, she emphasizes that her friend still provides her with a sense of security and support: "I can always turn to her and she will always understand me." In addition to this relationship with an American, Sushila also developed friendships with Indians which accommodated changes occurring to them, and she continued to maintain a few of the older friendships which had changed little from what they had been in India.

The Fourth and Fifth Years

At the end of their third year in the United States, Sushila and her husband bought a house in a suburban neighborhood not far from their previous apartment. The move provided the opportunity for new kinds of relationships with Americans, and was the prelude to a new

[11] There is probably a connection between this tendency to give compliments and the "trial balloon" mentioned by Suttles (1970) (see p. 423). Compliments convey approval of the other's "style," and may be employed by Americans partly to bridge differences resulting from a diversity of styles in the society.

stage in Sushila's life. She found that her new neighbors were friendly, which was a surprise to her after the reserve and distance of her previous neighbors. As soon as she and her husband moved in, they were invited to a welcoming party. Since then, her relations with the neighboring women have progressively intensified. She discovered that she could ask them for help, and that she was free to borrow things from them. She mentions the easy moving back and forth as particularly important to her, and says it reminds her of India—in the summer, there is a good deal of chatting and visiting among the neighbors as they do their outdoor chores. She has begun to give parties for her neighbors, and for these parties she makes an effort to cook American food and provide alcoholic drinks. She has begun to exchange visits with her neighbors for coffee or lunch. She and her husband have made deliberate attempts to increase their participation in the life around them. For instance, she is taking bridge lessons, because her neighbors like to play bridge at parties. She recently became "hooked" on a TV soap opera, and watches it with great concern for whether it depicts American life in general.

Sushila's relationship with these new American friends is close because they live nearby and can visit and help each other easily. But even under these circumstances she keeps herself and her family separate and less "merged" with these friends than with her Indian friends. In spite of physical proximity, she does not exchange baby-sitting with them in a casual, impromptu way. She plans her meetings with them ahead of time, and she does not ask them for family-oriented favors like rides to the airport or doctor. She also does not talk with them about her relationship with her husband. She does not invite American and Indian friends to her house at the same time, for fear that they will not "mix."

Although Sushila now describes some of her neighbors as "good friends," she also says of them that she experiences "a vacuum, or barrier, with Americans, and this is something I can't break through."[12] This vacuum is in contrast to her feeling of "I can be so free" with her old friends, and refers, I think, to the lack of "close interdependence" of which Coelho speaks. In the process of getting to know her neighbors she has experienced several misunderstandings and painful incidents

[12] Mehta (1970), who studied Indian women in India, gives this quote from one of her informants: "Even after three years in the United States, seeing the same people again and again, we never seemed to get any closer. There was always a distance; people were afraid to become involved. I never felt free to impose on them, and after a while it became a barren exchange" (p. 123). This quote makes particularly clear the connection between being able to "impose" and the meaningfulness of the relationship.

of rejection. Misunderstandings arise out of her tendency to "overextend." For instance, she baked a cake and cleaned the house when a neighbor planned to visit her; when the woman casually canceled, Sushila felt that she had overdone her anticipation. She finds that she assumes that a friendship has been established, on a basis of reciprocal favors, before her neighbors do. As she says, "When I try to get close to somebody, I get kicked back. I keep taking too much trouble to please them." By deliberately learning about the interests and pastimes of her friends (for instance, bridge) she is hoping to build a base of "common interests"; she senses that this, rather than "trying to please," is the foundation of American relationships.

Sushila has developed a vocabulary of several opposing categories to describe herself. On the one hand, she says that she is "the same underneath," that she hasn't really changed inside, but only superficially. That is, she has developed new definitions for relationships while keeping her initial definition of friendship, and of herself within the friendship relation, intact. She now has "acquaintances" and "casual friends" of whom she does not expect permanence, mutuality, or closeness to her family. She does not expect them to share her style of doing things, and she is careful not to depend on them too much.

On the other hand, Sushila also says that she has "changed *so* much." This seems to have begun, not with the change in outward appearance, but with her realization, in her relationship with her "local guide," that different styles of life were acceptable within a close friendship. The corollary of this acceptance was that she had more independence and freedom available to her in this country than she had at first suspected. "Here, I can just do anything I want, without worrying what anyone will say." One way in which Sushila expresses a sense of freedom from being told what to do is by organizing her time and housework carefully, and being very conscious of interruptions to her daily schedule.[13]

Sushila uses two other words to describe the alternatives that she feels are available to her. On the one hand, there is "submissiveness," which means subordinating oneself to others both in the family and with friends. She calls it "stepping on your ego a little bit," and she points out that in India, where it is difficult to get a divorce and one must live in close proximity with family and friends, this is a necessity. Her ambivalence about her husband's doing the dishes is partly an expression of her feeling that she is still "submissive."

Sushila contrasts this submissiveness with what she calls "inde-

[13] This is in contrast to Kamala, who is quite casual about the organization of her daily life.

pendence." Independence means doing things on one's own and speaking one's mind in family arguments. Americans, she says, "will not keep quiet. Why should they, right?" The contrast between submissiveness and independence corresponds to another opposition, mentioned by many of my informants, between being and doing. Americans, they feel, emphasize *what they do,* their interests, and use these as a basis for relationships both in and outside the family. Sushila feels that with Indians what is important is *what one is,* rather than what one does. She contrasts just being, and being submissive, with the imagined independence of an American woman who is not afraid of separating from family or friends, and always has the option of supporting herself and living alone. The American, she thinks, can do things outside the family context which will give her a separate identity. Although Sushila does not think that she has this separate identity, she has begun, in the context of her new neighborhood, to differentiate emotionally her family from her friends. She sometimes leaves her children with a baby-sitter in order to "have some privacy" when visiting friends. In attempting to develop separate interests, she has begun to loosen what another woman called the "bondage" to close friends and family.

Comparison of the Two Friendship Patterns

Kamala and Sushila came to America with similar expectations and values around friendship. For both women, friendship presupposed an intense mutuality, and rested on the assumption that close friends could not be "imposed on" or "lost." In friendships as they knew them, it was possible just "to be" with the other person. Large areas of their intimate life within the home were open to their close friends, and they did not sense a boundary or barrier protecting their "private" family life from intrusion by these friends.[14]

Although they evolve from common core values, the friendship patterns which the two women have developed in this country are quite different. Kamala's "boundary-maintaining" friendship style illustrates the use of friendship to form a small circle of immigrants who keep themselves separate from the surrounding social environ-

[14] It should be emphasized that this view of friendship in India is reported (both in this study and in Coelho's) after an immigrant has been away from India; it is likely that casual and expedient relationships are the first to be forgotten by someone who has left home. Nevertheless, it is clear that immigrants feel a real difference between remembered relationships and those that they observe in the United States.

ment. Kamala devotes her social energies to the creation of a correspondence between the friendship values with which she came and the social acts which give meaning to her current friendships. For her, "friends" are by definition those who affirm the values with which she came to this country, and she cultivates these friends by maintaining a structure in which these values continue their validity.

Kamala's view of the world outside her network remains, after five years, a very simple one, in which the most important thing she has learned about America is where to shop; by stressing the notion of threat and danger "outside" the known social environment, she is able to keep her interactions with this outside world on an entirely formal basis. The inner world of her friends is, in contrast, small, complex, and close. The intimate social sphere which is important to her includes both friends and family; she is not concerned to separate them or to maintain "privacy" for her relations with her immediate family. In this inner sphere, daily acts of friendship take shape as affirmations of the values of intimacy and permanence. For instance, the exchange of baby-sitting, which is of minor importance in India, assumes prominence as an expression of reciprocity. Similarly, cooking for friends, which is not so personal when there are servants in the kitchen, becomes expressive of solidarity and shared experience in this context.

Thus Kamala defends and maintains her values around friendship through daily acts which define and perpetuate an "inside" and an "outside" social world. New social rituals are invested with these values, and become a way of maintaining them even in the face of contradictions. By thus transforming her present social situation into a model or "mirror" of the one left behind, Kamala experiences security, reciprocity, and continuity in a new environment. On the other hand, she is vulnerable to changes in her present social situation, for her adjustment depends largely on the manipulation of the external social world. She still has the same expectations of friendship which she had when she came; in more isolated circumstances these expectations might be disappointed. It is likely, in the socially mobile atmosphere of the United States, that she will be exposed to renewed discontinuity in her relationships as her friends move away or change their own ways of living here.

Sushila has experienced from the beginning of her stay in this country a discontinuity between her values and expectations and her social environment. Her "outward-reaching" style of friendship involves a gradual change in her own definitions and expectations, enabling her to form new kinds of social relationships. Her energies are employed, not in trying to obtain a "fit" between her initial values and her experience, but in looking at the discontinuities between her old and new situations and sorting out new understandings which will apply to both.

Sushila no longer regards acts of friendship as automatically emblematic of reciprocity and permanence. She has had the experience she calls "getting kicked back" when she presumes too much on someone; as a result, she has created categories for less intimate friends of whom less is expected. She now sees friendly overtures and exchanges, especially with Americans, as problematic and open to manipulation and redefinition. One result of Sushila's redefinition of friendship is that she regards outsiders and acquaintances as potential friends, and is also open to the possibility that close friends may change or lose contact with her. She still tries to maintain her original expectations of her closest friends, but even with them she has shifted in her behavior toward separating them from her family and creating a tighter, less easily breached core of intimate and private family relationships.

One result of Sushila's approach to friendship is a much larger and more fluid group of friends and acquaintances; where Kamala knows well only five or six women, Sushila counts perhaps twenty among her "close friends." At the same time, Sushila has become far more self-conscious about friendship, so that acts which are for Kamala natural symbols of relationship become charged with uncertainty for her. Sushila has experienced more stress, more disappointment, and less security than Kamala over the past five years. On the other hand, she has developed a way of looking at things which is likely to see her through any further changes in her circumstances.

In looking at the difference between Kamala and Sushila we can see that Coelho's value of "permanence" means something more than the simple expectation that friendship will endure. It seems to stand for a crossing over into the "private" realm of the person. It implies a social reality which is so thoroughly shared that attachment becomes almost a merging of identities. For Kamala, and initially for Sushila, this crossing is expected to link people in a permanent way; they feel that a certain level of intimacy transforms a relationship so that it is "like kinship," and not subject to further change. What is confusing to Sushila in observing Americans is that they appear to cross this barrier into the other person's "privacy," but the crossing does not imply the same kind of reciprocity and commitment which she expects of such intimacy.[15] It is this discrepancy which is disorienting to the

[15] Lewin's (1948) discussion of German/American differences includes much that is helpful in this context.

newcomer, and which, in Sushila's case, has forced a redefinition of her situation.

Conclusion

The two women discussed in this paper illustrate two styles of friendship formation: (1) a maintenance of relationships modeled on those left behind in the home country, and (2) a movement toward new kinds of friendship, and a corresponding change in self-definition. In the first style, illustrated by Kamala, the discontinuity in relationships which is created by immigration is overcome by an emphasis on maintaining the boundary separating friends from outsiders. In the second, illustrated by Sushila, discontinuity is recognized and resolved through the formation of new kinds of friendships.

Although Kamala and Sushila came to this country with a culturally shared point of view about friendship, they arrived at divergent solutions to the problem of reconstructing a social network in a new environment. I want to conclude by suggesting four areas in which these differences have implications for further work on the question of friendship patterns among Asian immigrants.

1. The case histories presented here suggest a possible correlation between the initial living situation of an immigrant and her coping style with regard to friendship. This correlation holds true as well for the other women studied. In the case of Kamala, the fact that she moved immediately into an urban enclave of immigrants with similar backgrounds seems to be related to many aspects of her response to life here. Her knowledge of her new environment is largely filtered through the perception of other people in similar circumstances. She has no need to reach beyond this community for social contact, and within the community there is a high level of agreement as to the symbolic meaning of specific acts of friendship. Kamala illustrates the influence which this kind of early contact with co-migrants may have on an immigrant's approach to social relations in the new country.

Sushila, on the other hand, began her life here further away from other immigrants than did Kamala. From her description, she was much like Kamala when she first arrived (that is, "traditional"). But without the same proximity to other Indians her initial loneliness was much greater. It was this loneliness and sense of isolation that provided an incentive to reach out to new friends; at the same time freedom from the intimacy and gossip of a closed community allows her to redefine specific friendships according to her new circumstances. Thus, for Sushila the isolation caused by living in the suburbs was one impetus for her outward-reaching style.

Many immigrants to the United States have begun their lives here in city neighborhoods composed largely of others from the same country. Often one or two generations have passed before a move to the suburbs could be accomplished. South Asians who come to America as students or professionals often go directly to academic communities or to well-paid positions which allow them to buy a house within a few years; thus many of them bypass ths earlier pattern. Sushila's experience is, therefore, not an uncommon one for this group of immigrants. Further research with a larger group of Asian immigrants might focus on whether the outward-reaching style is consistently correlated with this type of initial living experience.

2. Another difference between Kamala and Sushila emerges when one looks at their orientation toward remaining in this country. Both women say exactly the same thing about going home. They want to go back, they say, "in about five years." Dahya (1974), talking about Pakistani immigrants in London, terms their attitude toward going back the "myth of the return" (p. 83). Most of the immigrants he studied said that they were going back, no matter how extensive their commitment to their host country. Very few of the South Asian women I talked with said that they were never going back; the five-year projection is a common one. It seems to serve as a symbolic tie expressing a continued connectedness with the home country.

Although Kamala and Sushila say the same thing about going back, there is clearly a difference in their actual commitment to staying in America. Kamala has done very little to commit herself to this country. She lives in an apartment, stays entirely within her own home, and buys only things which can be taken back to India with her. She does not look forward in a definite way to moving out of the city.

Sushila, however, has made serious commitments to staying here. She and her husband have bought a house and outfitted it with expensive and long-lasting furniture. They have many plans about ways to decorate the house, bring up their children in America, and learn more about American life. Her husband is committed to his work here and to the expense of a trip to India every few years. Therefore, although Sushila does not actually say that she is staying, she has a fairly large investment in her new country. Her new friendships with Americans after buying the house are not merely the result of proximity, but also of her own recognition of the beginning of a permanent connection with a specific neighborhood. In her relations with other Indians, her commitment to staying seems to allow her to see changes and differences in them which Kamala does not recognize.

This difference in orientation of the two women suggests another factor in their lives which may be correlated with their styles of friendship formation. Sushila's willingness to examine her own assumptions about friendship when these are challenged by new experiences may be related to an orientation toward staying in the United States which allows her to project a new identity for herself onto the future. Kamala's commitment to her group of friends may arise, in part, from a sense that the experience of living in the United States is simply a temporary interlude within the larger continuity of life lived in India. Again, further research might investigate the existence of this correlation for larger numbers of immigrants.

3. I have tried to show here that women are limited to the home and to the events of daily life in dealing with the discontinuities in social relationships caused by immigration. The arrangements of everyday social life are the medium through which a woman's approach to friendship is expressed. For instance, I have pointed out that, whether children are taken everywhere with the mother, left with friends on a casual basis, or left with baby-sitters, can be indicative of assumptions about privacy and reciprocity within friendship itself.

In further investigating this topic, it is important to keep in mind the centrality of these details of home life. Decisions about social life which are not easily detected from conversation or from a woman's activities outside the home are likely to be embedded in such matters as baby-sitting, the keeping of schedules, or the practice of dropping in on friends. Kamala's case suggests that, where the concern is with the maintenance of a homelike network, these acts will become almost ritualized as symbols of solidarity and reciprocity; Sushila illustrates the possibility that, as a more outward-reaching style is established, the same daily acts of friendship become problematic—markers, instead, of changes in attitude toward, for instance, privacy and timing in relationships.

4. The case histories have been presented in phases suggestive of a relationship between the passage of time and the development of coping styles. What happens at certain stages in an immigrant's stay may have an influence both in terms of opportunities available to her and in terms of her attitude toward her surroundings. Kamala is typical of the woman whose initial experience provides familiarity and security. The boundary-emphasizing style which she shares with her friends is relatively static through time, with the passage of time unmarked by sharp personal changes. However, a move away from this community of immigrants might produce a very sharp discontinuity for Kamala, not unlike the original discontinuity experienced on immigration.

Sushila shows the influence of isolation, in the early stages, in bringing an immigrant into an active involvement with the new environment which entails greater frustration and greater individual change. Here, the years since emigration become markers for changes in attitude and belief, and finally for changes in self-concept as well. At the five-year mark, Sushila's life is beginning to take on the more stable, community-oriented flavor which Kamala's has had since the beginning.

Although Sushila's case suggests a kind of "progression," it would be a mistake to see this kind of change either as inevitable or as the only kind of change possible; more work is needed to map out the various "routes" taken by immigrants entering new social territory. It does seem likely that the two "styles" delineated here extend into other areas of life, such as work and family relations. One question raised by this study is how the experience of change over time in the area of social relationships influences an individual's experience of the passage of time itself; Sushila seems more conscious of change in all areas of life than does Kamala.

The histories of Kamala and Sushila indicate that initial values are not in themselves indicative of the type of social adjustment an immigrant will make; each woman's style is the result of a complex mix of circumstances and personal choice. Both women developed friendships as a primary way of dealing with discontinuity and change, and each evolved a pattern of friendship formation which fills, at least partially, the gaps in social relationships created by emigration.

Acknowledgment

The author is indebted to David Jacobson, Elliot Mishler, Stuart Hauser, Ellen Dissanayake, and George Coelho for their comments on earlier versions of this work.

References

Coelho, G. W. Changing Patterns of friendship in modern India. In C. DuBois (Ed.), *Studies of Friendship*. Unpublished manuscript, 1955, the Tozzer Library, Harvard University.

Cohen, Y. A. *Social structure and personality: A casebook*. New York: Holt, Rinehart & Winston, 1961.

Dahya, B. The nature of Pakistani ethnicity in industrial cities in Britain. In A. Cohen (Ed.), *Urban Ethnicity*, ASA Monograph #12. London: Tavistock, 1974. (Distr., Harper & Row).

DuBois, C. The gratuitous act: An introduction to the comparative study of friendship patterns. In E. Leyton (Ed.), *The compact: Selected dimensions of friendship*. Newfoundland Social and Economic Papers #3, Institute of Social and Economic Research. St. John's: Memorial University of Newfoundland, 1974.

Eisenstadt, S. N. Friendship and the structure of trust and solidarity in society. In E. Leyton (Ed.), *The compact: Selected dimensions of friendship*. Newfoundland Social and Economic Papers #3, Institute of Social and Economic Research. St. John's: Memorial University of Newfoundland, 1974.

Jacobson, D. Fair-weather friend: Label and context in middle-class friendships. *Journal of Anthropological Research*, 1975, *31*, 225–234.

Kurth, S. B. Friendship and friendly relations. In G. McCall (Ed.), *Social relationships*. Chicago: Aldine, 1970.

Lewin, K. Some socio-psychological differences between the United States and Germany. In K. Lewin, *Resolving social conflicts: Selected papers on group dynamics*. New York: Harper, 1948.

Leyton, E. Irish friends and "friends." In E. Leyton (Ed.), *The compact: Selected dimensions of friendship*. Newfoundland Social and Economic Papers #3, Institute of Social and Economic Research. St. John's: Memorial University of Newfoundland, 1974.

Mehta, R. *The western-educated Hindu woman*. New York: Asia Publishing House, 1970.

Paine, R. In search of friendship: An exploratory analysis in middle class culture. *Man*, 1969, *4*, 505–524.

Paine, R. Anthropological approaches to friendship. *Humanitas*, 1970, *VI*, 139–159.

Saran, P. Cosmopolitans from India. *Transaction: Social Science and Modern Society*, 1977, *14*, 65–69.

Suttles, G. D. Friendship as a social institution. In G. McCall (Ed.), *Social relationships*. Chicago: Aldine, 1970.

United States Department of Justice. *Immigration and Naturalization Service Annual Reports*, 1961–1975.

VI

Stressful Situations of Uprooted Communities: The Role of Public and Government Bodies

Introduction

Population groups most at risk include those who are uprooted by natural catastrophes or who are forced to migrate owing to war, or political or cultural oppression. Similarly, members of ethnic minorities, intellectual exiles, and certain disadvantaged populations forcibly relocated by land-use developers are vulnerable to emotional disability and health hazards. A major theme of the papers in this concluding section is that, as the individual's freedom of action is reduced by lack of access to valued resources, the role of social and political institutions and bureaucracies tends to expand. Freedom of action implies both the capacity and the authority to determine the direction of one's strivings, and to make effectual self-motivated decisions. Thus, freedom of action can be restricted when individual authority is not recognized, as, for example, when entrepreneurs decided to exploit the native lands of the Amazon Indians (Chapter 20), or when Jews were the victims of discrimination and persecution (Chapter 21). But freedom of action can also be limited by (1) lack of preparation owing to the abruptness of the uprooting, as in the case of many Vietnamese refugees (Chapter 22), (2) lack of necessary language or economic skills (Chapter 20), or (3) lack of relevant information (Chapter 23). The crucial policy issue is to recognize the stressful situations and to identify the groups most in need of services and institutional intervention. It is essential to distinguish these situations from those in which public intervention may itself hinder the development of coping efforts on the part of the individual. For example, Kolker and Ahmed in Chapter 21 argue that increased dependency on government institutions may hinder the formation of primary and secondary affiliations. On the other hand, the lack of knowledgeable intervention by social agencies contributed to the development of "uprooting neurosis" in the case of Anna Maurer (Chapter 8).

Forced migration is particularly stressful for stable, indigenous, and traditional communities, because in many cases they have to cope with change at short notice and often without their informed consent, with little preparation and few resources. In Chapter 20, Trimble examines four cases of groups displaced by outside agencies which wanted to explore and exploit natural resources or potential energy sources. In these situations, the stressful circumstances were aggravated by the fact that relocation was enforced by an external agent under the following conditions: (a) the relocated groups were denied decision-making power, and (b) the goals of the agencies were incompatible with the goals of those being relocated. Participation in the decision-making process (a recurrent theme in coping) was absent in planning for the relocation. Thus the conflict between the native population and the outside agencies was aggravated. These communities not only failed to benefit from the development of their traditional lands, but also experienced environmental and emotional stresses and hardships as a result of resettlement. In each case the traditional economic patterns were destroyed, and the native population did not have the psychological skills or the economic techniques necessary for adapting to the new environment. Coping strategies appropriate in their native ecological settings proved ineffective in the new environment. This presents a challenge to community mental-health practitioners who increasingly are called upon to share their insights into the dynamics of uprooting and to help organize support systems essential to the maintenance of a vital sense of community life.

A further issue in the assimilating of immigrants into a new culture is the importance of national policy and of institutional priorities. In Chapter 21, Kolker and Ahmed trace the variations in the Israeli government's policy toward the resettlement of immigrants. From a policy of geographic dispersion and occupational redirection, the government shifted in the 1960s to a policy of accommodating immigrants' residential and occupational preferences whenever possible. Also bureaucratic intervention has lessened considerably over the years, and immigrants have been given greater opportunities for self-reliance.

Personality and motivational characteristics of individuals facilitate or inhibit integration. Kolker and Ahmed hypothesize that those who are alienated from their home society may be able to adjust more easily to the new society, since they have an internal readiness to embrace the new political ideology and social values. In spite of the stresses, these refugees are predisposed to make determined efforts to integrate themselves into the new culture. Hence cultural similarity or difficulty of initial assimilation cannot alone predict the success or failure of personal integration in a new society; internal readiness for change must be taken into account. This analysis may also be applied to people who are displaced without their consent. Since such populations retain their emotional ties to the former culture, integration into a new environment is expressed in resentment and ego–alien reactions.

The authors recommend that host institutions mitigate the degree and rapidity of change required of immigrants by recognizing the values of cultural pluralism in the integration of the diverse groups of immigrants. They further suggest that, where social and cultural supports are available and used by immigrants, reduced government intervention would foster informal primary group contacts which are necessary for successful social adaptation.

In Chapter 22, the degree of governmental intervention is studied in the case of the United States response to the uprooting of Asian refugees after the fall of the South Vietnamese Government in 1975. Ahmed, Tims, and Kolker trace the role of the American Government after the decision was made to limit its active participation to a short duration. Although several governmental departments and agencies provided assistance, the major burdens of finding employment, sponsorship, resettlement facilities, etc. were delegated to voluntary agencies. In the second phase of resettlement, the Health, Education, and Welfare (HEW) Refugee Task Force undertook to coordinate a wide range of resettlement-support programs, which included language and vocational training, mental-health counseling, and assistance with employment problems. The authors maintain that, though the settlement process has proceeded smoothly, many problems continue to face the refugees. Some of the issues considered in this chapter have been noted elsewhere in the volume. These include (1) underemployment owing to lack of marketable skills or deficiencies in language and professional qualifications, (2) conflicts between the social patterns learned in the socialization process in Vietnam and those prevalent in the United States, (3) the loss of key members in the extended family network, and (4) difficulties in finding acceptable counseling from culturally sensitive mental-health workers.

The effects of federal policies on relocation are increasingly important as new energy sources are being explored and developed, often at the cost of uprooting indigenous populations. Chapter 23, by

Thomson, reports on two cases of dislocation of semirural populations in the state of Washington. Thomson finds that, in spite of special legislation designed to assist residents in relocating, a great deal of tension and ill will existed. Lack of adequate information resulted in failure to create a comprehensive plan acceptable both to local residents and to the Army Corps of Engineers. In the second case, although elected officials obtained special legislation which enabled the county in question to receive federal impact assistance, funding levels proved inadequate. Thus, the limitations of federal, state, and local agencies in assisting local populations are delineated.

The author closes with recommendations which would reduce the threat of destruction to small communities. She argues that accurate and trustworthy information must be made available to all segments of the population, in order that they may understand and react intelligently to the proposed changes. She argues further that technical assistance and outside consultants are necessary to aid local officials in making crucial decisions. Communities undergoing radical changes in their way of life must attempt to form comprehensive planning programs which can propose and evaluate methods of coping with change at the community level.

20

Forced Migration: Its Impact on Shaping Coping Strategies

Joseph E. Trimble

Small settlements of tribal and indigenous native groups scattered in various countries are being forced to leave their ancestral homes; for many, if not for all such groups, the move is permanent. Many of the group's members are confronted with new, complex circumstances that demand attention and require the use of a variety of known coping and adaptive mechanisms. New forces stemming from the move, the settlement, and the new environment impinge upon individuals, compelling them to deal with the stressors in the best way they know how. Many of the forces create stress, tension, anxiety, and related life strains. Sadly, many relocated individuals are unable to prevent, avoid, or control the distress, and succumb in a number of tragic ways.

Native and cultural groups subject to the conditions and circumstances of forced migration have a great deal in common, especially the factors that lead to the resettlement. Basically, one can identify these factors to include at least one or a number of the following: (1) the movement is initiated by an agent, industrial, and/or governmental, external to the group; (2) migration is implemented for reasons that go beyond economic and/or political concerns; (3) resettlement programs are conducted in such a manner as to give the appearance that the external agent holds the safety and cultural preservation of the group as its primary objective; (4) the movement is exerted for the purpose

Joseph E. Trimble • Department of Psychology, Western Washington University, Bellingham, Washington 98225.

of exploitation of land resources; and (5) little or no effort is exerted on the part of external agents to prevent deleterious psychophysiological and psychosocial effects generated by the movement (Trimble, 1978). The most outstanding commonality, however, is the finding that each of these groups is forced to move against their will; that is, group leaders and members must move, as there is no alternative.

A discussion of experiences created by the forced movement of four native groups form the major focus of this chapter. A case-study approach is used; relocation experiences will serve to identify the effects of forced migration on the disruption of developmental life cycles, the efficient and effective use of coping and adaptation strategies, and the theoretical and substantive mental-health and policy implications for groups about to be a party to forced migration. An emphasis will be placed on identifying the impact of forced migration on the shaping of coping strategies, both at the individual and at the group level.

Social and Cultural Change

Few people enjoy being coerced into doing something they don't want to do. Under these conditions, people tend to cope with the circumstances in a variety of ways. Individuals may resist by withdrawing and refusing to do anything. They may engage in the activity with vigor, and convince themselves that what they're doing isn't so bad after all, or they may half-heartedly engage in the activity, hoping that someone will notice their ineptitude and reverse the decision. Then, again, individuals may randomly test new problem-solving methods to meet new dilemmas with which they have not had previous experience. Also, people sometimes emulate the problem-solving behaviors that they observe employed by the members of a new place or new society to which they are exposed. Nonetheless, the point is that individuals forced into actions not of their own choosing tend to experience initial difficulty and distress, which lead to the development and subsequent use of coping strategies.

A definite relationship exists between change and use of coping strategies. One usually occurs with the other. Change is a fundamental component of physical and social life. Some human groups are evolving more rapidly than others, and some seem to be standing still. In this connection, rate of change, level of technological and industrial development, and degree of accumulated knowledge vary dramatically from one culture group to another. Whether change is accelerating rapidly or seemingly creeping along, members of respective societal groups respond in ways appropriate for dealing with the impact and effect of change itself.

The social science study of change follows a number of specific orientations. Anthropologists concentrate on acculturation and assimilation; sociologists study modernization and social change; psychologists view change as a process of individual adaptation. Social and behavioral scientists attempt to identify and isolate the characteristics of change agents, the process of change occurring within and between groups, individual reactions to change, and short- and long-term effects evoked by the change process (cf. Berry, 1977).

A number of behavioral and social scientists take the position that

> Social change is accompanied by the intensification of social and cultural sources of psychological conflict, by new stresses and new adaptation requirements in new milieus, and by the loss of the stabilizing effect of old cultural patterns. (Kiev, 1972, p. 9)

Some social scientists have examined specific phenomena that result from sudden changes, such as the effects of bereavement among widows (Marris, 1974), the loss of one's home through urban renewal (Fried, 1963), effects of confinement and persecution (Bettelheim, 1943; Ostwald & Bittner, 1968), the impact of natural disasters on quality of life and emotional stability (Baker & Chapman, 1962), inequality of family status and stress in marriage (Pearlin, 1975), and changing social status and psychological disorder (Dohrenwend & Dohrenwend, 1969). Thus, some social-science research tends to concentrate on the negative impact of social-change agents on individuals, families, small communities, and even large cultural and religious groups. Human adaptation and coping with life crises have formed the theme of a collection of articles devoted to various forms of situational and personal actions that evoke sudden change (Moos, 1976).

Yet it would be absurd to suggest that change inevitably results in negative outcomes. Certainly there is enough anthropological and biological evidence to suggest that the ability of the human and lower phylogenetic life forms to adapt successfully to change is the key to survival and to the evolution of life itself. Nomadic tribal groups were constantly subjecting themselves to change in their seemingly never-ending search for sustenance and safety, as well as responding to a changing environment which they were unable to alter. We have to recognize, though, that nomadic groups expected change and hardships; they were an integral part of daily living (cf. Honigmann, 1959, pp. 323–326). Social changes therefore may not be hazardous and unduly stressful. Pastoral individuals expect changes, and internalize them as a normative process.

Coping and Adapting to Social Change

All changes require use of coping strategies. Sudden and very rapid social change, however, are likely to be especially disruptive for individuals.

The abruptness of a change, some argue, is a central source of individual stress (Kiev, 1972; Marris, 1974) leading, in many cases, to psychiatric illness, shock, and death. Others argue that the sudden change produced by the coming together of different cultural traditions produces "acculturative stress" (Berry, 1977) leading to the use of destructive coping mechanisms such as alcohol and drug abuse at the individual level, and massive disruption of cultural values and norms at the group level. Abrupt change can be disruptive especially when individuals and groups are not totally informed of the conditions and the consequences produced by the change. Under these conditions, people typically do not have the opportunity to assess and implement coping and adaptive mechanisms to deal with stress.

Abrupt social change can create situations that produce individual stress leading to a systematic disruption of life-style and possibly quality of life. Coping, or responding to external life strains in such a way that emotional distress is prevented, reduced, or avoided (Pearlin & Schooler, 1978), can occur in many forms. Occasionally, individuals demonstrate the inability to produce or evolve coping mechanisms to deal effectively with the strains and stresses. Efforts to utilize known defense, coping, and adaptive mechanisms seem to become dysfunctional in situations created by the abrupt presence of change. Unfortunately, recognition of the ineffectiveness and inappropriateness of mechanisms does not come about during or immediately after a sudden change. Instead, it typically occurs after a vicarious effort to use known or socially acceptable mechanisms to deal with the new conditions created by the change. As each effort to deal with strange and unfamiliar situations meets with failure, stress, anxiety, and destructive forms of behavior seem to increase. Many persons, through self-initiated efforts or intervention by others, manage to pull out of the crisis. Many of these individuals are strengthened by the experience. To the contrary, others succumb to their never-ending search for structure and meaning created by the new circumstances. Individual responses to rapid change are somewhat like the effect of fire on material objects—some are consumed, and some are strengthened.

Although I have emphasized coping as one strategy, social change can also be dealt with through adaptation, mastery, and defense (White, 1974). Each form of adjustment requires a particular kind of behavior typically learned within one's cultural environment as a result of past experiences and shared information from group members. Thus, four main strategies can be used to describe the behavior used by individuals to deal with change. For our purposes, however, coping behavior is of central concern, as it is a form of adaptation evoked "under relatively difficult conditions or circumstances" (White, 1974, p. 49).

Social Change and Stress

Like coping and adaptation, stress is another dramatic response to the circumstances created by social change. Similarly, stress reactions can be intimately linked to coping and adaptation efforts. The stress experience may be sparked by a generalized stressful situation, and lead to attempts to cope effectively with the circumstances. However, if coping efforts fail, the stressful experience can be intensified. Viewed in this manner, a systematic relationship exists between change, coping, and stress and, as suggested by Manderscheid, Silbergeld, & Dager, (1975), can generate negative cybernetic feedback loops.

As suggested by Leighton (1949), basic stress types such as threats to life and health, loss of means of subsistence, enforced idleness, restriction of movement, and capricious and unpredictable behavior on the part of those in authority upon whom one's welfare depends can create additional disturbances. These may include:

1. Frustration of expectations, desires, needs, intentions or goals.
2. "Ambivalence" or "multivalence."
3. Circumstances creating confusion and uncertainty as to what is happening in the present and what can be expected in the future (Leighton, 1949, pp. 76–77).

Smelser (1968) suggests that the above can constitute independent variables, where the reactions to stress form the unit of analysis, and per Leighton's suggestion result in patterns of behavior that may consist of:

1. Constructive activity directed toward overcoming the source of stress.
2. Vicarious activity evoked in succession.
3. Suspiciousness, hatred, hostility, and destructive action sometimes redirected toward surrogate causes instead of the actual ones.
4. Withdrawal and resignation leading to a state of hopelessness and futility (cf. Leighton, 1949, pp. 77–78).

Certainly these four response patterns are but a few of the many reactions that can occur at a psychological and physiobiochemical level.

In considering stress and stress reactions, attention should be given to what Smelser (1968) calls the: (1) differential vulnerability to stress and (2) differential availability of responses. Each of these affects the reactive style of individuals experiencing stress and, indeed, stems from one's own cultural milieu and experiential background.

Other factors are known to relate closely with stress reactions. Studies over the past two decades have led investigators to conclude that minority and social status (Dohrenwend, 1973; Dohrenwend & Dohrenwend, 1969; Vaughn, Lin, & Kuo, 1974) and age and sex (Dupuy, Engel, Devine, Scanlon, & Querec, 1970; Traxler & Linksvayer, 1973) are strong predictors of stress vulnerability. Similarly, some investigators have isolated certain personality characteristics as potent contributors to stress reactions (Coelho, Hamburg, & Adams, 1974). For example, rigidity, tolerance, and adaptability were found to foster effective use of coping mechanisms in the face of stressful situations (Kocowski, 1971; Opton & Lazarus, 1967; Wolff, Friedman, Hofer, & Mason, 1964). Manderscheid *et al.* (1975) theorized that alienation is linked "to the dynamics of social structure through the mechanism of stress and to behavior through perceptual style" (p. 101). Stress can be linked to a number of causal factors; however, little is known about stress factors among culturally unique groups.

To summarize, stress and stress reactions are systematically related to social change. And as with coping and adaptation, clearly identifiable stress reactions can be identified. Nonetheless, coping, adaptation, and stress reactions are so closely interwined that it can sometimes distort the understanding and investigation processes. An individual may be engaging in the use of coping mechanisms and subsequently meet with failure. Stress ensues, leading to the use of coping mechanisms to deal with the state. Negative cybernetic feedback loops can occur and amplify the circumstances, potentially locking the individual in a never-ending search for a way out of the difficulty. Observers may not be able to detect the shift in emphasis, and possibly distort the explanation of the behavior. Understanding the nature of the reaction can be complicated further when investigators know very little about a society's vulnerability to stress, and the availability of response repertoires in that society.

Forced Migration as an Abrupt Form of Social Change

Forced migration is a highly specialized form of migration that can arouse some confusion among researchers as to its implications and meaning (Micklin, 1978). It is a highly specialized form of social change, and therefore needs clarification. All forms of migration can be forced; that is, the causal impetus can be interpreted as the push or shove that initiates movement of people. Group members can be denied the decision to move by an external agent; economic conditions may provide no other option but to leave an area and search for new sources of income; environmental conditions may be such that continued residence spells disaster; or, one leaves an area because of some primary drive to search for and explore new lands. Whatever the motive, and motives may be mixed and not entirely conscious, all forms of migration may be forced. Nonetheless, forced migration has unique characteristics that set it apart from processes such as migration in general and voluntary migration specifically. By its very sound the word "forced" seems to connote something caustic and harsh. Attributing the motive for migration to some primal need for movement seems benign. Having to move when you have been denied the choice not to move seems less benign and perhaps harsh in contrast. Petersen (1958) refers to forced migration as an instance where migrants do not retain the power to decide on the move. Similarly, Heberle (1955) prefers to call forced migration involuntary, again appealing to a group's lack of decision-making power. It is safe to assume that when an individual is consciously and openly denied the choice to do something especially contrary to his better judgment, his actions are under the control of that agent. Right of due process is denied, alternative choices are almost nonexistent, and few relish the eventual departure and move. As will be illustrated later, outcomes of forced migration, the type where people are compelled to leave an area when their presence is not desired by an external agent, are disruptive, occasionally tragic, and in many cases generate irreversible problems. In this context, forced migration is viewed as migration induced by an external agent in a situation where groups are denied decision-making, and carried out to fulfill goals inconsistent with those of those who are forced to leave their ancestral or native homelands.

Students of history are generally aware of the hundreds of cases where groups have been forced to leave their native soil. In many situations, overriding interests of outside groups intrude and deny groups their rights to remain. Thus, forced migration as described above is not new. However, there is ample evidence to indicate that forced migration is very much a part of the contemporary policies of many governments. Communities and groups, whether they exist in developed or less developed countries, are subject to the circumstances surrounding forced migration.

A number of pressing issues and concerns seem to be contributing to the incidence of forced migration. For example, rapid population growth and the need for energy sources are two forces having a dramatic impact on contemporary society. Land speculators are looking for and developing areas for housing to meet population expansion. A large part of the land speculation and development is occurring in rural, agricultural areas. In many countries, such as Brazil, Colombia, and sections of Africa, native communities are forced to leave and find refuge in more remote areas. In related instances, community residents are forced to move to large urban areas to seek new forms of livelihood. In the United States, many farmers and ranchers are abandoning traditional life-styles and taking up residence in suburban and rural towns, largely owing to economic pressures and the demands placed upon them by land developers.

Use of land for the development of energy resources such as fossil fuels presents a slightly different set of circumstances for the native residents. If small communities exist within an area defined by the presence of a fuel source, minerals or timber, they usually are asked to leave. In some cases, residents are informed that development will occur around them. A "boom town" emerges, and tends to have total disruptive effects on village life-style and quality of life. Depending on the residents' standard of living, means of socioeconomic support, and attachments to their native soil, relocated residents can expect to experience stress, coping, and adaptation problems. This is especially probable for groups who are forced to leave the area and settle in nearby communities. Adjustments can be particularly stressful for individuals who have a limited repertoire of coping mechanisms, limited experience with "outsiders" and with the effects of migration.

Sole responsibility for developments occurring in the form of either forced migration or the emergence of "boom towns" does not rest solely with government and industry. Occasionally, some group members may cooperate or collude in activities that eventually lead to developments, such as migration. For example, the Thai Dam and Catholic Vietnamese in North Vietnam and the Hmong in Laos committed themselves initially to the French, and later to the Americans, in the struggle against the Viet Minh and later the communist North Vietnamese. In Paraguay, "acculturated" Aché Indians were used to hunt and track their own kind and resettle them on reservations (Münzel, 1973). This collusion or cooperation in a native group's own eventual undoing can lead to considerable self-anger and consequent depression.

In a somewhat related circumstance, modern industrialized societies may use tribal and small countries for exploitative purposes that are not directly related to immediate economic gains. Tribes or tribal leaders may be asked to serve in a military or political capacity to swing and effect a direction desired by certain developers. Part of the direction may eventually require the wholesale resettlement of rural, agrarian villages. Such hegemony may lead eventually to economic advantage, of course, but the economic aspects and the consequences for village and tribal members may not be evident initially.

To conclude, as population numbers increase, the need for land will also increase. Land use will be determined by a number of factors, such as suitability for habitation, presence of natural resources, suitability of land for producing food, distance from large population centers, and accessibility to trade and transportation routes. Certainly we can expect the search for land to spread rapidly and increasingly to remote rural areas. Residents of these areas indeed cannot expect to remain isolated and insulated. Changes will come, and will have a dramatic effect on life-styles.

Forced migration as a form of social change is likely to create new problems for rural, indigenous native populations. Geographically isolated native and tribal communities occupying land containing valued natural resources are most susceptible. Encroachment of developers, pressing needs of dominant societies, and increasing scarcity of certain natural resources are likely to effect irreversible social and psychological changes. If the land contains valued natural resources, we can expect residents to be relocated.

Forced migration and social change are closely related. Forced migration can come about through the need for change; change can emerge as a consequence of forced migration. Whatever the case, forced migrants experience stress as they attempt to cope and adapt to circumstances presented by a new environment together with an altered life-style.

Typically, the social change associated with forced migration is abrupt, as events are carried out in rapid succession. When native and tribal communities are forced to migrate, planners ordinarily attend to material needs, such as construction of homes and subsistence provisions. Unfortunately, little attention has been given to psychosocial needs associated with the move and the resettlement process (Trimble, 1978). Forced migration, as a process, has a definite impact on use of coping and adaptive strategies. In many cases, the forced movement and resettlement create a great deal of stress, that is manifested in a number of socially deviant ways. The remainder of this chapter focuses on four native groups who experienced negative consequences resulting directly from forced migration.

Forced Migration and the Experiences of Four Communities

This section sets out to describe in summary form the resettlement experiences of four native communities. The communities selected for the discussion have a number of factors in common, as follows: (1) they are native, indigenous groups whose cultural orientation and life-style is different from those responsible for their resettlement; (2) they were relocated so that their ancestral lands could be explored and developed to benefit a larger, more dominant population; (3) none of the groups volunteered to resettle; in fact, they were convinced that resettlement was an absolute necessity; (4) after resettlement none of the groups benefited directly or indirectly from the use or development of their ancestral lands; and (5) each of the groups experienced hardships as a result of the resettlement. More importantly, the hardships and related outcomes of earlier experiences have continued up to the present.

Material used in the preparation of the descriptions was compiled from a variety of sources, most of which are first-hand accounts prepared by journalists, anthropologists, and historians. Apart from the section on the Marshall Island community, the material came from personal letters, newspaper accounts, unpublished manuscripts, and personal interviews with observers who witnessed events surrounding the resettlement processes. Extreme efforts were taken to establish the authenticity of the accounts, including but not limited to checking and rechecking information with different sources. The point should be made that future accounts of communities subject to forced migration should be more carefully documented and researched if we are to have a better understanding of the phenomena. For the moment, unfortunately, reliance must be placed on unofficial observers to provide us with a data base.

The Kreen-Akrore: The Near Extinction of a Brazilian Tribe

Within the past few decades, South American governments coupled with foreign interests in timber, oil, uranium, and other energy-related resources have begun massive explorations in that continent's vast, unexplored territories. To permit exploration, South American governments have engaged in a rigorous program of relocating indigenous groups.

In Brazil a program was initiated to relocate native tribes from their ancestral homes to over two dozen federal reserves. Xingu Park is the largest reserve, and the only area exclusively reserved for the indigenous population (Jungueira, 1973). The relocation of tribes to Xingu and management of internal affairs in the park is carried out primarily by Claudio and Orlando Villas Boas, two brothers committed to the safety and welfare of Brazil's Indian population. Over the years the Villas Boas brothers have assisted thousands of Brazilian Indians in making the transition from their aboriginal life-styles to life at Xingu.[1]

The Kreen-Akrore (Kréen-Akaróre) was a small tribe of approximately 600 people who lived in a remote region near the Peixoto de Azevedo river in the northern Mato Grosso State in Brazil. Prior to 1973 numerous attempts were taken to contact the Kreen-Akrore, but none were successful (cf. Cowell, 1974).

Numerous skirmishes with government soldiers, the encroachment of settlers, and the development of a highway which was to pass through Kreen-Akrore territory convinced the Villas Boas brothers that contact and resettlement had to occur. Otherwise the tribe was doomed to extinction. Their efforts were initiated in 1970; however, it wasn't until 1973 that direct contact occurred. On February 5, 1973, about 30 Kreen-Akrore men walked into the Villas Boas brothers' camp, exchanged gifts, and through an interpreter were eventually informed of the need for their resettlement (Davis, 1977). After months of negotiation and subsequent preparation, the entire tribe was removed from their forest home to a new location.

Initially, FUNAI (*Fundacio Nacional do Indio*), Brazil's agency for controlling and regulating Indian affairs, wanted the Kreen-Akrore relocated in the Xingu reserve. Claudio and Orlando Villas Boas vehemently protested the choice, and urged the selection of a site within the Kreen-Akrore's ancestral home, but away from projected development areas. Despite the protest and the obvious suitability of the Villas Boas brothers' site choice, the Kreen-Akrore were settled on a reserve that included the Santarém-Cuibá highway as one of its boundaries. The decision to relocate the tribe close to a major highway led to the eventual devastation of the tribe.

Within eight months after the tribe's resettlement, their numbers were reduced substantially. Contact with *civilizados* also meant contact with contagious diseases to which many succumbed. As if death through disease were not enough, the tribe fell into a state of despair and "were in a state of sickness . . . and hunger" (Davis, 1977, p. 72). More than that, use of traditional customs had dissipated quickly, and been replaced with begging and the use of alcohol and tobacco.

Along with others, the Villas Boas brothers were fearful for the future of the Kreen-Akrore. To remain at the first reserve meant certain destruction. In October, 1974, after much debate with FUNAI officials,

[1] The Villas Boas brothers established a permanent camp in Kreen-Akrore territory in 1970. Their experiences associated with their attempts to contact the Kreen-Akrore, together with their feelings about the venture, are beautifully described in Adrian Cowell's prize-winning film, *The Tribe that Hides from Man*. Copies of the film are readily available from major film libraries.

the remaining 135 members of the Kreen-Akrore tribe were scheduled to be removed to Xingu. Davis (1977) notes, "A Brazilian doctor aboard (the) airplane reported that the Kreen-Akrore women were purposely aborting their children, rather than produce offspring who would face the new conditions of the tribe" (p. 73). At the time of the second removal, it was reported that several members of the tribe were suffering from another disease from which many would die.

By the last reports, the tribe numbered some 80 men, women, and children. The Kreen-Akrore continue to struggle to adapt to their new surroundings. Relocation for the tribe has not been a pleasant experience. In three short years an entire tribe is on the verge of extinction created for the most part by their forced movement to new lands.

Underneath the tragic circumstances surrounding the current status of the tribe lies a strange paradox. The Kreen-Akrore can reach a physical height of a little over 2 meters, which is at least 31 centimeters taller than the average Amazonian native. Many of the Kreen-Akrore's enemies referred to them as "giants." Moreover, one observer noted that the Kreen-Akrore were immune to disease and sickness in their ancestral homes; death was caused either by old age or combat. The centuries required for them to evolve their physical height and immunities have been wasted by their relocation, a move required to protect them from destruction. Yet their ability to cope with and adapt to the new environment, complete with all of its trappings, was limited. Whether or not they will accept assistance remains to be seen. A handful of Kreen-Akrore will probably survive, but they will never again experience life as it was in their original home.

The Kreen-Akrore were well equipped with the skills to live and survive in their native homes. However, the thought of developing coping and adaptation mechanisms for dealing with relocation probably never occurred to them. It's possible that they probably didn't know where to begin, if the thought did occur. Although the Villas Boas brothers took great care in preparing the Kreen-Akrore for the move, some very important steps were overlooked, not understood, or ignored. The ability to generalize coping skills from one environment to another was limited if not nonexistent. Perhaps the move was done too quickly. Whatever the explanation, circumstances surrounding the status of the Kreen-Akrore serve as an example of what can happen to a group that is unprepared for a move they did not want to make in the first place.

Mazatecs and Chinantecs of Southern Mexico

Our next short case study involves two tribes who reside in the state of Oaxaca in southern Mexico, the Mazatec and the Chinantec.

Unlike the Kreen-Akrore, the Mazatec and the Chinantec have been in contact with outsiders for hundreds of years. Despite these contacts, the two tribes have managed to retain many of their traditional customs, while borrowing and internalizing some of the skills and technologies of modern-day Mexico. Like the Kreen-Akrore, the two agrarian tribes were subjected to a massive relocation that resulted in tragic consequences.

In the late 1940s, the Mexican government, under advice from the Secretariat of Hydraulic Resources (SRH), began construction of two large dams in the Paploapan Basin in southern Mexico. Construction of the dams resulted in the permanent relocation of approximately 80,000 Mazatec and Chinantec people (Barabas & Bartolomé, 1973).

Prior to resettlement, fertile lands immediately surrounding the eventual dam were distributed to sugar refineries, lumber and paper factories, and industrial concerns which rely heavily on hydroelectric power. Native groups were given a chance to move. Many Mazatecs refused to move, so the SRH "provided a taste of its power by opening the dam's floodgates" (Barabas & Bartolomé, 1973, p. 7).

According to accounts, readjustment experiences of relocated Mazatecs were comparable to, and perhaps worse than, those of tribesmen who resisted removal. Irrigation, electricity, and safe, passable roadways were denied. Many have no deeds to their lands. Barabas and Bartolomé (1973) add:

> For others the transfer meant death. At least 200 simply died of depression (*tristeza*); the removal was especially hard as the aged, who grieved upon leaving lands where ancestors were buried and their sacred objects secure. (p. 7)[2]

Obligations were placed on many that eventually put most in debt. By the last reports, alcoholism and other forms of social deviancy have increased considerably, and traditional ceremonial life has faded rapidly.

In addition, the Mexican government resettled Mazatecs, Chinantecs, and mestizos in the same communities in hopes of promoting assimilation. As a result, violence and intergroup hostilities emerged, adding to the already destructive elements associated with the forced resettlement of the two tribes.

[2] By itself, depression is not usually thought of as a state that can produce death. I was reminded by professor Juris Draguns, Pennsylvania State University, and Dr. Joseph Westermeyer, University of Minnesota, that instances of psychogenic death have been recorded; however, circumstances are unclear. The observation by Barabas and Bartolemé is not contested; nonetheless, it should be recognized that depression can lead to self-imposed starvation, dehydration, suicide, and intercurrent infections. Therefore, it is possible that the depressive state led to death-producing behavior instead of being solely causative.

The resettlement of Mazatecs and Chinantecs has added another segment to the growing number of Mexico's impoverished groups. Prior to relocation, both groups were self-sufficient, and had developed an economy and a means of subsistence that provided for the needs of community members. In a matter of a decade, their stable life-styles have changed, largely owing to the exploitative character of industrial concerns and government policy. Moreover, both tribal groups are attempting to rebuild what remains of their culture in an atmosphere of violence, confusion, bewilderment, and inequity.

Removal of natives from traditional lands is a policy that is likely to continue, according to Beltran, Mexico's leading authority on indigenous affairs. He states, "It is necessary to change their [the natives] cast-like position to a class position where the possibility exists that the Indian may enter a group, in this case the proletariat" (Barabas & Bartolomé, 1973, p. 16).

Integration of native groups such as the Mazatec and the Chinantec has been a common practice of many countries. Typically, groups are forced to integrate. We see that the Oaxaca tribes afforded resistance to forced integration. At the same time the people had to cope and attempt to adapt to a new environment and life-style. Both the resistance to the move and the efforts to adapt appear to have been ineffective. The fact that many experienced serious depression suggests that a sense of futility had set in while they were attempting to survive and adapt. The individual need to search for and identify new coping skills seemed useless to those who succumbed to a depressive state. Many apparently recognized that they "found no practical manner to exercise (their) rights" (Barabas & Bartolomé, 1973, p. 15). The loss of native lands, accelerated contact with outside groups, an abrupt change in life-style, and the apparent loss of the right of due process contributed to the decline of the Mazatec and the Chinantec people. In this context, one cannot help but wonder what the effects generated by forced integration and migration will be on the youth of the two cultural groups.

Relocation of the Native People of Chemahawin, Manitoba

Development of lands for use as a source of hydroelectric power has affected the life-styles of other native communities. The situation occurring among the Mazatec and the Chinantec is not unique in the Western Hemisphere. In the winter of 1960, the Grand Rapids Hydro Development project was announced by the Canadian government. The native community residing in Chemahawin, Manitoba, would have to be moved to another site in order for construction to begin on the dam. Flooding of the native community was inevitable, and steps had to be taken to assure that the Indians would be settled safely in a nearby area.

Approximately 400 Cree Indians lived on an island in Cedar Lake at the mouth of the Saskatchewan River. The main source of livelihood for the Cree at Chemahawin was commercial fishing. Some of the men earned their living by trapping, hunting, forestry, and small construction work. The Cree had a high level of internal social organization, in part owing to their relative isolation from non-Indians and metis communities (Matthiasson, n.d.). Metis (mixed-blood Indians) participated in Chemahawin community activities on a limited basis. In short, the relative isolation of the Chemahawin Cree permitted tribal cohesion, prevented major forms of conflict and social deviancy, and enabled them to maintain an economy that provided opportunities and sustenance for residents. With some exceptions, the Chemahawin Cree had managed to maintain a traditional Cree life-style with a minimum of internal strife and disruption.

Construction of the dam began in 1961. During this period, representatives of the Canadian government met with Manitoba Hydro and Indian and metis communities to identify and select the new community. Eventually an area located on the southeast shore of Cedar Lake was selected as the site. The site eventually was named Easterville. According to Matthiasson and Landa (1969), the site decision did not reflect the sympathies of the total Chemahawin community. Many residents felt that they had not "been intimately involved in either discussions concerning the move or the selection of a site" (Matthiasson, n.d., p. 8). Despite protests and efforts to reverse the site-selection decision, the Easterville site was to become the new home of the Chemahawin Cree.

Troubles began for the Chemahawin Cree when they relocated at Easterville in 1964. The physical layout of new housing prevented the once-cherished value of privacy; the rocky soil prevented the development of family gardens; the unsheltered waters of the harbor were too rough for moving fishing boats; and the absence of suitable grazing lands meant that livestock had to be sold off. The flooding of Cedar Lake placed restrictions on hunting and trapping, once a major source of livelihood. The new community was more compact in size, bringing the Cree Indians into closer proximity with the metis. Intergroup friction was accelerated as a result, and led to problems in community organization and interpersonal relationships.

The relocation of the Chemahawin Cree from an island to a mainland community brought the people closer to the town of Grand Rapids. Thus, although the move brought the people closer to the outside world, it also provided them with an easier access to alcohol. As a consequence, heavy drinking became almost normative, rather than sporadic or incidental as it had been on the island. Significant increases in juvenile delinquency, marital problems, and unemployment also have been observed (Matthiasson, n.d.). Although the Cree had faced adversity and hardship in the past, the relocation presented the community with problems that they knew existed among other tribes, but thought would never happen to them.

The relocation of the Chemahawin people was not successful. The move was forced and abrupt. Even though the physical distance between the old and the new communities was slight, conditions in Easterville were enough to create an assortment of problems likely to endure for decades. Most of the Cree expected the relocation to be a smooth one, where life would continue much as it had on the island. Clearly, that was not the case. The Cree are now faced with problems for which there are no immediate answers. Most importantly, the move has created a sense of bitterness and frustration toward outsiders, the Canadian government, and among themselves. We can expect the socialization of youth to be even more disruptive and less structured than it was before. The effects of social deviancy and social disorganization are likely to feed on themselves. The future for the Chemahawin Cree is no longer so promising as it might have been if they had not been forced to move.

The Bikinians and Their Desperate Struggle to Return Home

Our fourth and final case study has a slightly different twist. This example involves the one-time residents of the Bikini Atoll, located in the northern Marshall Islands in the South Pacific. In 1946 they were forced to relocate on other islands and atolls in the Marshalls, to permit testing of nuclear weapons. However, after nearly 30 years, they are struggling to return to their ancestral homes. Both the circumstances surrounding the suitability of the Bikini Atoll for rehabilitation and the experiences of relocation make this case study another interesting example of the effects of forced migration on an indigenous group.[3]

In 1946, the small community on the Bikini Atoll were relocated when their ancestral home was selected as the United States' first postwar nuclear test site. The Bikinians, together with a few other Marshallese communities, have attempted to adjust to circumstances

[3] Dr. Robert Kiste, University of Hawaii, was especially helpful in providing information for this case description (cf. Kiste, 1974).

different from their normal life-style. Considerable problems have been posed for the administration of the United States Trust Territory of the Pacific and the Department of Energy (an agency succeeding the former Atomic Energy Commission). Prior to relocation, Bikinians were marine-oriented, with a long tradition of sailing. They relied heavily on fishing and the capture of marine fauna; horticultural activities were minimal. They were an isolated people who resided on a remote atoll, and were among the least acculturated of the Marshallese.

The Bikinians were initially moved to an uninhabited atoll, also located in the northern Marshalls. It did not have sufficient resources; in less than two years, the Bikinians faced near starvation. They were evacuated to a military base, and for eight months all their needs were provided for. In late 1948 they were once again resettled, this time on Kili, a small, single island in the southern Marshalls.

Kili represented a major alteration in the Bikinians' life-style. The southern Marshalls are in a wet climatic zone, as opposed to the dry northern atolls, and agriculture is a much more important subsistence activity. The Bikinians did not have agricultural skills appropriate for their new home, and, compounded with the difficulties of adjustment, the conditions at Kili did not permit the people to rely on marine fauna as extensively as in the past. Kili is a single island, not an atoll. It has no lagoon or sheltered fishing grounds. Rough seas and fringing reef prohibit the use of sailing canoes, and bar access to the island for several months of each year. The Bikinians have never adjusted to Kili, and their efforts to carve out an existence have been hampered by a burgeoning population. In 1946 the Bikinians numbered about 160; they are now well over 400. The years on Kili have been years of physical deprivation, mental frustration, and anguish.

Since relocation, Bikinians have had a burning desire to have their ancestral lands restored. Officials usually asserted that a return home was impossible because of radiological contamination. In 1967, a change occurred. A radiological survey report about Bikini indicated that it could be made safe for rehabitation. A clean-up program began. The two largest islands of the atoll, Bikini and Eniu, were replanted with coconuts and other crops. A village was constructed on Bikini Island, the traditional residential site.

Recent surveys of Bikini Island have reversed the optimism that prevailed a few years ago. It had been determined that Eniu, not Bikini Island, was safe for resettlement. Nonetheless, in March, 1977, about 80 people returned to Bikini Island.

The decision to allow the Bikinians to return has met with tragedy. The recent plants on the island were absorbing radiological elements from the soil. The relocated Bikinians were told not to eat edible flora; nonetheless, they did. As a consequence, the ingestion of the contaminated flora, when added to a higher than normal radiation from the damaged environment, subjected the people to exposure levels that exceeded the maximum amount recommended by the Environmental Protection Agency. In October, 1977, an official of the Department of Energy informed the Bikinians that it was not advisable for them to remain, because of the health hazards involved. Then the body burdens of radiological elements increased so dramatically that government officials decided to remove the Bikinians. In the fall of 1978, 112 Bikinians were returned to Kili with the impression that the Bikini Atoll would be forever unsafe for resettlement.

Several problems are apparent. First, what measures can be taken to alleviate the plight of the Bikinians? Hopes for a full-scale return to their ancestral home have all but disappeared. They are to continue residing on Kili, an island to which they have not made a satisfactory adjustment in 30 years. Second, resettlement of Eniu is still a possibility; however, it is much smaller in size than Bikini Island, and, with the dramatic increase in population, total resettlement is out of the question. Third, it is not at all certain that Bikinians comprehend the hazards of exposure to radiologically contaminated environments. It is not clear just how they perceive the risks that are involved. Obviously, the people needed an accurate notion of such risks if they were to refrain from eating crops on Bikini and avoid the "hot" islands of their home atolls. The nature of "unseen" radiological hazards is difficult to communicate effectively to a people whose culturally accumulated knowledge is based on notions foreign to the principles of Western science.

Although the people may not have a technical understanding of radiation, they are fearful and anxious about its consequences. They have good reason for concern. In 1954, one nuclear test resulted in fallout over at least two and perhaps several other atolls. The people of the two atolls are now experiencing cancerous thyroid disorders and leukemia. Death has occurred from the latter, and the "coconut telegraph" fans rumors of the "poisons" contaminating the northern Marshalls.

Sentiments toward relocation are mixed. Many older islanders want to return home to spend their last years, and the risks to them are minimal. Younger people are hesitant about resettlement on an atoll they have never known, especially when the threats to their own and their children's health are substantial. Yet other sentiments and beliefs, ones that are shared by all, filter the intent and content of communications with federal representatives. The people feel that they have suffered a great injustice, and anger exists. Bikinians have become accustomed to the food subsidies and other welfare measures, and they believe that such assistance is only partial retribution for past wrongs. In an anxious and fearful atmosphere created by past deprivation and relocations, they are seeking security in a state of permanent dependency.

Bikinians experienced near tragedy and deprivation from the forced relocation. Hope was inspired when the first group was allowed to return to Bikini. That hope has all but disappeared. Over the years their life-style has been disrupted and changes have occurred in traditions and customs. Like the other groups discussed in the preceding case descriptions, the Bikinians are bitter and uncertain about their future. Their once stable culture is in an unsettled state. Roles, values, beliefs, and quality of life are disrupted to the point that youth and adults alike will struggle to form strong new supportive coping and adaptation styles. Nonetheless, the likelihood of this coming about is questionable; indeed, personal costs and societal consequences are certain to emerge in the course of the effort.

An Analysis and Developmental Challenges

The relocation experiences of the Kreen-Akrore, Mazatec, Chinantec, Chemahawin Cree, and Bikinians raise some profound questions and concerns that are linked closely with the ability of cultural groups to accommodate and adapt to new surroundings. In retrospect, each group was subjected to abrupt social and cultural changes. They responded to the circumstances in similar ways but, to the contrary, the end result was tragic for some. None of the groups wanted to move, and none of them seemed comfortable in their new environments. The experiences of the groups may be extreme examples of the effect of forced migration. Yet, it is doubtful that any group subjected to similar cirucmstances would respond much differently.

In this section, attention is given to stress reactions and coping behavior induced by circumstances generated by forced migration. Against the backdrop of social change, discussion focuses briefly on stress types, stress reactions, and coping strategies. Emphasis is placed on formulating new descriptive and predictive models for handling the particular circumstances generated by forced migration. This section concludes with a series of fundamental challenges formed around research needs and policy implications. Where possible and reasonable, examples will be drawn from the experiences of the groups described earlier.[4]

Coping with Forced Migration: A Speculative Analysis

Forced migration, as a type of migration and a process initiated by an agent external to the group in question, inevitably contributes to psychological and sociological losses among group members. It also results in the loss of traditional, ancestral dwelling areas around which an entire cultural orientation and way of life was developed. Experiences generated by the relocation introduce an assortment of stressors, some of which are psychologically disturbing, but most of which are at least mildly disruptive. Stress reactions emerge and lead to the use of coping strategies to deal with them. As a consequence, the total relocation process is disruptive at a sociological level—familial patterns, role relationships and inherent responsibilities, forms of social control, subsistence patterns, and organizational structure are altered. Disruption at a psychological level occurs, too, affecting routine behavior, attitudes, motivation and emotion, and cognitive-perceptual styles. The interplay of the forces and structures at both levels therefore produces a complex state that may never stabilize to the form it once was.

At present there is no adequate social-change model that can handle the complex interplay and interaction of variables generated by forced migration (Micklin, 1978; Trimble, 1978). However, like many social and psychological phenomena, the forced migration process can be sliced up and assessed with a variety of molecular-type paradigms. For example, stress theory can be used to investigate both stressors and stress reactions generated by the resettlement, or attribution theory can assist in defining the way relocatees perceive the stimuli connected with certain segments of the resettlement process. At a fundamental psychiatric level of analysis, physiological reactions produced during and after the resettlement can be assessed and linked to

[4] It should be clear that empirically derived information is almost totally lacking on the four groups. To the best of my knowledge, the circumstances surrounding the forced migration of two of the groups were not thoroughly investigated. Matthiasson (n.d.) and Landas (1969) investigated a few variables among the Cree along with their detailed chronicle of events. Mason (1954) and Kiste (1974) documented the events surrounding the forced migration of the Bikinians. None of the authors, however, systematically assessed individual behavioral patterns and cognitive perceptions of community members.

one or a number of stress theory tenets (Burdock & Zubin, 1956). On this note, Kiev (1972) asserts:

> Detection of differences in mental illness in different cultures . . . requires a greater reliance on physiological indicators, which are more likely to get at the basic biological vulnerability, than on social-performance indicators. (p. 161)

In furthering an understanding of the forced migration of native and tribal groups, emphasis must be placed on two major points: (1) forced migration is a type of abrupt social change, and (2) relocatees are reluctant if not unwilling to move. These factors are interrelated. Abrupt change can be disruptive, and upset an established ecological equilibrium (Moos, 1976). Furthermore, the change-inducing decisions do not come about through active participation by those affected. Change, therefore, is induced and imposed upon the relocatees, forcing them to do something that they do not want to do. Thus, the abruptness of the change, and the absence of commitment on the part of the relocatees, can be potent sources of stress.

Abrupt Change and Stress

To lend some credence to the "abrupt change" argument, Coelho and Stein (1977) hypothesize that: (1) "individuals experience psychological stress reactions when there is unexpected or unusual change in their social or physical environment" and (2) "emotional stress is likely to increase when environmental change occurs at the same time as a severe life crisis that involves the disruption of emotional bonds" (p. 38). The disruption of emotional bonds plus the abrupt and unusual nature of change sets up stressors, usually the independent variables in stress studies. When the stress types combine to produce further disturbances, additional stress variables emerge and lead to stress reactions, typically the dependent variables.

Within a cross-cultural perspective, conceptualizing stress types and stress reactions emerging from forced migration can create an indeterminant framework. As yet no systematic attempt has been taken to organize respective physiological and psychological stressors and stress reactions among native and tribal groups. Furthermore, relationships between independent and dependent variables have not been carefully established. And finally, it seems plausible to suggest that a number of intervening variables should be identified and specified to strengthen relationships between certain independent and dependent variables. For example, Smelser (1968) suggests the addition of two broad classes of intervening variables to stress models: (1) differential vulnerability to stress and (2) differential availability of responses. Including these two classes of intervening variables would enable investigators to assess particularly stressful situations or classes and categorize the repertoire of stress reaction. Knowledge of this information as it exists in the prerelocation environment can be extremely useful in predicting outcomes during migration and upon settlement.

While living in their ancestral lands, the Kreen-Akrore and Bikinians had maintained well-established patterns of responding to stress situations. These patterns were deeply ingrained, and to an extent overlearned. The patterns were also an integral part of the folk history and folk psychiatry of the culture. In fact, the patterns were very much a part of their total ecological perspective. Sudden change, brought on by forced migration, exposed individuals to new environments. Under these conditions, traditional responses for dealing with stress became ineffective for dealing with the realities that follow migration (Reusch, Jacobson, & Loeb, 1948). On the surface, it would appear that the relocation experiences of the four groups described earlier introduced new forms of stress. Traditional forms of responding to the stress appeared to be ineffective; if they had been effective, the communities would be in far better position than they are today. Westermeyer (1978) suggests that "some aspects of culture might be more ecology sensitive while others are nonecology resistant" (p.121). It may be that stress reactions are least resistant especially among cultural groups with fairly rigid ways of responding to stressful situations.

Decision Making and Stress

Despite the plausibility of the argument that migration, as a form of social change, creates stressful situations and emotional difficulties, some have questioned it (Fried, 1963). Berry (1977) argues that "earlier assumptions that social change and acculturation tended to be psychologically devastating are no longer maintained" (p. 97). However, Berry goes on to add that "there is fairly persistent evidence for somewhat higher levels of psychosomatic stress among [native] populations than among nonnative samples" (p. 98). Furthering this point, Berry and Annis (1974) argue that individual preparedness has a bearing on accommodating and responding effectively to acculturative processes. Those who are less prepared for acculturation tend to experience more stress than those who are better prepared. The process of acculturation, though a form of social change, is not an abrupt process. People have time to think it through and decide how much of another culture they want to internalize. Therefore, it seems that denying an individual a decision-making opportunity is an important variable in the understanding of the relationship between migration and stress, one deserving of a great deal more attention.

The experiences of the Mazatec and Chinantec are especially illustrative. Not only were the groups denied the decision to move, they were also denied the decision where they were to be relocated. As it turned out, intercultural contact was thrust upon them. Along with mourning the loss of their homelands and acclimating themselves to a new environment, they had to contend with acculturation through contact with lower-class metis populations. At each step of the relocation process the people were forced to accept circumstances not of their own choosing. Certainly the absence of decision making contributed to many of their stress-related experiences.

Communal Experiences with Migration and Stress

Communal experiences with migration are important considerations in the understanding of the relationships between stress and forced migration. If, as we have pointed out earlier, a tribe or native culture is nomadic, they probably will experience less stress than a tribe or group that is not. The four native groups discussed earlier are not nomadic; each has resided in essentially the same environmental setting for centuries. Although each group roamed their respective areas in search of food, a permanent settlement was always present; this was especially true for the Oaxaca tribes, the Bikinians, and the Cree. Short-term mobility therefore was customary and necessary. However, these groups were never subject to constant movement involving exposure to a multitude of ecological settings. The available flora and fauna were always sufficient to sustain life. Therefore, having limited experience with migration and associated stresses implies that relocatees can expect to experience increased difficulties in adapting with the new environment. Knowledge of this characteristic is essential if we are to predict and prevent future tragedies generated by forced migration.

Coping Strategies and Forced Migration

Pearlin and Schooler (1978) refer to coping as "behavior that protects people from being psychologically harmed by problematic social experience" (p. 2). Essentially, coping is a process for minimizing the impact of life strains, stresses, and problematic situations on the psychological well-being of the individual. As a source of life strains, forced migration can require individuals to cope with (1) the process of adapting to new ecological conditions and (2) transferring technologies and skills. Each of these can create stress, necessitating the use of coping strategies for dealing effectively with them. Some strategies will be effective, and others will fail.

At a general level "we know relatively little of the nature and substance of people's coping repertoires and even less of the relative effectiveness of different ways of coping" (Pearlin & Schooler, 1978, p. 2). Of course, psychoanalysis provides useful descriptions of coping through the broad rubric of defense mechanisms. Similarly, White (1974) suggests use of adaptation, defense, and mastery as interchangeable concepts for coping. Nonetheless, current conceptualizations of coping tend to be broad, and to be discussed in omniscient and universal terms.

If any cultural group experiences stress, it can be certain that coping strategies have been developed for dealing with them. However, just as there may be a limited number of stressors, so there will also be a limited number of coping mechanisms. As we have pointed out earlier, forced migration can create new stresses. Will the prevailing repertoire of coping mechanisms be sufficient for dealing effectively with the new stresses? Will the failure of coping strategies contribute to increases in stress? In situations where coping strategies fail, what alternatives are available for dealing with the stress? These are not mere academic questions, but ones that have profound implications for the welfare and stability of a community, native or otherwise.

Stresses that accompany forced migration require more than identifying a coping strategy and putting it to immediate use. There are additional considerations. Pearlin and Schooler (1978) list four points of concern: (1) situations must be recognized as problematic before coping efforts can be implemented; (2) people may lack knowledge or experience required to deal with identifiable problematic situations; (3) strategies implemented in one situation can produce unwanted and unexpected outcomes; and (4) many problematic situations are impervious to coping interventions. These conditions can indeed affect the way native and tribal groups respond to new circumstances; certainly they are potential sources of difficulty, that compound the hardships associated with the resettlement. Like some many other points raised in this chapter, the concerns demand attention.

In summary, forced migration of native and tribal groups creates stress, produces stress reactions, and requires use of coping strategies. The process of uprooting and resettlement is foreign to most native communities, thus compounding relocation problems. More important, native groups respond to identifiable problematic situations with coping strategies particular to the ecological settings in which they were developed; that is, coping strategies are developed to deal with lifestrains posed by the specific environment in which one resides. Sudden and abrupt change in life-styles, resettlement to a new and sometimes unfamiliar ecological setting, and individual efforts to adapt can produce irreconcilable problems, attributable largely to one's inability effectively to reduce or eliminate strains with "known" coping strategies. Strains produced by forced migration can be so overpowering that many individuals have been known to give up, refusing to respond to the strains and to meet the basic requirements of daily subsistence. Forced migration's impact on the effective use of coping strategies is a subject worthy of a great deal more systematic attention.

Challenges for the Future

It is reasonable to assume that, as nations accelerate acquisition efforts for energy development and exploration of natural resources, more and more rural groups will be forced to relocate. It is also reasonable to expect these groups to experience complex problems in adjusting to new surroundings. Native and tribal communities are particularly vulnerable to the psychological and sociological strains associated with forced migration. Their cultural orientation may be geared to a specific ecological setting. Some of the elements within their sociocultural milieu may be resistent to the effects of change; others will not, and these raise the distinct possibility that tragic losses can occur. Within this framework, social science should mount a concerted effort to increase awareness and understanding for the eventual purpose of identifying potent workable mitigation strategies.

Future research on the effects of the forced migration of native and tribal groups will have profound implications for increasing theoretical and substantive knowledge about social change, stress reactions, and the effective use of coping strategies. At present, social science is painfully ignorant of these factors. Moreover, for groups like the Kreen-Akrore there is a complete absence of ethnographic data, further compounding the understanding process.

Seminal efforts should be taken to identify areas and native communities most likely to be relocated in the near future. Industry together with federal and state governments have targeted certain regions for resource development, especially in Alaska and parts of the western United States. Similarly, multinational corporations, working closely with foreign governments, are exploring large tracts of land in remote regions, especially in South America, Canada, and Southeast Asia. Native and tribal groups, most of whom have limited understanding of resource development and technology, reside in many of the regions. In all likelihood the presence of these groups is known to developers and concerned social scientists. For example, according to Davis (1977) presently there are about 118 indigenous groups in Brazil, and about 60 mineral exploration projects planned or under way, with six national parks and 15 reserves, or 21 areas, where tribes have been or are scheduled to be relocated. Moreover, there are countless groups residing in remote areas of the Amazon basin on whom there is very limited information. Further surveys and accountings are encouraged. Similarly, the status of resource development should be identified for each region, and where imminent, efforts should be taken to assess those psychological and sociological factors most likely affected by the impact of development.

The identification of cultural-specific stressors, stress reactions, and coping strategies is a logical starting point, especially for groups targeted for relocation. Similarly, efforts should be undertaken to document thoroughly the experiences of native groups recently subjected to forced migration. As part of this effort, research should focus on identifying the stressors, stress reactions, and coping strategies employed by relocatees. Investigators may find that some have managed effectively to cope and adapt to their new surroundings. Knowledge of the successful attempts will provide an important data base for devising mitigation strategies for marginal individuals, that is, those still struggling to cope and adapt, and those who will be subjected to relocation in the future.

Research efforts should concentrate studies on youth and the aged, two key stages in the life-cycle process. On the one hand, existence of social-science research on socialization occurring among culturally different adolescents and the elderly is lacking (Dinges, Trimble, & Hollenbeck, 1979). Future studies must focus on the developmental tasks that occur during these crucial and significant life cycles, especially as they occur in traditional settings. We can never be in a position to understand the nature of adaptation and coping behavior in new environments until we understand traditional patterns of socialization and the disruption of these patterns through abrupt rapid forced relocation.

Future studies of forced migration should include systematic efforts to monitor the relocation experiences (Micklin & Leon, 1978). Although such studies may be costly, the costs incurred may be far less than those required to implement and sustain future mental-health programs. And certainly the costs in human life and life-satisfaction cannot be equated with the costs in money.

Apart from the theoretical gains, future studies of forced migration can and should be useful for planners and policy formation. Data can be useful for eliminating or mitigating identifiable sources of stress and implementing a program for interpersonal support. Crisis-intervention programs could be established and managed in close cooperation with native healers and tribal leaders.

Planners should be more concerned about the effects of abrupt social change on a community. Traditional subsistence patterns and daily living habits may not coincide with circumstances existing in a new ecological setting. Manderscheid *et al.* (1975) suggest that "specific directives for change would be little more than mere speculation" (p. 100) if planners systematically assessed the total impact of the planned change. Planning should take on an interdisciplinary focus and research results applied to eliminate identifiable stressors. By itself, removal from one's ancestral home creates grieving, bereavement, and stress. Relocatees should not have to contend with additional stresses. Prevention of stress and mitigation strategies must become integral components of large development efforts where the lives and way of life of a cultural group are at stake.

Finally, research and development should take account of the importance of the cultural traditions of any group about to be relocated. The traditions are an integral part of the ecological fabric of any cultural group. Appell (1977) argues that "if a people have access to their cultural traditions and are able to evaluate them positively, they have the resources to cope creatively with the social change and move into the future without apprehension" (p. 16). He suggests that planners implement a "salvage archeology" model, or a program to help relocatees regain control over the use and management of their cultural traditions, history, and language. Maintenance of the cultural heritage, though no simple task in the face of the psychological and sociological losses created by relocation, can provide a source of strength.

Forced relocation of culturally diverse groups is a topic that has received far too little interest. Interest in the land which groups occupy seems to take clear precedence over the needs of the occupants. Yet, the need for responsible and culturally sensitive planning is apparent, as evidenced by the experiences of the four groups herein discussed. Few will doubt the need for natural resources, but few have questioned the need for systematic research on the adaptation and coping of groups forced to relocate to meet the energy needs of millions. Prevention of conflicts, stress, and adaptation produced by forced migration not only makes practical sense, it can also lessen the likelihood of governments' having to invest millions of dollars in the future to assist groups in recovering from the misfortunes of relocation.

Acknowledgment

I am grateful to the following persons for their thoughtful assistance in helping me prepare this chapter: Dr. Joseph Westermeyer, University of Minnesota; Dr. Christopher Cluett and Dr. Marvin Olsen, Battelle Human Affairs Research Centers; and Dr. Robert Kiste, University of Hawaii.

References

Appell, G. N. The plight of indigenous peoples: Issues and dilemmas. *Survival International Review*, 1977, *3*, 11–16.

Baker, G., & Chapman, D., (Eds.). *Man and society in disaster.* New York: Basic Books, 1962.

Barabas, A., & Bartolomé, M. *Hydraulic development and ethnocide: The Mazatec and Chinantec people of Oaxaca, Mexico.* International Work Group for Indigenous Affairs Document Series (Vol. 15). Copenhagen: International Work Group for Indigenous Affairs, 1973.

Berry, J. Social and cultural change. In H. Triandis & R. Brislin (Eds.), *Handbook of Cross-Cultural Psychology* (Vol. 4). New York: Allyn & Bacon, 1977.

Berry, J. W., & Annis, R. C. Acculturative stress: The role of ecology, culture and differentiation. *Journal of Cross-Cultural Psychology,* 1974, *5*, 382–405.

Bettelheim, B. Individual and mass behavior in extreme situations. *Journal of Abnormal and Social Psychology,* 1943, *38*, 417–452.

Burdock, E. I., & Zubin, J. A rationale for the classification of experimental techniques in abnormal psychology. *Journal of General Psychology,* 1956, *55*, 35–49.

Coelho, G. V., & Stein, J. J. Coping with stresses of an urban planet: Impacts of uprooting and overcrowding. *Habitat*, 1977, *2*, 379–390.

Coelho, G. V., Hamburg, D. A., & Adams, J. E. (Eds.), *Coping and adaptation.* New York: Basic Books, 1974.

Crowell, A. *The tribe that hides from man.* New York: Stein & Day, 1974.

Davis, S. H. *Victims of the miracle: Development and the Indians of Brazil.* Cambridge: Cambridge University Press, 1977.

Dinges, N., Trimble, J. E., & Hollenbeck, A. American Indian adolescent socialization: A review of the literature. *Journal of Adolescence*, 1979, *2*, 259–296.

Dohrenwend, B. S. Social status and stressful life events. *Journal of Personality and Social Psychology,* 1973, *28*, 225–235.

Dohrenwend, B. P., & Dohrenwend, B. S. *Social status and psychological disorder: A causal inquiry.* New York: Wiley, 1969.

Dupuy, H. J., Engel, A., Devine, B. K., Scanlon, J., & Querec, L. *Selected symptoms of psychological stress: United States.* Washington, D. C.: Government Printing Office, 1970.

Fried, M. Grieving for a lost home. In L. J. Duhl (Ed.), *Urban condition.* New York: Basic Books, 1963.

Heberle, R. Types of migration. *Southwestern Social Science Quarterly,* 1955, *36*, 65–70.

Honigmann, J. J. *The world of man.* New York: Harper & Row, 1959.

Jungueira, C. *The Brazilian indigenous problem and policy: The example of the Xingu National Park.* International Work Group for Indigenous Affairs Document Series (Vol. 13). Copenhagen: International Work Group for Indigenous Affairs, 1973.
Kiev, A. *Transcultural psychiatry.* New York: Free Press, 1972.
Kiste, R. C. *The Bikinians: A study in forced migration.* Menlo Park: Cummings, 1974.
Kocowski, T. Resistance to stress as a personality factor. *Polish Psychological Bulletin,* 1971, *1,* 25–30.
Landa, M. J. *Easterville: A case study in the relocation of a Manitoba Native Community.* Unpublished master's thesis, University of Manitoba, 1969.
Leighton, A. H. *Human relations in a changing world: Observations on the use of the social sciences.* New York: 1949.
Manderscheid, R. W., Silbergeld, S., & Dager, E. Z. Alienation: A response to stress. *Journal of Cybernetics,* 1975, *5,* 91–105.
Marris, P. *Loss and change.* New York: Pantheon, 1974.
Mason, L. *Relocation of the Bikini Marshallese: A study in group migration.* Unpublished doctoral dissertation, Yale University, 1954.
Matthiasson, J. S. *Forced relocation: An evaluative case study.* Unpublished manuscript, University of Manitoba, n.d.
Micklin, M. Psychosocial aspects of forced migration. Unpublished manuscript, 1978.
Micklin, M., & Leon, C. A. Life change and psychiatric disturbance in a South American city: The effects of geographic and social mobility. *Journal of Health and Social Behavior,* 1978, *19,* 92–107.
Moos, R. H. (Ed.). *Human adaptation: Coping with life crises.* Lexington, Mass.: Heath, 1976.
Münzel, M. *The Aché Indians: Genocide in Paraguay.* International Work Group for Indigenous Affairs Document Series (Vol. 13). Copenhagen: International Work Group for Indigenous Affairs, 1973.
Ostwald, P., & Bittner, E. Life adjustment after severe persecution. *American Journal of Psychiatry,* 1968, *10,* 87–94.
Opton, E. M., & Lazarus, R. S. Personality determinants of psychophysiological response to stress: A theoretical analysis and an experiment. *Journal of Personality and Social Psychology,* 1967, *6,* 291–303.
Pearlin, L. I. Status inequality and stress in marriage. *American Sociological Review,* 1975, *40,* 344–357.
Pearlin, L. I., & Schooler, C. The structure of coping. *Journal of Health and Social Behavior,* 1978, *19,* 2–21.
Petersen, W. A general typology of migration. *American Sociological Review,* 1958, *23,* 256–266.
Reusch, J., Jacobson, A., & Loeb, M. B. Acculturation and illness. *Psychological Monographs',* 1948, *62,* 1–40.
Smelser, N. *Essays in sociological explanation.* Englewood Cliffs, N. J.: Prentice-Hall, 1968.
Traxler, A. J., & Linksvayer, R. D. Attitudes and age-related stress periods in adulthood. *Proceedings of the 81st Annual Convention of the American Psychological Association,* 1973, *8,* 783–784 (summary).
Trimble, J. E. Issues of forced relocation and migration of cultural groups. In *Overview of intercultural education training and research,* (vol. III): *Special research areas.* Washington: Society for Intercultural Education, Training and Research, Georgetown University, 1978.
Vaughn, J., Lin, N., & Kuo, W. *Some major determinants of stress among Chinese Americans:*

A test of two theories. Paper presented at the meeting of the American Sociological Association, Montreal, August 1974.

Westermeyer, J. Ecological sensitivity and resistance of cultures in Asia. *Behavior Science Research*, 1978, *13*, 109–123.

White, R. W. Strategies of adaptation: An attempt at systematic description. In G. V. Coelho, D. A. Hamburg, & J. E. Adams (Eds.), *Coping and adaptation*. New York: Basic Books, 1974.

Wolff, C. T., Friedman, S. B., Hofer, M. A., & Mason, J. W. Relationship between psychological defenses and mean urinary 17–hydroxycorticosteroid excretion rates. I: A predictive study of parents of fatally ill children. *Psychosomatic Medicine*, 1964, *26*, 576–591.

21

Integration of Immigrants: The Israeli Case

Aliza Kolker and Paul I. Ahmed

Of all modern immigration societies, none has absorbed so large a proportion of immigrants or placed so emphatic a stress on immigration as the State of Israel. In the first three years of the state's existence the Jewish population doubled; in the first 25 years it quadrupled.

Much can be learned from the Israeli experience about the integration of immigrants. Although a great deal has been written about it, the major concern has been with issues of economic adjustment. In fact, the Israeli government's policy concentrates exclusively on the economic adjustment of immigrants, particularly in the areas of housing and employment. Much less is known—or done—about the psychological aspects of adjustment, which are much more elusive, yet no less critical. In this chapter we shall analyze factors contributing to the psychological adjustment of immigrants in Israel, and draw some implications for policy.

Immigration as an Ideological Challenge

In this section we shall review briefly the facts about immigration to Israel, and explain its significance in the context of Israeli society

Aliza Kolker • Department of Sociology, George Mason University, Fairfax, Virginia 22030. **Paul I. Ahmed** • Office of International Health, Department of Health and Human Services, Rockville, Maryland 20857.

(for more complete accounts, see Mishal, 1971; Government of Israel, 1973, 1977; Jewish Agency, 1964).

Immigration is viewed as no less than the fulfillment of the State of Israel's historic mission in the context of the Jewish people's 4,000-year history. It is viewed as the "repatriation" of Jews to their true "homeland" from which they were exiled by the Romans 2,000 years ago and for which they have never ceased to long. In modern times (since 1881) the history of the Jewish community in Palestine has been characterized by a constant struggle to win the right to unlimited immigration in the face of the hostility of the Arab inhabitants and of the colonial governments. The plight of Jews in many countries, and particularly the extermination of six million Jews by Hitler, lent a desperate urgency to this struggle.

It is not surprising, then, that when the State of Israel was founded in 1948 free immigration was declared the *raison d'etre* of the Jewish state, the cardinal principle upon which it was founded. The Law of Return, adopted in 1950 and later amended to exclude "those who constitute a danger to public health or to national security," grants any Jew a "natural right" to "return" to Israel.

The Jewish population in Israel in 1948 numbered about 650,000, of whom 90% were of European birth. During the first three years of the state (1948–1951) 700,000 immigrants arrived, doubling the population of the country. Among these, half came from Europe, the battered survivors of World War II and the Holocaust. They arrived from displaced persons' camps in Italy, Germany, and Austria, from detention camps in Cyprus, and from newly Communist Eastern European countries. The remainder came from the Middle East. During those years entire Jewish communities were transported to Israel by massive air and sea lifts. Virtually the entire Jewish communities of Yemen, Iraq, Lybia, Syria, and Bulgaria, as well as substantial portions of the Jewish communities of Poland, Rumania, Hungary, Iran, Egypt, and Turkey emigrated to Israel.

From the beginning the only limitations on immigration were to be the Jews' readiness to leave their countries of residence, those countries' readiness to let them leave, and the availability of means of transportation. Israel's capacity to absorb immigrants was taken for granted. Yet when the tides came, surpassing all estimates, Israel was woefully unprepared. The most critical needs were housing and employment. Abandoned Arab housing was filled quickly. Many immigrants were sent to immigrant camps where they were taken care of by the government, and where enforced idleness and bureaucratically induced passivity contributed to a massive demoralization. In 1950 the government's policy was changed: from then on immigrants were sent to "transitional camps" *(ma'abarot)* composed of tents and tin shacks, where they looked after themselves. Many of the camp dwellers were employed in special relief-work projects, such as land reclamation. By 1952 a quarter of a million people were living in over 100 transitional camps.

A lull in immigration set in in 1952–1954, when only 50,000 Jews arrived, and a substantial number left the country. The lull was caused partly by economic recession, which compelled the government to impose a strict austerity regime and created mass unemployment, and partly by the clampdown on further Jewish emigration on the part of Eastern European countries. The lull was used by the government to reassess and overhaul its absorption policy. Henceforth new immigrants were to be sent directly to Spartan, hastily constructed permanent houses in new agricultural and urban settlements, bypassing the transitional camps. Two major purposes of this policy were to decentralize the Jewish population, thitherto concentrated in the urban center of the country, and to facilitate the agricultural development of the country. The transitional camps were progressively abolished, and by the 1960s nearly all their inhabitants had been resettled in permanent housing.

Immigration increased again in 1955–1957, when 100,000 Jews came. As in previous years, the majority of immigrants came from countries undergoing political upheaval, or where the existence of Jews was threatened. But the composition of the immigration had changed: the majority were now North African, primarily from Morocco. With 120,000 in Israel by 1957, Moroccan Jews now comprised the largest single ethnic community in the country. During this period, immigrants were sent directly "from ship to settlement," founding many villages and towns throughout the country. Frequently they remained in those remote and inhospitable locations only for lack of an alternative. Others left for the slums of the big cities.

Immigration slowed down again in 1958–1960, when only 70,000 came. Another wave of 200,000 arrived during the recovery years of 1961–1964, only to be followed by another lull during the recession of 1965–1967, when the total number reached only 70,000, and was frequently equaled by the rate of emigration from the country.

The Six-Day War in 1967 sent shock waves throughout the Jewish world, and motivated many Jews in Western countries to translate their emotional and financial support for Israel into actual immigration. To these were added 100,000 Russian Jews who were allowed to leave after strong internal and international pressures on the Soviet government. Altogether the years 1968–1973 brought 200,000 Jews to Israel. However, after the Yom Kippur War in 1973 immigration slowed down, now comprising an average of 20,000 a year, and roughly equaled by emigration.

In the 1960s the character of immigration changed significantly. By this time most of the Jewish communities in the "countries of distress" (with the exception of the Soviet Union) had been transplanted to Israel. Since the 1960s the majority of immigrants have come by choice from countries where their existence was relatively secure. These people, predominantly white-collar and professional, have immigrated out of Zionist motives, the "pull" of Israel rather than the "push" of their original countries, and have come as individuals rather than as groups or communities.

The changed nature of immigration forced the government to change its policy drastically. It would no longer do merely to "process" the immigrants and "fit" them into the needs and goals of the country. Now the government, convinced of the need to entice Jews to immigrate and to prevent their reemigration, tried to accommodate their needs and wishes with respect to housing and employment. Professionally trained immigrants could choose to settle in large metropolitan areas near sources of employment. Before making permanent plans they could spend up to six months in an "absorption center," learning the language and exploring opportunities. In addition, all new immigrants were entitled to many tax, customs, and housing privileges. This tended to irk many older immigrants who had received no such privileges.

The rate of immigration and the successful integration of immigrants into society are seen as the ultimate indices of the viability of the Jewish state. The failure of immigrants to integrate is perceived as both a national disgrace and an ideological threat, and those old-timers or new immigrants who, having failed to adjust, reemigrate, are often denounced. In reality, of course, some failure is inevitable. In the United States it is estimated that during some periods the percentage of immigrants who returned home approached 50%. In Israel it is estimated that 300,000 Jews, or 10% of the present Jewish population, have left the country since its establishment. There is no way to estimate how many more would leave if they could, or how many potential immigrants were deterred by the discouraging stories of those who preceded them.

Theoretical Framework

We will now examine several factors that affect the social-psychological adjustment of immigrants in Israel. But first we must clarify the meaning of adjustment or integration. Of what dimensions does this process consist?

Earlier American students of immigration often spoke of "assimilation" or "amalgamation," rather then "integration." They implicitly assumed that the desired, indeed inevitable, end result was the complete disappearance of differences—cultural, social and biological—which distinguished the original immigrants. Thus, Fairchild (1944) defined assimilation "as the process by which different cultures, or individuals or groups representing different cultures, are merged into a homogeneous unit" (p. 276). Cuber (1955) defined assimilation as "the gradual process whereby cultural differences (and rivalries) tend to disappear" (p. 609). And Rose (1956) characterized assimilation as the process of acquiring "the culture of another social group to such a complete extent that the person or group no longer has any characteristics identifying him with his former culture and no longer has any particular loyalties to his former culture" (p. 557–558).

More recently, American sociologists have viewed assimilation as a process of mutual accommodation, rather than an eradication of differences. Horton and Hunt (1976), for example, define assimilation as the "process of mutual cultural diffusion through which persons and groups come to share a common culture" (p. 304). Others have shifted the focus of their discussion of intergroup relations to acculturation rather than assimilation. Gordon (1964), though recognizing that complete assimilation exists only as a theoretical limiting case, depicts several interrelated dimensions that together may be said to constitute assimilation:

1. Cultural or behavioral assimilation. This, usually the first step in the process, refers to the adoption of the everyday behavior patterns of the host culture, including language, dress, and manner. This entails first a cognitive learning of new roles and then an enactment of them; it does not necessarily entail internal identification with them.

2. Structural assimilation. This consists of full integration into the primary and secondary groups of the host society, that is, into economic and social institutions, and into formal and informal face-to-face groups. This is the keystone of the process of integration, a necessary precondition to marital and identificational assimilation.

3. Marital assimilation. Consisting of intermarriage and interbreeding, this is probably the process evoking the strongest resistance on the part of both the host society and the immigrant group, though to some extent it is an inevitable result of structural assimilation.

4. Identificational assimilation. This consists of developing "a sense of peoplehood" based on the host society. Park and Burgess (1921) described this process, the hallmark of integration, as acquiring

"a sense of national solidarity" based on internalizing the memories, sentiments, and attitudes and on sharing the historical experience of the host society. This cannot occur fully unless structural integration into social and kinship (primary) groups has already taken place.

5. Acceptance by the host society without prejudice and discrimination. The onus of "receptional assimilation" lies with the host society; it is a necessary precondition for structural and marital assimilation, though it may also be seen as a result of these processes.

Gordon goes on to show that the policy of "cultural pluralism" has replaced the "melting-pot" or "assimilation" approach in the United States. The *degree* of assimilation expected of immigrants has decreased; indeed, they are expected (or at least permitted) to retain some cultural uniqueness. In Israel, too, the "melting-pot" policy has been replaced by an approach which tolerates, and sometimes encourages, cultural pluralism. In this chapter, however, our focus will be not on the ethnographic aspects of subcultures, but on the social-psychological process of adjustment to the new country. Whether or not subcultural differences continue to exist (and the indications are that they indeed will), adjustment has occurred when the immigrant can call Israel "home."

Three of the above-mentioned processes—behavioral, structural, and identificational integration—will be used as a conceptual framework for understanding the Israeli experience. Following Eisenstadt (1954) we may further distinguish between two spheres in which structural assimilation may take place, the economic and the social (Eisenstadt refers to these as instrumental and expressive integration, respectively.) Only in the case of the latter and invariably more resistant sphere may we witness the formation of those intense, intimate, primary relationships which eventually facilitate the complete transferring of identificational loyalties to the host society. Paradoxically, Israeli immigration policy largely ignores the painful long-term psychological and social processes that lead to a full integration, focusing instead on economic integration, which is more easily measured and manipulated by bureaucratic intervention.

Eisenstadt (1954) and Bar-Yosef (1969), focusing on the social-psychological dimension, consider the adjustment of immigrants a process of resocialization, similar to the process of socialization that takes place in childhood. The immigrant's resocialization involves the learning of new roles, the acquisition of new reference groups, and the transferring of identification and solidarity to the new society. This is always a traumatic process, because it must be preceded by a partial desocialization—a stripping of former roles, reference groups, values, and identity—before any resocialization can take place. Sociologists (e.g., Bar-Yosef, 1969 and Goffman, 1961) describe desocialization as involving cognitive disorientation, disintegration of the role set, impairment of ability to function in many roles, and temporary degradation of status. In layman's terms, the process of immigration is typically accompanied by a painful change to an unfamiliar locale, a new language, and new norms for conducting everyday life. It is also accompanied by a disaffiliation from previous ties of friendship, kinship, and ideological loyalties. These realities complicate the immigrant's adjustment to new roles and to new ways of performing old roles, for the requirements of job, parenthood, political participation, etc., are likely to be bafflingly different in the new country. All this constitutes a loss of the immigrant's previous social identity, a process which Goffman (1961), in his analysis of total institutions, calls a mortification of the self. The process may stop here, creating only apathy, alienation, and anomie. The successful formation of a new, "integrated" identity requires the mobilization of all the inner resources of the immigrating individual and group, as well as an enlightened absorption policy aimed at easing the transition.

In the rest of this chapter we shall analyze several factors affecting the rate and thoroughness of the integration of immigrants into Israeli society. We shall first consider the impact of individual characteristics: socioeconomic level, alienation from the country of origin, ideological identification with the new country, and the persistence of the old culture as a salient reference group. We shall then analyze the impact of conditions in the host country on the process of adjustment. These conditions include those over which Isreal has no control, such as war, as well as those which fall within the purview of the government, such as the overly bureaucratic nature of the integration process. This discussion will lead us to formulate some conclusions for policy vis-à-vis immigrants.

Individual Characteristics: Socioeconomic Level

The immigrant's educational and occupational level have an obvious impact on his successful adjustment to Israeli society to the extent that they determine his place in the country's stratification system. In view of the government's policy of perceiving economic adjustment as both the goal and the criterion of successful integration, the immigrant's socioeconomic characteristics are obviously crucial. Even more notably, the government actually discriminates *in favor* of highly educated and skilled immigrants by entitling them to preferential rights for apartments in desirable urban locations, and by enabling them to learn Hebrew in residential "absorption centers" at government expense. This policy in effect creates a sheltered formal transition period, a "moratorium" from normal role obligations, to which only those classified as "professionals" are entitled.

A closer look, however, reveals that it is not really the immigrant's absolute educational and occupational level, but the compatibility between his qualifications and the country's needs and opportunities, that determines successful adjustment. The same is true with respect to cultural patterns in general: the greater the similarity between the original country's culture and that of Israel, the easier will be the process of adjustment. This stems from the theory of resocialization: the greater the continuity between the old roles and identity and the new ones, the easier and more successful will be the resocialization, for the need for radical and painful changes will be minimized. Thus, during Israel's harsh early years of austere economic conditions, highly educated and professionally trained Jews avoided immigrating to Israel if they had a choice, or emigrated shortly after arriving if alternatives existed. Among Moroccan Jews, for example, the poorer and less educated came to Israel, whereas the more affluent and highly educated, having learned of the hardships of life in Israel, went to France. Today immigrants from the Soviet Union face a particularly difficult adjustment, because the cultural patterns to which they are accustomed differ from those of Israeli society more than do those of American or Western European immigrants (Russian immigrants, for example, expect the government to provide them with a job, and are often appalled by the "chaotic" appearance of Israeli political and civic life; see Gitelman, 1973).

Educational attainment, however, has a further significance for the process of adjustment. As we know from many studies (e.g., Adorno, Frenkel-Brunswick, Levinson, & Sanford, 1950; Lipset, 1960), education is directly correlated with a psychological openness to change, a tolerance for ambiguity and cultural relativity, and a capacity for self-detachment. These attributes, in addition to facilitating the learning of new roles and the internalization of new values, should also enable the individual to preserve his self-image in the face of the temporary degradation of status which accompanies immigration. Of course, education alone does not lead to a greater capacity for detachment and empathy, any more than a lack of education precludes it. Yet it is safe to assume that a high level of education, by increasing the individual's exposure to new experiences and value systems, increases the chances of a successful psychological adjustment.

Individual Characteristics: Alienation and Ideology

Eisenstadt (1967) speaks of the immigrant's "internal readiness for change" as contributing greatly to his economic adjustment. The Israeli society into which early immigrants stepped had preconceived notions about the economic roles they were expected to fill (in the early years, primarily agricultural roles), but even today immigration often entails occupational change, sometimes involving downward mobility. As Eisenstadt (1954, 1967) points out, the individual's internal readiness for change is largely a function of the degree of social-psychological support forthcoming from his primary groups. The solidarity and flexibility of the family structure and of the immigrant group's leadership greatly contribute to the individual's adaptation.

Internal readiness for change may be seen as consisting of two dimensions: alienation from the previous society (desocialization prior to immigration), and ideological identification with the new society. Desocialization, as we have seen above, consists of a stripping of old roles, identities, and loyalties. Goffman (1961), in analyzing this process, focuses only on its most extreme manifestation—after admission to a total institution. In his eagerness to establish insightful analytic uniformities, Goffman deemphasizes the possibility of a voluntary partial desocialization from the old identity and identification with the new one *before* admission. There is reason to assume, however, that resocialization will be less painful and more effective where the authorities have the cooperation of the individual who is pursuing a *desired* new identity, for example, at West Point, in a religious order, or in medical school. The same will hold true for the voluntary immigrant.

Alienation from Country of Origin

Until modern times, alienation from their countries of origin has been so integral a part of the Jewish experience that social scientists have often considered Jews as the very prototype of the concepts "pariah," "marginal man," "sojourner," or even "minority." (Weber, 1963). Jews as a people were never considered full-fledged members of society in the Middle East, in Eastern Europe until the Russian Revolution, or in Central and Western Europe until the 19th century. In our own times, of course, the persecution of Jews reached its culmination under Hitler. Short of physical violence, persecution has taken the forms of denial of civil, legal, and political rights, residential segregation, economic persecution, occasional cultural repression, and an invariable prohibition against any kind of primary group contacts with Gentiles.

This unrelenting rejection on the part of their countries of residence was complemented by the Jews' own powerful collective identification as a separate people, with their own separate frame of reference anchored in their religious institutions, in their preexilic past, and in a messianic future. Psychologically as well as physically Jews were always ready to leave on short notice to "follow the call of the Messiah to Palestine." The most pathetic but by no means the only example of this readiness occurred in the 17th century, when hundreds of thousands of European Jews followed the call of a misguided "Messiah" named Shabetai Zevi into oblivion. In modern times the religious collective identification of Jews was easily transformed into a political identification, Zionism. Jews emigrating from countries where they had always been barely tolerated were relatively easily socialized into the new Israeli society. For many, the opportunity to become full-fledged members of a society in which Jewishness was not stigmatized, but rather proudly affirmed, overshadowed the early difficulties of adjustment. The clearest example of this orientation was manifested by the Yemenite Jewish community, which suffered through the harshest processes of desocialization and resocialization uncomplainingly. At the other extreme we find the German Jewish community, so thoroughly integrated into their former society that they have yet to dissociate themselves fully from its memories, or to become fully integrated into Israeli society, or even to learn Hebrew.

A more common form of ethnic alienation in modern times is individual rather than collective alienation. Following their emancipation in Western Europe, Jews were integrated behaviorally and to some extent structurally into their countries of residence. They adopted the general population's dress, language, occupations, everyday manners, outer forms of religious worship, and political loyalties. They were usually not accepted into primary groups, however. Friendship patterns, club memberships, and other nonwork activities were usually segregated. The same situation was largely true in the United States and Western Europe until World War II, and in Middle Eastern countries, in the case of well-to-do Europeanized Jews, until the 1950s; it holds true in Latin America and in the Soviet Union to this day. As we saw above, the primary group is considered by sociologists as the crucible of identity development. Nothwithstanding their outward indistinguishability from the general population and their desire for complete acceptance, therefore, Jews continued to feel alien; identificational integration was denied to them. Immigrants from these societies are motivated to seek in Israel an end to their personal malaise, a fuller acceptance into the general society. The task of resocialization is eased somewhat in that some desocialization has taken place before migration, to the extent that the immigrant's goals and values are already partly dissociated from those of his original society, and his network of role relationships is already truncated.

Finally, a third, more recent form of alienation in the West and in the Soviet Union is ideological (nonethnic) alienation. Along with many non-Jews, some Jews have experienced a loss of identification with the goals and values of the surrounding society. In this they have acted not primarily as Jews, but as youths, radicals, intellectuals, or dissidents. Here the issue is not Jews' inadequate structural integration or lack of acceptance by the larger society, but an ideological rejection of that society couched in universalistic terms. Although this phenomenon is on the wane in the United States, it has motivated many American Jews to seek an alternative, more ideologically acceptable life-style in Israel. The same motivation has prompted many Russian Jews to immigrate to Israel. In these cases, however, Israel's attractiveness does not lie in her Jewishness, but in her image as more ideologically progressive or democratic, or at least less repressive, than the original country. This immigration wave, for whom particularistic considerations of Jewishness were irrelevant, has turned out to be unstable: many of its members reemigrated shortly after arrival.

All forms of alienation (ethnic-collective, individual, and ideological) contribute to the desocialization of the immigrant from his original country even before migration. The alienated individual has few role relationships which bind him to the surrounding society, and little identification with it. His adjustment to Israel, though still painful, of course, will be more effective because the old role relationships, values, and frames of reference will not continually threaten or compete with the new allegiances.

Ideology

Identification with the goals and values of a group which the individual hopes to join or anticipates joining may be said to constitute anticipatory socialization. Whether the individual is preparing to enter a profession, is moving up to a higher social class, or is planning to immigrate to another country, prior identification will facilitate adjustment in many ways. It will facilitate cognitive learning and assimilation of relevant information about the new system; it will spur the formation of new primary-group relationships which should help ease the transition; and it will accelerate the internalization of the values, goals, and attitudes of the new system.

As we saw above, Goffman in his study of resocialization in total institutions does not deal with anticipatory socialization. Merton (1968) discusses the significance of the nonmembership reference group, that is, of adopting the attitudes and success criteria of the group which one aspires to join. This is a part of the process of anticipatory socialization into a new profession, military rank, social class, or national culture.

Anticipatory socialization refers to the internalization of a group's value system or ideology, as well as of some of its modes of behavior, prior to joining that group. The ideological context of much of the emigration to Israel before and after the establishment of the state was Zionism. Does adherence to the ideology of Zionism make a difference in the rate of adjustment of immigrants in Israel? Shuval (1959), examining Zionism not as a set of beliefs but as formal membership in the movement, concludes that it does indeed. She finds that identification with Zionism as a "predisposing frame of reference" accounts for both a greater cognitive familiarity with Israel's culture and role expectations and a greater readiness to reorient personal plans in accordance with these expectations. Among today's immigrants, especially from the West, identification with Zionism is often strongest among religious or orthodox Jews. This helps to explain the fact that American immigrants who are orthodox tend to stay in Israel at twice the rate of the nonorthodox.

The significance of ideological identification with Zionism, religious or secular, is crucial. First, such identification facilitates the learning of relevant factual information about Israel, both before and immediately after immigration. As Bar-Yosef (1969) points out, one of the major aspects of the crisis of immigration is the immigrant's temporary cognitive disorientation, that is, his initial ignorance of the appropriate definitions of new situations, his lack of knowledge of the rules of social interaction, his unawareness of available opportunities and expectations, and his unfamiliarity with the language and institutions of the country. Although it might seem at first that all immigrants (especially if they come voluntarily in a planned fashion) would be motivated to learn as much as possible about the country, there are wide variations in the amount of learning that actually takes place. Many immigrants feel too baffled by the avalanche of new information, or too pressed by other immediate concerns such as job, schooling, and housing, to devote extensive resources to learning the language and the culture. Furthermore, many do not know how to obtain such information, for they lack an effective communications network. If the prospective immigrant is an orthodox Jew or a member of a Zionist organization, however, he may have learned the language before immigrating, and in any case he has been exposed to written and oral information about the country. It will be easier for him to absorb and interpret new information upon arrival, not only because of his prior partial understanding of Israel, but also because of his broader familiarity with the Jewish context into which the new information can be integrated. Herman (1969) confirms that learning and using Hebrew rather than the mother tongue is an indication of desire to identify with Israel and Israelis, and that potential immigrants who so identify, such as Zionists, are often motivated to learn Hebrew before immigrating.

The Salience of the Old Reference Group

Once the physical transition has taken place, the degree of persistence of the old culture as a viable frame of reference for behavior and identification will be inversely related to the immigrant's adjustment to the new culture. Adjustment to Israel requires accepting the new cultural patterns as "natural" or inevitable. As Schuetz (1944) explains, those born or reared in a culture will make "of course" assumptions about it: they will accept its patterns unquestioningly as "recipes" for interpreting the world and for getting things done. The stranger, however, cannot make any "of course" assumptions about the validity of the new culture, because he is acutely conscious of the discrepancies between it and the old culture which also claims a matter-of-fact validity, even when he has cognitively apprehended the new behaviors and expectations. The persistence of the old culture as a source of interpretations, attitudes, and possible action plans will constantly get in the way of his becoming an integral, unselfconscious member of the new culture. In short, his resocialization will be completely effective only when old cultural patterns are no longer salient.

The old culture may retain its salience in one of two ways: the physical possibility (or likelihood) of returning to the original country and the persistence of the immigrant group's old cultural patterns in the new country. Broadly speaking, only those immigration waves were successful (that is, stayed in Israel) whose members had nowhere else to go: those who were forcibly expelled from their countries or whose original communities were devastated by the Holocaust. Many immigrants from Iraq or Czechoslovakia, for example, report that in the early years of hardship they would have gone back if they could have; in the long run they adjusted because they had no choice. In fact, notwithstanding the vehement denunciation of emigrants, emigration from Israel has always been substantial, as is the case for any immigrant society. Thus, it is estimated that of the famous Second Wave of Immigration (1903–1914), the wave that brought to Palestine David Ben Gurion and Yizhak Ben Zevi and which laid the foundations to the country's political, social, and economic institutions, between 80% and 90% left the country.

The "burning of the bridges" does not have to be physically coercive; it may be psychological and voluntary. Those who firmly make up their minds that the new Israeli reality is the only viable alternative, and who expunge the old memories of perhaps a higher standard of living, different life-styles, and discrepant cultural patterns as salient alternatives, will adjust more rapidly. They will seek out primary group relationships that will serve as a bridge between them and the new society; they may throw themselves into social and political activism; and by thus becoming structurally integrated (provided the Israeli society allows it), they will eventually transfer their identificational loyalties to the new society.

The Impact of Internal Conditions

Among the factors aiding the process of nation building in Israel are the common religious background and the state-of-siege mentality of the country. Overriding the ethnic differences and individual frustrations is a feeling of shared solidarity and of a common fate: "We are all in this together," and "The whole world is against us." We need but remember Durkheim's contention that in times of war or major crisis national cohesion increases and alienation and deviance decline, as well as Sumner's point that out-group hostility enhances in-group cohesion, to understand the impact of war and the threat of war on the identificational integration of immigrants. Whether the cohesion-building effect of Arab hostility counterbalances the disintegrating effect of recurrent loss of life, prolonged military service, and economic hardships is a difficult question to answer. Perhaps the only clue is the fact that repeated surveys of emigrants *from* Israel cite economic hardship rather than the threat of war as motives for leaving. War *per se* is perceived as a unifying factor (see, for example, Levy & Guttman, 1975).

The impact of economic conditions on immigration is more complex. Overall, lulls in immigration have coincided with recessions (1952–1955, 1965–1966, 1974–1977), peaks in immigration with prosperity (1956–1957, 1968–1973). Although recession no doubt discourages potential immigrants, and prosperity encourages them, there is no question but that immigration contributes to the country's economic forces by increasing the labor force and by spurring consumption and capital investment (see Berger, 1975). To a large extent, then, fluctuations in the Israeli economy are a product of highs and lows in immigration as well as a contributing factor to them.

The human variable in integration is even more complex. Whereas in the United States much of the resettlement is done by informal groups and families in their role as sponsors, in Israel the process is handled by bureaucratic authorities—the government and the Jewish Agency. By contrast, before 1948 "absorption" was handled informally by the primary groups in the Jewish community— kibbutzim, political and ideological movements, etc. As we know from both the theoretical literature and empirical experience, dependence on a bureaucracy for a large share of one's role functions tends to induce apathy, demoralization, and alienation. Bar-Yosef (1969) and Eisenstadt (1954) point out the components of the bureaucratic interaction which contribute to the client's status degradation. These components include (a) the categorization into universalistic categories (e.g., "professionals") for the purpose of allocating benefits, thus denying any individuality of needs and situations, (b) the loss of the power to make important decisions affecting one's life, (c) the invasion of spheres usually left private and not exposed to strangers, (d) the loss of age status (clients become infantilized, the bureaucrats being the decision-making, powerful "adults"), (e) the loss of time perspective, (f) the added difficulties of communicating across linguistic and cultural barriers. We know from the experience of Japanese resettlement camps in World War II that the process may be dehumanizing and alienating. In the Israeli experience, similar situations with similar unfortunate consequence occurred in Jewish colonies in Palestine administered by Baron Rothschild's officials in the 1890s, in the administered immigrant camps in the early 1950s, and in the administered farms of the Jewish Agency in the late 1950s. The average immigrant today is subject to much less bureaucratic administration, of course; he alone takes care of a large portion of the resettlement process. Yet the dependence on bureaucracy for the allocation of benefits and resources hinders the process of resocialization, because it hinders effective role performance in adult roles. A rapid integration into informal primary groups rather than bureaucratic relationships may, on the other hand, facilitate resocialization.

Conclusions and Recommendations

We have considered immigration as a process of resocialization leading to the formation of a new social identity. We have seen that major components of successful resocialization include a prior desocialization or a stripping of previous roles and values, as well as an integration into the primary (face-to-face) groups of the new society. This is invariably a traumatic process. Much of its success depends on the "internal readiness for change" of the immigrant, on his willingness to undergo considerable dislocation, to unlearn familiar patterns and to learn strange ones, and to suffer a temporary degradation of status.

A large portion of the process, however, depends on the facilitating efforts of the host society. Probably the most important thing that the host society can do for the immigrant is mitigate the degree and rapidity of change required of him. A great deal of personal change is inevitable, of course, in due time. Yet, the more lenient the host society is in requiring immigrants to adopt its own identificational loyalties and cultural patterns, and the more opportunities for occupational continuity it extends to immigrants, the easier will be their readjustment. The person undergoing the very real shock of immigration needs to cling to fragments of familiar roles and values in order to salvage his self-image. Social pressures to behave and think like an Israeli, to speak only Hebrew, to live among and associate with Israelis or immigrants from other countries, to change occupations in accordance with the needs and ideology of Israel, and to drop the previous national or ethnic identification as a focus for self-image, intensify the strains to a point where they may seriously impede adjustment. It is important to strengthen the immigrant's self and community, not weaken them. Israeli society has become undeniably more tolerant of cultural, linguistic, and identificational differences than in the early years, largely by necessity. The "melting-pot" approach had to be replaced by a pluralistic approach because immigrants refused to "melt." Thus, residential integration within new villages had to give in to a regional approach where ethnically homogenous villages are served by a single-service town. The policy of occupational redirection to farming has been abandoned since agricultural development has ceased to be a national and ideological goal. In fact the country's economic development today is directed by the future needs of potential immigrants. Yet more efforts must be made to keep the immigrant's self-image and immediate environment relatively intact by respecting his needs for residential and institutional segregation and cultural continuity where such needs are expressed (for example, by Georgian Jews). This is difficult to follow both for pragmatic reasons (such as the lack of sufficient apartments and service institutions within the same neighborhood to accommodate an entire community) and for ideological ones, for the pressure for identificational assimilation is intense. In the long run, however, the "Israelification" of immigrants (or at least of their children) is inevitable. By minimizing the strains involved in this process Israeli society can significantly reduce disillusionment, stress, and failure to adjust.

The positive aspects of resocialization must be equally emphasized. We have seen that structural integration (integration into primary groups) is a keystone to successful adjustment. Typically the immigration process involves the breakdown of primary groups, and the bureaucratic nature of the resettlement process removes incentives for the formation of new ones. Unlike the United States, Israel views the provision of immigrant's needs for housing, employment and schooling as the responsibility of the government, not of private groups and individuals. Yet Israel may do well to adopt the noneconomic aspects of the American "sponsor system" by fostering informal links between immigrants and Israeli families, kibbutzim, work places, community organizations, etc. Side by side with strengthening existing primary-group relations within the immigrant community, the building of "expressive" bridges into Israeli society will reduce the frustration, anomie, and role-disintegration of immigrants, and facilitate their resocialization into functioning, integrated members of society.

References

Adorno, T. W., Frenkel-Brunswick, E., Levinson, D. J., & Sanford, R. N. *The authoritarian personality.* New York, Harper, 1950.

Bar-Yosef, R. W., Desocialization and resocialization: The Adjustment process of immigrants. In M. M. Lissak, B. Mizraki, O. Ben-David (eds.), *Immigrants in Israel: A reader.* Jerusalem: Hebrew University Press, 1969. (Hebrew)

Berger, L. Immigration motivations and the structural changes in Israel. In *Israel Yearbook, 1975.* Tel Aviv: Israel Yearbook Publications, 1975.

Cuber, J. F., *Sociology: A synopsis of principles,* New York: Appleton-Century-Crofts, 1955.

Eisenstadt, S. N. *The absorption of immigrants.* Westport, Conn.: Greenwood Press, 1954.

Eisenstadt, S. N. *Israeli society.* Jerusalem: Magnes Publishing Co., 1967.

Encyclopedia Judaica. Jerusalem: Macmillan, 1971.

Fairchild, H. P. (Ed.). *Dictionary of sociology.* New York: Philosophical Library, 1944.

Gitelman, Z. *Absorption of soviet immigrants.* In M. Curtis & M. Chertoff (Eds.), *Israel: Social structure and change.* New Brunswick, N. J.: Transaction Books, 1973.

Goffman, E. *Asylums.* Garden City, N.Y.: Doubleday, 1961.

Gordon, M. M. *Assimilation in American life.* New York: Oxford University Press, 1964.

Government of Israel, Central Bureau of Statistics. *Immigration to Israel 1948–1972.* Special Publications Series No. 416, Jerusalem, 1973.

Government of Israel, Central Bureau of Statistics: *Israel statistical abstract.* Jerusalem: Sivan Press, 1977.

Herman, S. N. Explorations in the social psychology of language choice in Israel. In M. Lissak, B. Mizraki, & O. Ben-David (Eds.), *Immigrants in Israel: A reader.* Jerusalem: Hebrew University Press, 1969.

Horton, P. B., & Hunt, C. L. *Sociology.* New York: McGraw-Hill, 1976.
The Jewish Agency: *16 years of immigration.* Jerusalem, 1964 (Hebrew).
Levy S., Guttman, L: The desire to stay: Analysis of findings from the current survey of the urban Israeli public, April, 1974. Jerusalem, Bureau of Applied Social Research and Bureau of Communication, Hebrew University. In *Israel Yearbook, 1975.* Tel Aviv: Israel Yearbook Publications, 1975.
Lipset, S. M. *Political man.* New York: Doubleday, 1960.
Merton, R. K., *Social theory and social structure.* New York: Free Press, 1968.
Mishal, S. *Israel: Immigration & absorption.* Jerusalem: Office of Education and Culture, 1971. (Hebrew)
Park, R. E., & Burgess, E. W. *Introduction to the science of sociology.* Chicago: University of Chicago, Press, 1921.
Rose, A. M. *Sociology: The study of human relations.* New York: Knopf, 1956.
Schuetz, A. The stranger: An essay in social psychology. *American Journal of Sociology,* 1944, *49,* 499–507.
Shuval, J. T. The role of ideology as a predisposing frame of reference for immigrants. Human Relations, 1959, *12,* 51–63.
Weber, M. *Sociology of religion.* Boston: Beacon Press, 1963. (Originally published in 1922.)

22

After the Fall: Indochinese Refugees in the United States

Paul I. Ahmed, Frank Tims, and Aliza Kolker

Refugees have been a major phenomenon of the twentieth century. Since World War II, the massive numbers of refugees and displaced persons have occasioned large scale, programmatic responses by governments and international agencies. Aside from the resettlement efforts growing out of the Second World War, the United States has opened its doors four times to refugees since 1945—to the Hungarians in 1956, the Cubans in 1960 and again in 1980, and those from Indochina in 1975. Each of these four cases has been markedly different in its circumstances, and the nation's response has been different each time.

The Hungarians who escaped in the wake of the 1956 October uprising were received warmly as refugees from Soviet oppression. Their absorption into American society was facilitated by the fact that large numbers of ethnic Hungarians were already citizens of the United States, and many of these families acted as sponsors for the refugees. Most of these newcomers were settled in the industrial northeastern United States, where large Eastern European ethnic groups made for similarity and the availability of community cultural supports. In these circumstances it was not necessary for the government to mount a large scale resettlement effort.

The Cuban situation was similar to that of the Hungarians, in that

Paul I. Ahmed • Office of International Health, U.S. Department of Health and Human Services, Rockville, Maryland 20857. **Frank Tims** • National Institute of Drug Abuse, Rockville, Maryland 20857. **Aliza Kolker** • Department of Sociology, George Mason University, Fairfax, Virginia 22030.

the refugees were viewed primarily as escaping from an oppressive, totalitarian regime. Some came individually by air or by sea and others in an airlift which was especially negotiated with the Cuban government. Many of those arriving in Miami during the airlift were middle-class families who found the Castro regime uncongenial. For the greater part they settled in the Miami area, where, during the past two decades, they have become a political power bloc decisive in local elections. Unlike the Hungarians, the Cubans had among their number many who expected to return in the not-too-distant future, presumably after the overthrow of the regime they fled. Unlike the Hungarians, who had a community waiting to absorb them, the Cubans formed their own, and clung to it because of its significance to them and to their aspirations.

In the Cuban case of 1960, the Cuban Refugee Agency was formed within the U. S. Department of Health, Education, and Welfare. The primary concern of this small agency was looking after the social welfare and employment concerns of the Cubans and seeking to assure that the large numbers of new immigrants did not unduly strain the resources of Dade County or of Miami.

The fall of governments backed by the United States in Laos, Cambodia, and South Vietnam took place in a short time in the spring of 1975. Congress authorized the conduct of military operations in support of evacuating friendly local nationals from Cambodia and Vietnam. Large numbers of refugees escaped on their own, by sea, by air, or by crossing the Thai border. A total of 144,748 refugees from Indochina, most of them Vietnamese, had been admitted to the United States as of March 1, 1977, according to the HEW Refugee Task Force *Report to Congress* of March 21, 1977.

As of this writing (1980) many thousands more refugees have been admitted into the United States, often after spending harrowing weeks or months in squalid refugee camps or on unseaworthy boats. Emigration from Indochina and Cuba will probably continue for some time, as conditions there are still unsettled. With respect to American governmental policy, however, this immigration is very different from the one that followed the fall of Cambodia, South Vietnam, and Laos in 1975. First, it is not a wave but a trickle. It does not necessitate major new organizational efforts, but can be handled by existing institutions. Second, the substantial Indochinese and Cuban communities now existing in many places in the United States can handle a large share of the responsibility for resettlement, thus further reducing the need for governmental action.

This paper concentrates only on the wave of Indochinese refugees entering the United States in 1975–1976, and on the unique problems it presented for American policy makers. We will examine the major policy assumptions that guided the resettlement of those refugees and the mechanisms by which policy was developed and implemented. Of key concern will be the policy issues of cultural adjustment, service delivery, and long-term mental-health questions. Although the situation continues to change, it is important to examine the initial policy responses, because the United States will probably be confronted with similar waves of refugees in the future, following political upheavals in other countries. The lessons of 1975, like those of 1956 and of 1961, should prove instructive for future crises.

The Indochinese refugee situation presented a set of problems more complex than those dealt with in the Hungarian or Cuban operations. The refugees were of cultural backgrounds not familiar to most Americans. To the extent that Asian-American resources could be called upon, they were themselves largely foreign to the Indochinese refugees. It must also be said that some in the United States were less than enthusiastic about allowing the refugees to settle in the country. Some members of the press depicted this wave of new immigrants as comprised of less desirable members of their own societies—bar girls, war profiteers, and politicians with numbered Swiss bank accounts. The data compiled by the Interagency Task Force for Indochina Refugees provides quite a different picture of these newcomers. Still, the political problem remained—they were the reminders of our humiliation and failure in Southeast Asia. Whereas the government allowed the Cubans to settle primarily in Miami—one might even say that we encouraged it—the early policy pronouncements on the Indochinese refugees pointed out that the United States wanted to avoid their settling in "Little Saigons." They were to be dispersed and, presumably, absorbed inconspicuously into American society.

The Uprooting

The political events leading to the fall of South Vietnam seemed to unfold inexorably. Some observers awaited the collapse with hope that it would spell an end to the agony of war, others with predictions of a bloodbath in Indochina and a backlash in the United States, and still others simply with a sense of resignation. Many viewed the events as part of a cynical strategy in which the disputed countries would be allowed to fall after a "decent interval." After the decision of the Congress in the spring of 1975 to withhold additional military assistance funds which had been requested, President Thieu announced his decision to withdraw troops from the central highlands of South

Vietnam. This decision shocked the military commanders, who were convinced that the central highlands were an essential keystone of defense for South Vietnam. Panic followed. South Vietnamese troops fled the conceded highlands or simply deserted their units. A succession of Viet Cong and North Vietnamese Army victories followed. The world watched the drama's progress on television screens: the frantic evacuation of Danang, the helicopter evacuation of the American Embassy compound in Phnom Penh, and finally Saigon.

The American response to these events was, as might have been expected under the circumstances, purely defensive. No air strikes were provided to slow the advancing forces. The prevailing view in the United States at that time was that, realistically, little could be done.

The result was predictable. It was impossible to evacuate all who wanted to leave Vietnam and Cambodia (the transition in Laos had been less dramatic and, on the whole, less traumatic). Space on evacuation craft was made available to those whose relationship to the United States or South Vietnamese government and personnel might place them in jeopardy, though it must be said that the circumstances and a lack of planning severely limited this operation. A considerable number of South Vietnamese and Cambodian military and other government personnel managed to flee on their own aircraft or naval vessels to neighboring states. Many civilians simply fled by sea in a variety of crafts, particularly fishing or commercial vessels.

As refugees were processed through Guam, Wake, and the Subic Bay naval base in the Philippines, and then sent to the four relocation centers in the United States, American public opinion was uncertain about the most appropriate course of action. Newspaper stories and editorials depicted them variously as patriots, corrupt officials, criminal elements, and ordinary people who had the misfortune to cast their lot with the losing side. Many had no connection with Americans or the government, such as the fishermen and their families who fled by boat. A consensus quickly developed that the United States and its people had an obligation to offer them a new life, and machinery, both public and private, was set up to accomplish this task.

The Policy Response

The immediate reality of the new situation was that the United States had not officially recognized the possibility of a precipitous collapse, and therefore had not invested adequate resources in planning for this eventuality. The lack of planning has been adequately documented elsewhere, and will not be further discussed here. The rapid pace of events forced an immediate response, and on April 18, after consultation with Congress, President Ford established the Special Interagency Task Force for Indochina Refugees. The Task Force was initially directed by Ambassador L. Dean Brown, and later (beginning May 22, 1975) by Julia V. Taft, Deputy Assistant Secretary of Health, Education, and Welfare. A total of twelve government agencies were involved, including the Departments of State, Defense, Health, Education, and Welfare, Justice, and Labor. The Task Force was initially charged with the responsibility of coordinating all government activities "concerning evacuation of United States citizens, Vietnamese citizens, and third country nationals from Vietnam and refugee and resettlement problems related to the Vietnam conflict," according to the Interagency Task Force for Indochina Refugees, Report to the Congress, December 15, 1975.

With the formation of the Task Force, a series of policy decisions were to be faced immediately. To examine the factors, political and bureaucratic, which resulted in a void where contingency plans for refugees were concerned would serve no purpose now. This lack of planning was manifested in the fact that much of the first months of the Task Force's operations were consumed in defining policy, with numerous decisions being made on an ad hoc basis—either by the Task Force itself, or by the organizations which were major recipients of funds for resettling the refugees. The major policy issues confronted at the start related to (1) caring for the refugees and their immediate needs, (2) admission and resettlement of a large number of the refugees into the United States, and (3) provision of services deemed necessary to facilitate the adjustment of the refugees to their new homeland, and to take care of their basic health and welfare needs. There were, of course, political, economic, and social aspects to these policy questions which were of pronounced concern to government at all levels.

Prior to the establishment of the Task Force, representatives of the U.S. State Department had consulted with appropriate committees of Congress regarding use of the Attorney General's "parole" authority as a basis for admitting evacuees to the United States. Under this authority, parole was authorized on April 14 for dependents of American citizens in Vietnam at that time, and was subsequently extended to others, including relatives of American citizens or resident aliens who had the status of petition holders, Cambodians in third countries, and up to 50,000 "high-risk" Vietnamese by April 25, 1975. Ultimately, more than 144,000 evacuees and others who escaped Indochina were resettled through the efforts of the Task Force.

On April 25, the Task Force requested that all United States missions overseas give consideration to the matter of refugee resettlement. On April 29, the American Embassy in Saigon closed, and the emergency airlift of Americans and Vietnamese who wished to escape came to an end. Further evacuation by sea continued.

The initial part of Phase One of the Task Force operation (evacuation and resettlement) involved establishing refugee staging areas at the Subic Bay naval base in the Philippines, on Wake Island, and on Guam. The airbase at Utapao, Thailand, was also used temporarily as a staging area. Four military bases in the United States were designated refugee reception centers: Camp Pendleton, California; Camp Chaffee, Arkansas; Eglin Air Force Base, Florida; and Fort Indiantown Gap, Pennsylvania.

The resettlement effort must be viewed in the context of a number of constraining pressures. Although the consensus that the United States had a clear obligation to the refugees was evident, the nation was experiencing the ravages of an economic recession, with an unemployment rate of 9.2%. Governors of the states were sensitive to the possibility that the refugees might strain local social-service resources if they were permitted to concentrate in particular areas. There were doubtless other political concerns over the possible implications of having large numbers of the refugees settle in a few cities, though these need not be of concern here. In any event, an early decision was made to disperse the refugees, through a national campaign to have refugees sponsored by individual citizens or organizations.

A second policy decision was to place emphasis on obtaining employment for those refugees who were employable, and to provide English-language and vocational training to enhance the employability of those who needed it.

A special appropriation of $405,000,000 was obtained to fund the resettlement program and provide programmatic support in connection with this, under the Indochina Migration and Refugee Assistance Act of 1975 (P. L. 94–23) in May. The legislation was requested by the President (the Administration had requested $507,000,000) and was to fund a wide range of social services and vocational programs. These efforts continued for several years.

The Supplemental Appropriation Act of 1977 (P. L. 95–26) provided $18.5 million for Title II, and $10.5 million for Title III for the implementation of the Indochinese Refugee Children's Assistance Act (P. L. 94–405). Title II provides funds for local school districts for the education of refugee children, and Title III provides funds for project grants to state and local education agencies for adult refugee education.

Another key policy decision was to utilize those resources and channels already in place to accomplish as much as possible of the resettlement, and to deliver services and cash assistance to the refugees. There was concern that the tasks be accomplished without the creation of a new bureaucracy, and that the resettlement effort be kept a short-term endeavor. Among the existing resources which played a key part in the resettlement were the "voluntary agencies" (VOLAGS), private nonprofit organizations experienced in international refugee matters. The VOLAGS presented numerous advantages to the Interagency Task Force. Aside from their experience with refugees, many of these organizations had worked under contract to the U. S. Agency for International Development, and had the capability for quickly organizing the necessary talent and structure for the resettlement. VOLAGS could also function as sponsors and employers for some of the refugees, and had the capability to provide "mass sponsorship," a significant fact when one considers the pressures experienced by the reception centers to close during calendar year 1975. Four VOLAGS (also termed "Voluntary Resettlement Agencies" by the Task Force) accomplished the resettlement of some 117,000 refugees.

The VOLAGS were provided with direct cash payments for the refugees they resettled. The terms of reference in allocating funds were

Table I. Voluntary Resettlement Agencies (VOLAGS)[1]

Agency	Approximate number of refugees resettled
United States Catholic Conference Migration and Refugee Services	60,000
International Rescue Committee	19,500
Church World Service Immigration & Refugee Program	19,000
Lutheran Immigration & Refugee Services	18,500
Hebrew Immigration Aid Society (HIAS), Inc.	3,900
Tolstoy Foundation, Inc.	3,600
American Council for Nationalities Service	4,830
American Fund for Czechoslovak Refugees	1,200
Travelers Aid International Social Service of America	530

[1] From HEW Refugee Task Force *Report to the Congress*, March 21, 1977.

somewhat permissive, and VOLAGS were allowed to determine for themselves how best to utilize the monies in the resettlement effort.

A major task confronting the Interagency Task Force was the effective coordination of the numerous federal, state, and private agencies involved in the resettlement. Several states entered into contracts as resettlement agencies. A total of fifteen departments and agencies were involved in the effort, either in terms of their statutory authority (such as in the case of the Immigration and Naturalization Service) or in supporting roles (for example, the U. S. Information Agency was pressed into service to provide print media for the refugees). Again the use of VOLAGS simplified this formidable task of coordinating the private agencies involved. Among the federal agencies participating in the task force were:

Office of Management and Budget
Department of State
Department of the Treasury
Department of Defense
Department of Justice
Department of the Interior
Department of Health, Education, and Welfare
Department of Housing and Urban Development
Department of Transportation
General Services Administration
U. S. Information Agency
Small Business Administration
  (Department of Commerce)
Agency for International Development
  (Department of State)
Immigration and Naturalization Service
  (Department of Justice)
Central Intelligence Agency

Key responsibility for the reception centers rested with the Department of Defense, which was charged with providing housing and support for the refugees as the tasks of clearance for parole, processing for immigration, and arranging sponsorship were carried out by the appropriate agencies. During this period, the Interagency Task Force provided necessary social and health services, and began preparing the refugees for life in the United States by means of English-language and orientation classes.

The reception-center stay was intended to be of short duration. There was, in fact, considerable pressure on the Task Force to limit this period, and to accomplish specific reductions in the refugee population at each center in the United States by specific dates. The staging areas at Subic Bay and Wake Island were both closed by August 1, 1975; Eglin Air Force Base's center closed the following September 15; Centers at Guam and Camp Pendleton closed October 31, and the remainder of the centers were closed before the end of 1975. Provision was made to repatriate those who wished to return to their native lands, including 1,546 Vietnamese who set sail from Guam on October 16, 1975.

The Interagency Task Force *Report to the Congress* on December 15, 1975, gave the following description of reception-center administration:

> The leadership at the staging areas and reception centers was shared equally by a civilian coordinator and a military officer. The military commander, who reported through Defense Department channels, was responsible for all facilities, and logistical support required by the Senior Civil Coordinator, whose authority came from the Director of the Task Force.
>
> Administration of the centers was largely in the hands of civilian Indochina experts, most of them employees of the Agency for International Development, the State Department, and the U. S. Information Agency—former members of the U. S. Mission in Vietnam and Cambodia. The experience and language skills of these officials was invaluable, not only in the narrow sense of communicating with the refugees but also in providing a sympathetic link between the refugees and some of the other Americans with whom they had to deal. In one instance, the Civil Coordinator came from a domestic agency, HEW. HEW also provided skilled specialists to staff the reception centers as did the Department of Labor, which provided efficient job search and job matching technical assistance and expansion to IATF operations, and INS, which was generally responsible for security checks for the refugees and for their parole into the United States. All the facilities were Defense Department installations, and the outstanding support of the military services was basic to whatever success the program has enjoyed.

With the closing of the reception centers, the evacuation and resettlement phase of the Indochina Refugee program was deemed completed, and the Interagency Task Force disbanded on December 31, 1975. On the following day the HEW Refugee Task Force within the Department of Health, Education, and Welfare was created to continue support of the resettlement. The new Task Force was made responsible for coordinating a wide range of resettlement support and social-service programs under P. L. 94–23 and other legislation related to the refugee resettlement.

During this second phase of the resettlement, which may be termed "consolidation," the Task Force worked closely with VOLAGS and developed federal strategies for dealing with the multitude of problems faced by the newcomers. Funds for English-language and vocational training were made available to the states and the federal regional offices. Monies were also provided to states and school districts through the Indochina Refugee Children Assistance Act of 1976 (P. L. 94–405). An additional appropriation of $50 million under P. L. 94–23 was passed by the Congress in October of 1976.

The HEW Refugee Task Force funded a range of services intended to be as comprehensive as possible in dealing with the critical problems of refugees during the resettlement period and their integration into American society.

There were factors peculiar to the situation of the refugees which made it necessary to pursue innovative strategies. The HEW Task Force provided financial support for demonstration projects to develop services. A case in point is in the mental-health area. Since Western psychotherapy is a foreign concept to the Vietnamese, Cambodians, and Laotians, other approaches to mental-health assessment and service delivery more compatible with Indochinese cultures had to be developed. Culture-specific approaches to problem solving were developed through refugee-assistance organizations, including over 100 Indochina refugee self-help associations.

A variety of mental-health programs, funded by HEW, have attempted to integrate the refugees' cultural perspectives by utilizing Indochinese professional and paraprofessional mental-health workers.

In addition to mental-health demonstration projects, states have established social services for Indochinese refugees. Examples of these services are:

Wisconsin: A $55,000 program for coordinated public education, information and referral effort, and printing and circulation of a bilingual newsletter.

Louisiana: Associated Catholic Charities, on behalf of the state, provide employment service and health-related services. The health services include medical arrangements, mental-health referrals, outreach, and health-education services. They also provide home-management services, housing-improvement services, and family-counseling services.

In summary, the policy response sought to quickly identify and make provision for the particular needs of the refugees in a climate of crisis and political unrest. The Interagency Task Force provided a reasonably efficient use of resources already in place to accomplish the reception and resettlement of those uprooted by the fall of the Cambodian and South Vietnamese regimes. The HEW Task Force assured continuing provision of services during the critical period of consolidation which followed. The policy of dispersal of refugees was in part a recognition of the political reality that acceptance of the newcomers would be facilitated if no disproportionate burden were placed on any community's resources or labor market, and perhaps also a move in the direction of promoting rapid assimilation of the refugees while the national consensus remained intact. The stress on employment was the most obvious and sensible policy, not only in terms of the political goal of integrating the refugees into the nation's economic life as self-sustaining individuals and families, but also from the perspective of enhancing their transition from an uprooted status to their new identities as productive, contributing members of American society. The provisions of English-language instruction and of cultural orientation were, of course, obvious necessities, although both must necessarily continue over the long term. Both the Task Force and Congress recognized the need for a range of social services and educational programs, and moved quickly to provide these, using both delivery systems already in place and new resources such as self-help organizations established by the refugees themselves.

The Refugees

This examination of the Indochina refugee phenomenon would be incomplete without a consideration of who the refugees were. The reception accorded them was, in the main, one of welcome and constructive assistance. There were exceptions to this, with some press allegations based on stereotypes of corrupt generals and war profiteers. An examination of the educational and occupational profiles of the refugees presents a different picture: one of a diverse population representing a cross section of social origin.

The great majority of the refugees were Vietnamese, with a small percentage being Cambodian and Laotian. There was considerable variation in the class background of the refugees, as reflected in such characteristics as education and occupation. Approximately equal numbers of males and females are found in the refugee population, reflecting in part the fact that many nuclear families escaped intact. Children under 18 accounted for about 42–43% as of January 1976, while only slightly more than 4% were over the age of 62. The relatively small numbers of older persons (only 11.6% were over the age of 44) may signal a special area of concern, given the importance of the extended family in traditional Asian cultures.

The refugees, though coming from a variety of occupational backgrounds, were generally a well-educated group. Of those evacuees 18 years of age or older, 37.9% were reported to have attended secondary school, 16.6% had attended a university, and an additional 2.9% had postgraduate education. Although no published data are available on the numbers of former military officers and civilian government employees, both press accounts of the evacuation and informal reports by participants in the debriefing and indexing project sponsored by the Department of Defense lead one to suspect their numbers to be substantial. The classification of occupational skills of heads of households (including one–person households) showed 31.2% to be in medical, professional, technical, and managerial occupations; 11.7% in clerical and sales occupations; 7.6% in service occupations; 4.9% in farming, fishery, forestry, and related occupations; and some 20% in occupations which could be classified as manual, construction, machine trades, or bench-work occupations. An additional 16.9% were classified as being in "miscellaneous" occupational fields, and 7.9% could not be classified.

In terms of composition of households, only about 14% of the refugees comprised single–person households, whereas 79% lived in households of three persons or more. In fact, 52% of the refugees lived in households of six persons or more. Of households which consisted of at least two persons, some 76% were headed by males and 24% by females. Among adult individuals living alone, 80% (13,502) were males and 20% (3,317) females.

Of the total of 145,000 refugees, 52,219 or 36% were receiving cash assistance as of July 1, 1977. Most of the recipients, though employed, had low enough incomes to qualify for cash assistance. It is interesting to note that only 17,656 refugees, or 12% of the total, were authorized medical assistance.

As previously stated, there was concern that refugees be resettled in a dispersed pattern rather than allowed to concentrate themselves in a few cities or states. There was especially keen concern on the part of the governors of states such as California, who felt that their states were likely to receive a disproportionate share of refugees, who might take jobs from their citizens or become a drain on social services. California, for example, had almost 25% of the total of all refugees at one point. The *Report to the Congress* dated March, 1977, indicated that only six states had more than 5,000 refugees resettled within their borders: California (30,495), Texas (11,136), Pennsylvania (8,187), Virginia (5,620), Florida (5,237), and Washington (5,205).

A number of reports, including some submitted by HEW Regional Offices, indicate that substantial migration of resettled refugees is taking place—notably to states with warmer climates, and, in particular, to California, where large numbers of Indochina refugees reside. This, of course, may well be a constructive development, and must be viewed in the context of the continuing adjustment of the refugees to their new life in the United States.

Refugees' Problems

As would be expected, the refugees arrived in the reception centers with a host of problems. The most apparent problems grew out of their sudden status as homeless persons suddenly thrust into an alien society, many of them arriving destitute and separated from members of their families. The immediate needs were dealt with by the Interagency Task Force on an emergency basis during the evacuation and resettlement, and on a continuing basis through numerous special programs funded by means of grants to states and localities. The most immediate set of problems, language and culture, were addressed at the reception centers by federal personnel and contractors who provided classes in "survival English" and problems of everyday life in the United States, as well as materials designed to assist the refugees in their resettlement. In addition to social services, the Task Force recognized the importance of providing vocational training to enhance employability, and of assisting those with skills in obtaining credentials. Numerous professionals in health fields, including hundreds of dentists and physicians, were confronted with problems of credentialing, and a special program was established to provide the necessary evaluation, training, and assistance in arranging for the necessary examinations. In the case of such professionals it was necessary to demonstrate a level of proficiency both in English and in professional knowledge. Similar deficiencies in both language and occupational qualifications were doubtless encountered by refugees of differing backgrounds, and many were faced with the difficult choice between lowered occupational status and unemployment. The presence of numerous former government officials and military officers in the refugee population implies that many of these may have been generalists without readily marketable skills, but who had been accustomed to prestige and power in Vietnamese society. Reports based on sample surveys as well as reports from the HEW Regional Offices indicate that underemployment is a serious problem, though one can only speculate as to whether the expectations of the individuals affected had been unrealistic. If so, having to adjust to lowered socioeconomic status might have implications for diminished self-esteem and associated problems of adjustment.

Other considerations that should be examined are the extent to which the dramatic uprooting under drastic circumstances might generate feelings of guilt, powerlessness, and betrayal among adults, and ambivalence toward parents on the part of children. The uprooting carried lingering physical, psychological, and social stress among parents and children. They experienced broad ranges of acute and chronic problems of identification and adjustment. The forced involuntary departure evoked the anxiety and insecurity inherent in the new and unknown. The sense of isolation and abandonment felt by many was exacerbated in the face of the unpredictable social, vocational, and economic problems, apathy and depression, disappointment and anger.

Being thrust into a strange and often confusing cultural environment generated a sense of aimlessness, powerlessness, and isolation which may well have led to a high risk of depressive states among some people. The sudden shift from a Confucian, tradition-oriented society to a modern Western culture with its attendant stress on individualism and achievement, and with its strong adolescent peer pressure, may have tended to undermine parental authority in ways not easily understandable by adults who have simultaneously to cope with a myriad of new pressures. These often subtle challenges to authority may have led to overreaction on the part of parents, leading them to feelings of inadequacy at a time of intense personal crisis. No systematic data exist for ascertaining the incidence of mental-health problems in this refugee population, though there are numerous reasons for addressing such questions in the immediate future, and over time. Refugee self-help organizations clearly can provide a bulwark against some of the continuing problems such as feelings of isolation and powerlessness, and may provide an acceptable context for mental-health counseling, though it is not now known to what extent this is being done.

A contributing cause to the Vietnamese adjustment problems was that they did not have locally established communities into which to migrate, where they could be assured of continuity of the Vietnamese culture, and economic and social support. They have not had established neighborhoods (unlike Hungarian refugees in 1956), churches, or familiar land makeup which would make them feel comfortable.

In summary, the chief causes of anxiety of the Vietnamese refugees were the following.

1. Although 94.6% of the male refugees and 86.4% of female refugees were employed, a large number of families remained underemployed or at less paying jobs. This can best be typified by the fact that, in spite of high employment rates, 36% of the refugee population received some kind of cash assistance.
2. The anxiety about adjusting their immigration status to Permanent Resident Alien remained high.
3. The frustration over family separation contributed to problems of resettlement and adjustment. There was growing frustration among refugees, individually and collectively.

4. The lack of psychological or financial preparations for the evacuation.
5. Decisions that were of necessity hastily made at the time of the evacuation.
6. The unlikelihood of being able to return safely, if at all, to their home land.
7. Cultural differences.
8. Inability to speak English and scarcity of persons in the American society speaking the Indochinese languages.
9. The lack of an Indochinese community.
10. Inadequate housing.

Conclusion

It is important that approaches be developed for assessing the magnitude of adjustment problems of the refugee population and for periodically following up those resettled under the program. Failure to do this may result in our concluding that the transition has been smooth, when in fact serious difficulties were masked by unique adaptations and geographic dispersion. Locator files were developed by the Interagency Task Force and by at least one Defense Department project which debriefed Vietnamese military officers and former government officials. These files could be put to good use in periodic assessments of mental-health service needs, or in doing longitudinal studies. Research studies are badly needed to record and evaluate the experience of these refugees. With our present knowledge, we cannot afford simply to declare the resettlement successful; though the government did perform an admirable job under very difficult circumstances.

23

Relocation and Rapid Growth: Case Studies of the Effects of Federal Policy on Life in Rural Communities

Margaret A. Thomson

Introduction

In our high-technology society, forced relocation and rapid population growth are often associated with shifts in public policy. These decisions are made in the public interest, but often have undesirable consequences for the people who live in or will be moving to the areas most affected.

Forced relocation is defined as moving from a dwelling against one's will, and includes, for example, relocation as a result of highway construction or transfer to a new job. Refusing a forced relocation often has consequences which are worse than accepting the move. Rapid population growth is defined as a population growth rate above ten percent per year, though rates of five percent can have the same effect

Ms. Thomson is affiliated with the United States House of Representatives. This article, however, was written by Ms. Thomson in her private capacity. No official support or endorsement by the legislative or executive branches of the federal government is intended or should be inferred.

Margaret A. Thomson • House of Representatives, U. S. Congress, Washington, D. C. 20015.

if the community has had a low rate of growth or no growth for several years.

The purpose of this chapter is to demonstrate the present level of responsiveness to the undesirable consequences of relocation and rapid growth; and to encourage innovative approaches to mitigating these consequences, using existing community resources rather than depending on resources outside the community, such as federal grants or loans. Increased awareness among local decision-makers should encourage communities to demand and work toward a truly comprehensive and equitable approach to mitigating the impacts of policy decisions made in the public interest.

Consequences of Relocation and Rapid Growth for Mental Health

In a growing community, new residents may quickly outnumber the old. Relatively homogeneous religious and political values may conflict with values shared by new residents who are often younger, better educated, affluent, and racially diverse. Social-interaction patterns shift and community institutions change, placing pressures on families to change too. If a community is growing rapidly, then the newcomer must not only adjust to a new community, but also do it while the community itself is undergoing change—an even greater challenge.

Adapting old behaviors or developing new behaviors to meet the stress of a new or changing environment is called coping. Although all coping strategies have positive and negative results, some are more effective than others. The effectiveness of coping strategies is influenced by the social, economic, cultural, and physical attributes of a community and the attributes of the individual (Monat & Lazarus, 1977).

Although we cannot predict with acceptable accuracy what changes will be perceived as stressful by particular individuals or communities, what stresses will result in psychomatic illnesses, or what stresses can be reduced by successful coping strategies, we do know that people who are already emotionally and physically vulnerable are likely to be more affected by the environmental changes associated with relocation and rapid population growth. These vulnerable groups include the elderly, female heads of households, children, and transients (Coelho & Stein, 1977).

For the elderly, rapid growth means a faster-paced life, a higher cost of living, fewer free recreational activities, and increased competition for low-cost social services. The elderly who are poor are the first to be affected by relocation or rapid growth.

In a community undergoing industrial growth, female heads of households are at the same disadvantage as the elderly. Among existing residents, these two groups are least likely to have the education and training to qualify for the new jobs, which will be taken by outside professionals. Industrial development may decrease the unemployment rate, but the number of unemployed will probably increase (Reiff, Schule, Nokkeo, Kolp, Regelson, & Williams, 1976).

A changing community also places demands on children. In Kitsap County, Washington, for example, the growth rate doubled when a military base was constructed. In one school district, two of four elementary schools were expanded, and four new ones built. As the new schools opened, service areas changed, and classmates were assigned to other schools. Children lost friends, and had to assimilate new members into their peer groups at a faster pace.

Transients, such as construction workers and the military, are also among the high-risk groups vulnerable to environmental stress. Construction workers are accustomed to inadequate facilities and services, but experience additional stress from living away from home, commuting long distances, and working long hours. Their presence in the community is temporary, but their impact is felt because they are not well integrated into the community social structure (Little, 1977). The presence of construction workers is correlated with drinking-while-driving and other alcohol-related arrests, suggesting that the adaptation to this transient life is not complete.

The presence of the military also makes community adaptation to rapid growth more difficult. Military families are at risk as a result of frequent forced relocation. Their presence is correlated with higher rates of teenage pregnancy and family abuse, and other indicators of environmental stress. The military services are aware of their responsibilities to help military families and communities adapt to forced relocation, but funds for social services are limited.

People for whom rapid growth means a drastic change in their way of life are also vulnerable. In Colstrip, Montana, for example, coal mines were reopened and electric generating plants built. Merchants and landowners profited, gaining in status and prestige, while ranchers lost their traditionally higher social and economic position. Newcomers to Colstrip were also at risk because the community lacked stability and failed to meet their higher expectations for community services and facilities. Failure of the community to meet their expectations increased absenteeism and turnover, which itself was a source of instability (Reiff, *et al.*, 1976).

For all of these high-risk groups—the young, the elderly, the poor, and the transient—the quality of life in growing communities is not conducive to good mental health. Those who overcome the adjustment period with successful coping strategies are rewarded with enriching experiences. Those who don't adjust contribute to increasing per capita rates of alcoholism, divorce, depression, and suicide (Little, 1977).

Public Policy Responses to Relocation and Rapid Growth

Communities with growth rates above seven to ten percent cannot absorb new residents without raising taxes to intolerable limits or lowering the quality of services. New residents contribute tax dollars for two years before they pay for the cost of facilities and services they need upon arrival.

To help pay for the "front-end" costs of capital construction, industry has been required to provide facilities and pay taxes, while state and federal governments have developed special revenue-sharing and impact-assistance programs (Staats, 1977; Stinson, 1977).

Industry is generally reluctant to provide assistance, and will do so only if the lack of facilities and services threatens the economic viability of investment. In Sweetwater County, Wyoming, for example, development of energy resources caused an annual growth rate of 19%. Employee turnover increased and productivity decreased. By providing housing and other community facilities, industry was able to help recruit and retain its employees, reduce equipment losses, and maintain productivity (Staats, 1977).

In Colstrip, Montana, Montana Power constructed houses, schools, shops, and recreational facilities. The company will also contribute to the maintenance of public facilities and services through taxes on the new coal-generating plant (Myrha, 1975). However, these revenues are not sufficient to cover increased school costs related to Colstrip's increased population, and no money will be available for other essential facilities and services (Hawks, 1974).

The federal government has been as reluctant as state governments to provide impact assistance for communities experiencing rapid growth. In 1976 the administration's policy was that the primary responsibility for meeting increased needs for community facilities and services belonged to state and local governments. However, during the same year, Congress enacted legislation which increased the state and local share of revenues from mineral leases on federal lands and provided payments in lieu of taxes to local governments having federal lands within their jurisdiction.

Legislation was also passed that assists coastal communities impacted by the development of offshore oil and gas resources. This Coastal Energy Impact Assistance program provides loans to impacted communities, but remains underutilized because lower interest rates are available on the private market.

A second problem is that projects are proposed on the basis of local priorities which usually rank community facilities above services such as community mental-health programs. Projects are funded at the discretion of the federal officials administering the program, who are aware of social impacts, but have limited expertise in helping communities identify the mental-health needs of newcomers and existing residents, and therefore have chosen not to influence community decisions in favor of social services.

The Inland Energy Impact Assistance Program is more responsive, but also emphasizes the economic over the social and cultural adjustments required by communities experiencing rapid growth. The Department of Energy recognizes that preventing and remedying social impacts is critical to the adjustment of the community, but no funding is made available. The only opportunities to comment on sociocultural impacts are the Environmental Impact Statement and the intergovernmental review of federal projects required under the Office of Management and Budget Circular A–95 (United States Department of Energy, 1978). No legal sanctions for or against any program can be exercised.

To illustrate the lack of awareness of social impacts by federal agencies and the ability to reduce the negative effects of public-policy decisions on community mental health, we shall describe the effect of relocation on North Bonneville, Washington, and the community-impact assistance programs for Kitsap County, Washington. Residents of North Bonneville have received national recognition for stopping the Army Corps of Engineers from dispersing their town, lobbying for and obtaining legislation which requires the Corps to relocate the community to a nearby site. Although North Bonneville residents have set a precedent, they have suffered both as individuals and as a community. In contrast, Kitsap County residents have enjoyed perhaps the finest community-impact assistance program in the nation, made possible again through special legislation. Nevertheless, the Kitsap County program has not been effective in responding to the increased demand for community mental-health services.

North Bonneville, Washington

North Bonneville, Washington, is a town of 500 located on the Columbia River which separates Oregon from Washington. The town is on a bluff, adjacent to the Bonneville Dam, and was selected as the site for the second Bonneville powerhouse by the Corps of Engineers. The town is located 45 miles east of Portland in a rural area where timber, tourists, and the Corps are the primary sources of employment. The town began when the Bonneville Dam was built, and consists of a few businesses and several hundred houses built along a highway.

In North Bonneville, the reactions to the proposed construction of the powerhouse and pending relocation were mixed. The residents had known for years that the town land might be needed. Some wanted to leave North Bonneville, and were willing to let the Corps pay them to move. Others felt that they had no choice, given the Congressional mandate to build the powerhouse and the authority of the Corps to implement the law (Johnston, 1975).

Although community reactions to the proposed powerhouse were mixed, the North Bonneville leaders opposed the planned relocation to other towns. The mayor and the townspeople asked the Corps to provide planning, engineering, and financial assistance to relocate the entire town at a site nearby. The request reflected the sense of community that the town derived from living in North Bonneville. The residents valued the familiarity with the town and surrounding land, and with their neighbors and friends. This feeling of community had been purchased with time spent in one place; they realized that these relationships were not generalizable to some other time, place, or group of people (Marris, 1977).

The Corps denied the town's request because under their legislative authority, the Uniform Relocation Assistance Act, they could not provide the utilities, roads, and schools that would be needed. Under the act, the Corps was only required to pay fair market value for property, and up to $15,000 to compensate for the increased cost of acquiring a comparable dwelling. Tenants could receive $4,000 and small businesses from $2,500 to $10,000. The Corps could also reimburse for moving expenses, but no other expenditures were authorized, certainly not the millions of dollars that would be necessary for constructing a new town and providing counseling services to help the displaced persons adjust to a new community.

When the Act was passed in 1970, documentation of the social and psychological effects of forced relocation was just beginning. The first detailed study was of urban renewal in Boston (Fried, 1973). The majority of the other early studies were concerned with slum clearance and the economic rather than the sociocultural effects of renewal (Cook, 1975).

The legislative history of the Uniform Relocation Act suggests that Congress was aware that relocation caused stress, but felt that financial assistance would be sufficient compensation.

In order that North Bonneville could be relocated in its entirety, legislation authorizing the Corps to provide relocation assistance to the town was introduced. With the help of Senators Magnuson and Jackson, the bill became law. The town hired a planning consultant, and an interagency relocation board was formed which served as a public forum for the negotiations still going on in 1978, almost four years later (Johnston, 1975).

While the bill moved through Congress, the Corps continued to implement the Uniform Relocation Assistance Act. By the time Congress passed the bill in March, 1974, the Corps had acquired about half the homes in North Bonneville. As people sold their homes and moved away, the population began to decline. Streets and buildings deteriorated, a highway was relocated, and business suffered. By 1977, when construction of the new town was finally under way, the people appeared similar to cultural groups forced by government to leave traditional lands. These groups exhibit a tendency toward destructive rather than constructive thinking. They are motivated to relocate by fear, do not have the right of due process, and are denied information (Trimble, 1974). North Bonneville residents were in a similar situation, because they lacked information about the relocation, felt that they had been denied due process in the appraisal of their homes and businesses, and respected the power of the Corps. In 1978, disagreements remain over interpretation of the contract between the Corps and North Bonneville, and the Congressional delegation is still involved in settling disputes. Both North Bonneville and the Corps underestimated the effort necessary to relocate the community, and neither side is satisfied with the present situation. The people have gained a new town, but their community is not the same as it was four years ago, nor will it be the same as it is now when the relocation is complete.

It should be noted that officials of the Corps recognized their agency's limited capability to carry out an effective relocation program. They indicated that they would have preferred to have another agency of the federal government responsible for the relocation, such as the Department of Housing and Urban Development, which provides over 80% of the relocation assistance awarded. HUD has recently reviewed the effectiveness of the program; however, no recommendations to amend the Act will be made, nor will the program be expanded to mitigate the social and psychological effects of forced relocation through mental-health counseling or other social-service programs.

Kitsap County, Washington

Kitsap County, Washington, is the site of the Navy's recently completed Trident Submarine Base at Bangor on Hood Canal. Kitsap

County is a semirural community of 125,000 located about an hour from Seattle by ferry. The main employer is Puget Sound Naval Shipyard, in Bremerton, a town of about 40,000. There are three other towns in Kitsap County, all about the size of Colstrip, Montana. Besides the Navy, the primary industries are timber, aquaculture, and tourism.

Construction of the Trident base in Kitsap County will double the rate of population growth, resulting in a population of 160,000 by 1985. A military installation is generally considered to be an impetus for economic development, and the business community supported the base. However, local officials and the state Congressional delegation, led by Senators Magnuson and Jackson, were concerned about the secondary impacts of the rapid growth. They realized that the existing federal programs lacked sufficient resources to provide community facilities and services for the projected population.

In return for accepting the base in Kitsap County, the delegation asked Congress to provide community-impact assistance to close the gap between costs of facilities and services and the revenues generated by the increase in population. Here again, it is the action of elected officials in obtaining special legislation which makes possible the local response to federal policy, allowing the community to exercise some control over its future.

The Trident community-impact assistance program was administered by the Department of Defense's Office of Economic Adjustment (OEA). The OEA's main function is to help the economy of a community adjust to changes in defense employment. The staff is also concerned about changes in community social structure related to defense realignments, and about the demand on social and health services made by families who relocate to new jobs, or those who stay behind after the jobs are gone. The Trident community-impact assistance program is somewhat unique, because it is the first time that OEA has assisted with a rapid *increase* in defense jobs.

Yet even this program is not adequate to provide the mental-health services required by persons relocating to a new environment or experiencing rapid population growth.

In a growing community, many social services can accommodate additional clients without expanding facilities, at least for a while. But after a certain period of time a major capital expenditure is necessary to expand or replace an existing facility. Because of rapid growth in Kitsap County, most social services will reach the threshold for capital improvements during the 1975–1985 period.

While the need for services is expanding, the funds available remain the same. Funding formulas for social services are often based on the decennial census data. Communities that grow at reasonable rates can anticipate a decrease in per-capita funding over the decade; however, rapidly growing communities cannot stretch 1970 program formulas far enough, and lack other resources to help meet the need.

Kitsap County, like other rural communities, is also handicapped by a social-service system which has developed over the years in response to specific needs. The services are dependent on private funding, and have a variety of sources of public funds. Other communities in rural areas are even less fortunate, and have no social-service system upon which to build. To help coordinate social services in Kitsap County, a county-wide board was formed for mental-health, drug-abuse, and developmental disability programs. Although these services do not duplicate each other, they do compete for limited sources of federal and state funding. In the past, competition and a lack of centralized decision-making authority has delayed applications for Trident community-impact assistance and the construction of a community mental-health center.

Social services in Kitsap County are having difficulty in coping with the increased demand for services; recently passed state law has made some program requirements more stringent.

In addition, many of these costs cannot be attributed to the Trident growth, and are therefore not eligible for community-impact assistance. The costs for services required by the existing population and by the normal growth in Kitsap County are also not eligible for Trident funds.

Of the $21.8 million needed to provide for social services and required facilities until 1985, only $3.8 million can be attributed to the Trident growth. Of this $3.8 million, $1.5 is for capital improvements, the only costs eligible for the front-end financing provided by the impact assistance. The operation and maintenance costs of the program will be paid for by the state and county. In addition to the $2.3 million in program costs, the county will also provide one-third, or $0.5 million, of the costs for capital improvements. These costs may not seem large, but are in addition to the funds needed for roads, schools, water, and sewer facilities.

For those who feel that the sums granted for social service programs are generous by today's standards (and in comparison to other localities, we agree that they are) the costs of social service programs funded by the Trident community impact assistance program should be compared to the total estimate of need for impact assistance ($60 million), the cost of base construction ($750 million), and the cost of developing and building the Trident submarine and missile ($20 billion, or $2 billion per submarine). The cost of the Environmental Impact Statement alone was $750,000, a sum more than adequate to construct a community mental-health center or provide special services to new residents or individuals at risk.

Kitsap County is further ahead than some local governments in that an effort was made to plan for the projected population increase. A grant was obtained from the Economic Development Administration, and task forces were formed to review the Navy's Draft Environmental Impact Statement on construction of the Trident base. Through the review of the DEIS, local officials and other key people in the community were familiarized with the scope and magnitude of projected impacts. The task-force members included representatives of the public agencies which would later have a role in helping new and existing residents adjust to the changes resulting from the base construction. In the early stages of planning for impacts the task-force members also helped funnel information about the base to the rest of the community.

In addition to the planning grant from the Economic Development Administration, a demonstration grant was obtained from the Department of Health, Education, and Welfare to plan for social services. Unfortunately, adminstrative and organizational difficulties limited the success of this demonstration project, and planning for social services could not be integrated into the overall county plan for land use, transportation, housing, and other urban services.

Perhaps the most positive result from the Trident population growth will be the realization that short-range ad hoc approaches to community mental health will not be adequate. Rapid growth increases an awareness of the lack of community policy in social services, and encourages the development of social-service planning. In another 10 or 20 years, perhaps, the final evaluation of the Trident community-impact assistance program can be made. Until then, many difficult and irrevocable decisions will be made. In most cases, these decisions will favor capital improvements, not social services.

Discussion

Some of the major psychosocial effects of relocation and rapid growth owing to changes in the built environment in two communities have been described, and the programs designed to reduce these efforts have been analyzed in terms of risks and costs to emotional health and social adaptation. The programs have been found to be less adequate in addressing the psychosocial needs of communities than their economic needs. On the basis of these two case studies, recommendations for other rural communities undergoing relocation and rapid growth will be made.

First, information is essential for successful adaptation. People living in rural communities are not generally well-informed about the effects of planned change on their communities, particularly the effects of this change on mental health. Their information about the planned change comes from the newspaper and conversations with friends, relatives, neighbors, and co-workers.

Community programs to inform people about the effects of a planned change are needed. Public hearings on development proposals and environmental-impact statements could be used; however, only a minority of the community will attend. The documents themselves may be more widely read, particularly if made available at local libraries. However, past evidence indicates that even these extensive studies do not provide adequate information on projected social impacts. Available data are repeated, without regard for quality or applicability to a specific situation.

Television and radio programs may be most effective in reaching the largest number of people on a regular basis, especially in sparsely-settled rural areas. Organization could also establish a home-visitation program for high-risk groups such as the elderly, the poor, and the young.

In providing information about a planned change it is important for those who have knowledge of the community to share this information. After the planned change has occurred, and the impacts are experienced, all changes in the community will be perceived as a result of the planned change, including the changes resulting from normal population growth. Knowing that not all impacts are the result of the planned change, and that some were inevitable, should help most people to accept the differences in their lives.

In North Bonneville, the relationship between the Corps of Engineers and the town residents could have been improved if the Corps had provided more information about the process of relocation and the proposed legislation authorizing construction of the new town. Some of the people who sold their homes to the Corps before the special legislation was passed knew nothing of the plans to relocate the town. They might not have sold their homes had they been aware of the relocation efforts.

Communication between the Corps and the town could have been facilitated through the relocation advisory service required under the Uniform Relocation Assistance Act. However, the main function of the service was to determine eligibility for relocation assistance and to locate a replacement dwelling. Relocation advisers employed by public-works agencies like the Corps are usually more capable of providing information about the real-estate market than ensuring that the residents of an entire community have received sufficient information to make fully informed decisions about their futures. Staff have little or no training in determining whether sufficient resources and coping strategies are present or can be developed in order that displaced persons can adjust more effectively.

The flow of information could have been improved if the Corps had trained people from the community to work as relocation advisers. The community advisers would have been better able to tap existing systems of communication, and would also have been more sensitive to the needs of people, since the advisers would themselves be in the same situation.

But even community advisers would not have been able to identify those persons who were not capable of coping with the relocation, and refer them to available community services for additional counseling and assistance. Licensed counselors trained in coping strategies are needed. These counselors would be able to provide the full range of services necessary (Coelho & Stein, 1977). At the very least, people should have more information about the availability of social services, the second recommendation.

People lack knowledge about the available social services because they learn about counseling agencies informally through close friends or relatives. They have less knowledge about the organizing and financing of social services because there is not a monthly bill or annual tax assessment; social services are generally financed through revenues, fees for services, or government grants.

Even local elected officials lack information about services, particularly mental-health services. In small rural communities, elected officials may be part-time volunteers who represent their communities to industry and other levels of government. Where once the elected official's role was largely ceremonial, officials in rapidly growing communities are being asked to make technical policy decisions in diverse areas. Where once road construction and law enforcement were the primary functions of local governments, cities and counties are now expected to provide social services, too.

In a rapid-growth situation the local government's responsibilities and budget may increase even more dramatically. When a city or county decision maker chooses between a community mental-health center and a storage tank for a public water system, he or she probably has more first-hand knowledge of the latter. The local official is also more likely to choose the storage tank, because water pressure affects the lives of more people than the mental-health program, and visibility of accomplishments means votes at the next election.

Knowledge of the effects of rapid growth on community mental health and the need for certain types of services to mitigate these effects can be provided through existing national or state groups, such as the National Association of Counties. These organizations have established credibility among their memberships for being responsive to the immediate practical needs of governments and the separate functions within governments when demands are made by higher levels of government without providing the resources to meet the demands. Recent examples include reporting requirements for affirmative action required of personnel officers, and environmental-impact statements required when significant actions are taken; and are analagous, although at a lesser scale, to the situation forced on a local government when the federal government decides to build a powerhouse, or establish a military base. These organizations of local governments or government officials can provide technical assistance, perhaps funded by a federal grant, and advise local officials concerning the effects of rapid growth on communities and the possible solutions to the problems created.

After the local officials have been made aware of the potential impacts of relocation or rapid growth on community mental health and related social services, then the remaining lead time must be maximized through comprehensive planning, the third recommendation.

Planning for change avoids decision-making in a crisis atmosphere. Armed with available information, and convinced of the need to begin preparations immediately, the local elected official may face opposition to planning from the local community. People resist planning for change, because it makes them acknowledge the inevitable, and implies a certain dissatisfaction with current living and working environments. Since most people are satisfied with the status quo, they become apprehensive when others plan for the inevitable reconstruction of their familiar working space and relationships.

Residents of rural America resist the interference with their lifestyle that they associate with effective planning. They believe that planning leads to big government and challenges cherished ideals, such as local control and home rule; when, in fact, planning often enhances these goals (Myrha, 1977).

People are more opposed to planning for social services than for more familiar services, such as water supply and sewage disposal. A local elected official is fortunate in gaining acceptance of the need for planning for services people know something about, and will have a more difficult time in explaining the need to plan for community mental health.

The fourth recommendation to local governments planning for relocation and rapid growth is to consider carefully the separate advantages and disadvantages of investing resources in community mental-health *services* versus community mental-health *facilities*.

In Kitsap County, the community-impact assistance will be used primarily for facilities and not for the services, because the impact assistance is a temporary source of revenue, available only for the period of rapid growth. This impact assistance will not be replaced by tax revenues from industry when the base is completed, and yet, if construction had not been provided, the military base would have created a permanent degradation in community facilities and services. Using impact assistance for buildings means fewer services during the period of rapid growth, but reduces long-term mortgage debt. As the personal property tax base grows, revenues can be used to provide services rather than to reduce bonded or other indebtedness.

However, if population growth will not be permanent—as is often the case in some energy-impacted communities, and some communities impacted by the military, too—providing services should have a higher priority than capital improvements. In temporary boom situations, needs for facilities should be met with temporary structures, even though they have a higher short-term cost than mortgage payments. Later, if the population decreases, the community has neither the long-term debt nor a monument to the growth in an oversized building that must be maintained. Another possibility would be to finance capital improvements with loans guaranteed by the federal government. If the population decreases, and the revenue to pay for the improvements is no longer available, then the loans can be forgiven. History has shown that approval for loan programs is easier to obtain from Congress than support for outright grants; these loan programs can be useful, provided the interest rates are below the market rates.

A fifth recommendation is for rural governments to obtain the services of outside consultants in the early stages of planning. Outside consultants can provide technical assistance consistent with local policy, obtaining the necessary background and historical perspective from officials and their staff. Consultants should have prior experience in preparing for rapid change and knowledge of applicable social-science literature, data, and theories, and should maximize the participation of community residents. Consultants can develop alternatives and identify consequences for elected officials who will then be better prepared to make the irreversible long-range decisions for the community. Although many local communities and their officials have met the challenges of planning for rapid growth and relocation without consultant services, they should not be expected to understand the effects of a rapid change, or to articulate community needs, while personally experiencing the changes.

A second advantage is that the consultant can be the scapegoat. If the community is opposed to change, and the projections do not match community expectations, then the consultant can bear the brunt of public criticism, leaving the community political system intact to prepare for the projected changes. In the early 1970s, many communities welcomed industrial development without knowledge of the rapid social changes that would occur. Similar communities, aware of the adverse impacts upon their neighbors, have become increasingly wary. Unfortunately, the legal authority of these small towns to reject the federal decisions which cause the impacts has not improved.

The need for technical assistance from outside consultants has been identified by the Western Governors' Conference, representing the Rocky Mountain states. Congress should provide funds for teams of researchers, administrators, and practitioners to be made available to local governments upon request, in much the same way as the Office of Economic Adjustment of the Department of Defense is available to communities impacted by changes in defense employment. Site visits could be followed by technical reports and recommendations, and, if conditions warranted, a team member or affiliate could remain in the community.

In addition to providing valuable technical advice to a community, the social-adjustment team could collect data from several communities. Evaluation of the data would facilitate development or implementation of new policies and action programs, encourage long-range analysis and implementation of new policies and programs, and perhaps even explore basic processes that govern the functioning and change of social and cultural systems. Establishment of such a research program would not be expensive, and would satisfy the concerns of those who wish to have better answers before increasing the money spent on community-impact programs. At the least, persons in charge of research programs in energy and behavioral sciences should encourage research on the social impacts of rapid growth and relocation.

In addition, hearings should be held on all issues of concern to communities experiencing planned change resulting from a shift in federal policy.

The experiences of communities in the United States could then be compared with the experience of foreign governments, the United Nations, and our own State Department to identify those universal human needs resulting from relocation or rapid growth which bridge political, cultural, and geographic boundaries.

Data and conclusions from a broad range of activities such as the above should be publicized by professional organizations representing behavioral scientists and government officials. The experience of the local elected official in the field, backed by empirical research, should be used to raise the priority of services mitigating social impacts.

Rural counties will continue to have difficulty in meeting the social- and health-service needs of new and existing residents during periods of rapid growth. Even if these communities are as fortunate as Kitsap County in having a source of adequate impact assistance, intergovernmental cooperation, and a commitment to minimize secondary impacts, the needs will not be adequately met.

Although improvements on the national and international scale are needed, one sufficiently concerned citizen can have an impact within a local jurisdiction. Bringing the problem to the attention of the service-delivery system, including the health-planning agency for the area, and then contacting local elected officials in general-purpose governments can significantly improve community response. Support of local citizen groups is especially effective at the local level in gaining elected officials' support for necessary improvements in social and community mental-health services.

With the support of an elected official and his or her constituency, one should then contact elected officials for larger jurisdictions, at each level of government requesting that an inquiry of the appropriate agencies be made. This process helps to increase knowledge and awareness of the problem, and, if a solution is found, all levels of government and elected officials can share the credit.

Contacting elected and appointed officials should be continued on at least an annual basis as long as necessary. If the federal policy toward the mitigation of psychosocial and emotional effects of planned change is to be altered, the modification will occur not because these changes are needed, but because they are politically expedient. It is up to the individual who values emotional and mental health to increase the political payoff for elected officials who have the power to change our national priorities.

References

Albrecht, S. L. *Sociological aspects of power plant siting.* Paper presented at a conference on Developing Utah's Energy Resources: Problems and Opportunities, Salt Lake City, 1972.

Boice, L. P. *Encountering a city: The spatial learning process of urban newcomers* (Exchange Bibliography No. 1264). Monticello, Ill.: Council of Planning Librarians, 1975. (Available from Mrs. Mary Vance, Editor, P. O. Box 229, Monticello, Illinois 61856.)

Brody, S. E. *Federal aid to energy impacted communities: A review of related programs and legislative proposals.* Cambridge: M. I. T. Press, 1977. (Available from Energy Impacts Project, Laboratory of Architecture and Urban Planning, Room 4–209, M.I.T., Cambridge, Massachusetts 02139. Energy Research and Development Administration Contract No. E 49–18.)

Coelho, G. V., & Stein, J. J. Coping with stress of an urban planet: Impacts of uprooting and overcrowding. *Habitat,* 1977, *2,* 379–390.

Coelho, G. V., Hamburg, D. A., & Adams, J. E. (Eds.). *Coping and adaptation.* New York: Basic Books, 1974.

Cook, E. *Bibliography on relocation of families as a result of government acquisition of property* (Exchange Bibliography No. 744). Council of Planning Librarians, 1975.

Donnermeyer, J. F. *Forced migration: A bibliography on the sociology of population displacement and resettlement.* (Exchange Bibliography No. 880). Council of Planning Librarians, 1977.

Fried, M. Grieving for a lost home. In L. J. Duhl (Ed.), *The urban condition.* New York: Basic Books, 1963.

Hawks, Units 3 and 4 at Colstrip. One Western Wildlands, 1974.

Johnston, R. W. Caught standing in the way of progress. *Sports Illustrated,* 1975, *24,* 50–52, 55–56, 61–62.

Little, R. L. Some social consequences of boom towns. *North Dakota Law Review,* 1977, *53,* 401–425.

Lovejoy, S. B., & Little, R. L. *Western energy development as a type of rural industrialization: A partially annotated bibliography.* (Exchange Bibliography No. 1298). Council of Planning Librarians, 1977.

Marris, P. *The uprooting of meaning.* Unpublished manuscript, 1977.

Micklin, M. *Psychosocial aspects of forced migration.* Paper presented at the meeting of the American Psychological Association, San Francisco, August 1977.

Monat, A., & Lazarus, R. (Eds.). *Stress and coping: An anthology.* New York: Columbia University Press, 1977.

Myrha, D. Colstrip, Montana: The modern company town. *Coal Age,* May, 1975, 54–57.

Myrha, D. *Factoring local interests into energy development planning.* Paper presented at the meeting of the Southern Governor's Conference, Nashville, March 1977.

Reiff, I. S., Schule, M., Nokkeo, S., Kolp, P., Regelson, K., & Williams, D. *Managing the social and economic impacts of energy developments.* Washington, D. C.: Energy Research and Development Administration, 1976. (Available from Energy Research and Development Administration, P. O. Box 62, Oak Ridge, Tennessee, 37830.)

Staats, E. B. *Rocky mountain energy resource development: Status, potential, and socioeconomic issues.* (Report No. EMD–77–23). Washington, D. C.: General Accounting Office, 1977. (Available from U. S. General Accounting Office, Distribution Section, P. O. Box 1020, Washington, D. C., 20013.)

Staats, E. B. *Changes needed in the relocation act to achieve more uniform treatment of persons displaced by federal programs.* Washington, D. C.: General Accounting Office, 1978.

Stinson, D. C. *Predicting the local impacts of energy development: A critical guide to forecasting methods and models.* Cambridge: M.I.T., Press, 1977. (Available from Energy Impacts Project, Energy Research and Development Administration, Contract No. # 49–18.)

Trident Fiscal Impact Analysis. Olympia: State of Washington, 1977. Available from Office of Economic Adjustment, Department of Defense, Pentagon, Washington, D. C., or Board of Commissioners, Kitsap County, Washington, 614 Division Street, Port Orchard, Washington 98366.

Trimble, J. E. To assimilate the native: A social psychological analysis of strategy and tactic. In G. Marin (Ed.), *Proceedings of the XVth interamerican congress of psychology.* Bogotá, Colombia: The Interamerican Society of Psychology, 1974.

Trimble, J. E. *Issues of forced relocation and migration of cultural groups.* Paper presented at the meeting of the American Psychological Association, San Francisco, August 1977.

United States Department of Energy. *Report to the President: Energy Impact Assistance.* Washington, D. C.: U. S. Department of Energy, 1978.

United States Department of Housing and Urban Development. *Rapid growth from energy projects: Ideas for state and local action: A program guide.* Washington, D. C.: U. S. Department of Housing and Urban Development, 1976.

Index

MIX
Papier aus verantwortungsvollen Quellen
Paper from responsible sources
FSC® C105338

If you have any concerns about our products,
you can contact us on
ProductSafety@springernature.com

In case Publisher is established outside the EU,
the EU authorized representative is:
Springer Nature Customer Service Center GmbH
Europaplatz 3, 69115 Heidelberg, Germany

Printed by Libri Plureos GmbH
in Hamburg, Germany